Deep Cognitive Modelling in Remote Sensing Image Processing

Sadique Ahmad
Prince Sultan University, Saudi Arabia

Muhammad Shahid Anwar
Gachon University, South Korea

Ala Saleh Alluhaidan
Princess Nourah bint Abdulrahman University, Saudi Arabia

Mohammed A. El-Affendi
Prince Sultan University, Saudi Arabia

IGI Global
PUBLISHER of TIMELY KNOWLEDGE

A volume in the Advances in
Geospatial Technologies (AGT)
Book Series

Published in the United States of America by
    IGI Global
    Engineering Science Reference (an imprint of IGI Global)
    701 E. Chocolate Avenue
    Hershey PA, USA 17033
    Tel: 717-533-8845
    Fax: 717-533-8661
    E-mail: cust@igi-global.com
    Web site: http://www.igi-global.com

Library of Congress Cataloging-in-Publication Data

CIP DATA PROCESSING

2024 Engineering Science Reference

ISBN(hc): 9798369329139
ISBN(sc): 9798369349151
eISBN: 9798369329146

This book is published in the IGI Global book series Advances in Geospatial Technologies (AGT) (ISSN: 2327-5715; eISSN: 2327-5723)

British Cataloguing in Publication Data
A Cataloguing in Publication record for this book is available from the British Library.

All work contributed to this book is new, previously-unpublished material.
The views expressed in this book are those of the authors, but not necessarily of the publisher.

For electronic access to this publication, please contact: eresources@igi-global.com.

Advances in Geospatial Technologies (AGT) Book Series

ISSN:2327-5715
EISSN:2327-5723

Editor-in-Chief: Nilanjan Dey, Techno India College of Technology, India

MISSION

The geospatial technology field is a fast-paced, high growth industry that is involved in a variety of fields, including military planning, public health, land use, environmental protection, and Google Earth mapping. With such a diverse body of applications, the research in geospatial technologies is always evolving and new theories, methodologies, tools, and applications are being developed. **Advances in Geospatial Technologies (AGT) Book Series** is a reference source and outlet for research that discusses all aspects of geographic information, including areas such as geomatics, geodesy, GIS, cartography, remote sensing, and other areas. Because geospatial technologies are so pervasive in such a variety of areas, AGT also includes books that address interdisciplinary applications of the technologies.

COVERAGE

- Cartography
- Public Health Applications of Geospatial Technologies
- Digital Terrain Modeling
- Remote Systems
- Geovisualization
- Military Security and Geospatial Technologies
- Public Sector Use of Geospatial Technologies
- Spatial Reference Systems
- Disaster Informatics
- Geospatial Analysis

IGI Global is currently accepting manuscripts for publication within this series. To submit a proposal for a volume in this series, please contact our Acquisition Editors at Acquisitions@igi-global.com or visit: http://www.igi-global.com/publish/.

Titles in this Series

For a list of additional titles in this series, please visit:
http://www.igi-global.com/book-series/advances-geospatial-technologies/73686

Geospatial Application Development Using Python Programming
Mohammad Gouse Galety (Samarkand International University of Technology, Uzbekistan)
Arul Kumar Natarajan (Samarkand International University of Technology, Uzbekistan)
Tesfaye Fufa Gedefa (Space Science and Geospatial Institute, Ethiopia) and Tsegaye Demsis
Lemma (Space Science and Geospatial Institute, Ethiopia)
Engineering Science Reference • copyright 2024 • 344pp • H/C (ISBN: 9798369317549)
• US $265.00 (our price)

Ethics, Machine Learning, and Python in Geospatial Analysis
Mohammad Gouse Galety (Samarkand International University of Technology, Uzbekistan)
Arul Kumar Natarajan (Samarkand International University of Technology, Uzbekistan)
Tesfaye Fufa Gedefa (Space Science and Geospatial Institute, Ethiopia) and Tsegaye Demsis
Lemma (Space Science and Geospatial Institute, Ethiopia)
Engineering Science Reference • copyright 2024 • 339pp • H/C (ISBN: 9798369363812)
• US $255.00 (our price)

Advanced Geospatial Practices in Natural Environment Resource Management
Rubeena Vohra (Bharati Vidyapeeth's College of Engineering, India) and Ashish Kumar
(Bennett University, India)
Engineering Science Reference • copyright 2024 • 292pp • H/C (ISBN: 9798369313961)
• US $275.00 (our price)

Emerging Trends, Techniques, and Applications in Geospatial Data Science
Loveleen Gaur (Amity University, India & Taylor's University, Malaysia & University of
the South Pacific, Fiji) and P.K. Garg (Indian Institute of Technology, Roorkee, India)
Engineering Science Reference • copyright 2023 • 309pp • H/C (ISBN: 9781668473191)
• US $240.00 (our price)

Addressing Environmental Challenges Through Spatial Planning

For an entire list of titles in this series, please visit:
http://www.igi-global.com/book-series/advances-geospatial-technologies/73686

IGI Global
PUBLISHER of TIMELY KNOWLEDGE

701 East Chocolate Avenue, Hershey, PA 17033, USA
Tel: 717-533-8845 x100 • Fax: 717-533-8661
E-Mail: cust@igi-global.com • www.igi-global.com

Table of Contents

Preface... xiv

Chapter 1
Deep Strategy of Object Detection in Remote Sensing Images: A Systematic
Review..1
> *Sadique Ahmad, Prince Sultan University, Saudi Arabia*
> *Mohamed A. El Affendi, Prince Sultan University, Saudi Arabia*
> *Ala Saleh D. Alluhaidan, Princess Nourah bint Abdulrahman University,*
> *   Saudi Arabia*
> *M. Shahid Anwar, Gachon University, South Korea*

Chapter 2
Deep Learning Classification of Empty and Full Container Ships in Satellite
Imagery: A Cognitive Modeling Approach for Maritime Applications 30
> *Serif Ali Sadik, Faculty of Engineering, Kutahya Dumlupinar University,*
> *   Turkey*

Chapter 3
Mapping Faces From Above: Exploring Face Recognition Algorithms and
Datasets for Aerial Drone Images 55
> *Sadique Ahmad, College of Computer and Information Sciences, Prince*
> *   Sultan University, Saudi Arabia*
> *Mahmood Ul Haq, University of Engineering and Technology,*
> *   Peshawar, Pakistan*
> *Muhammad Athar Javed Sethi, University of Engineering and*
> *   Technology, Peshawar, Pakistan*
> *Mohammed A. El Affendi, College of Computer and Information*
> *   Sciences, Prince Sultan University, Saudi Arabia*
> *Zahid Farid, Abasyn University, Pakistan*
> *Alaa Sal. Al Luhaidan, College of Computer and Information Sciences,*
> *   Princess Nourah bint Abdulrahman University, Pakistan*

Chapter 4

Ethical Considerations in Remote Sensing and Cognitive Modeling 70
　　Baseer Ul Haq, University of Malakand, Pakistan
　　Mohammad Faisal, University of Malakand, Pakistan
　　Mahmood Ul Haq, University of Engineering and Technology,
　　　Peshawar, Pakistan

Chapter 5

A Review of Capsule Network Limitations, Modifications, and Applications
in Object Recognition 88
　　Mahmood Ul Haq, University of Engineering and Technology,
　　　Peshawar, Pakistan
　　Muhammad Athar Javed Sethi, University of Engineering and
　　　Technology, Peshawar, Pakistan
　　Atiq Ur Rehman, Hamad Bin Khalifa University, Qatar

Chapter 6

Advancements in Machine Learning and Deep Learning..................................113
　　Dina Darwish, Ahram Canadian University, Egypt

Chapter 7

COMSATS Face: An Image Database of Faces With Various Poses, Its
Structure, and Features...151
　　Mahmood Ul Haq, University of Engineering and Technology,
　　　Peshawar, Pakistan
　　Muhammad Athar Javed Sethi, University of Engineering and
　　　Technology, Peshawar, Pakistan
　　Aamir Shahzad, COMSATS University Islamabad, Abbottabad campus,
　　　Pakistan
　　Muhammad Shahid Anwar, Gachon University, South Korea

Chapter 8

The Development, Applications, Challenges, and Analysis of a Cricket Player
Face Recognition Dataset...173
　　Mahmood Ul Haq, University of Engineering and Technology,
　　　Peshawar, Pakistan
　　Muhammad Athar Javed Sethi, University of Engineering and
　　　Technology, Peshawar, Pakistan
　　Subhan Ullah, University of Malakand, Pakistan
　　Abd Ullah, University of Malakand, Pakistan

Chapter 9
Exploring COVID-19 Classification and Object Detection Strategies: X-Rays
Image Processing 198
    Saifullah Jan, City University, Peshawar, Pakistan
    Aiman, City University, Peshawar, Pakistan
    Bilal Khan, City University, Peshawar, Pakistan
    Muhammad Arshad, City University, Peshawar, Pakistan

Compilation of References 219

Related References 253

About the Contributors 278

Index 282

Detailed Table of Contents

Preface.. xiv

Chapter 1
Deep Strategy of Object Detection in Remote Sensing Images: A Systematic
Review.. 1

    Sadique Ahmad, Prince Sultan University, Saudi Arabia
    Mohamed A. El Affendi, Prince Sultan University, Saudi Arabia
    Ala Saleh D. Alluhaidan, Princess Nourah bint Abdulrahman University,
        Saudi Arabia
    M. Shahid Anwar, Gachon University, South Korea

Recently, the demand for satellite image analysis increased with the recent advancement in various research areas (e.g., improvements in remote sensing image resolution, object recognition ideas, and deep learning techniques). Articles report new trends in object detection using remote sensing images, such as multi-temporal scene classifications, semantic segmentation, multiresolution, and large-scale optical image analysis. However, this study observed a black box of synchronization among the advancements in various technologies (i.e., new ideas in satellite image analysis) and advancements in deep learning techniques. Without investigating this black box, the optimization of object detection will be out of track. So, to keep the deep learning innovation on track, the primary goal of the current work is to explore the aforementioned black box to advance object detection in remote sensing images. In contrast, focusing on a specific technique, this study explores the black box while reviewing 150 articles from 2019 to 2023. First, the study highlights research methodologies and novel features of literature to achieve object detection in remote sensing images. Second, it evaluates effective deep learning models and assesses various featured studies to draw a clear picture of limitations and recent object detection trends to provide in-depth recommendations and future directions.

Chapter 2

Deep Learning Classification of Empty and Full Container Ships in Satellite
Imagery: A Cognitive Modeling Approach for Maritime Applications 30

*Serif Ali Sadik, Faculty of Engineering, Kutahya Dumlupinar University,
Turkey*

This chapter presents a novel approach to ship classification in optical remote sensing
(ORS) imagery, focusing on the distinction between empty and full container ships.
Leveraging deep cognitive modeling techniques, the study employs renowned
pre-trained deep learning models, including VGG-16, VGG-19, and InceptionV3,
with fine-tuning for enhanced performance. The investigation addresses the
challenges posed by class imbalance through strategic data augmentation. Results
demonstrate the efficacy of the proposed models, with InceptionV3 exhibiting
superior performance. Evaluation metrics encompassing accuracy, precision,
recall, F1-score, AUC-ROC, and AUC-PR are meticulously analyzed. These
findings contribute to the advancement of ship classification methodologies in
ORS imagery, with implications for maritime applications and decision-making
processes. The work underscores the importance of deep cognitive modeling in
addressing complex classification tasks and paves the way for future enhancements
and applications in the field.

Chapter 3

Mapping Faces From Above: Exploring Face Recognition Algorithms and
Datasets for Aerial Drone Images 55

*Sadique Ahmad, College of Computer and Information Sciences, Prince
Sultan University, Saudi Arabia*

*Mahmood Ul Haq, University of Engineering and Technology,
Peshawar, Pakistan*

*Muhammad Athar Javed Sethi, University of Engineering and
Technology, Peshawar, Pakistan*

*Mohammed A. El Affendi, College of Computer and Information
Sciences, Prince Sultan University, Saudi Arabia*

Zahid Farid, Abasyn University, Pakistan

*Alaa Sal. Al Luhaidan, College of Computer and Information Sciences,
Princess Nourah bint Abdulrahman University, Pakistan*

This chapter explains various facial identification algorithms used in drones. The
burgeoning field of face recognition technology makes use of image processing to
identify faces in people. Face recognition is becoming increasingly popular for a
variety of reasons, such as the growing population, which necessitates higher security
and surveillance systems, identity verification in the digital age, combat in rural
regions, disaster relief, and so forth. This research compares and contrasts various
face identification techniques, including neural networks, PCA, LBPH (local binary

pattern histogram), PAL, capsule network, and LDA (linear discriminant analysis). Additionally, a comparison of face datasets collected with drones is included in this chapter. The results of this study will help facial recognition technologists design a hybrid algorithm that meets the requirements of real-time applications.

Chapter 4

Ethical Considerations in Remote Sensing and Cognitive Modeling 70

    Baseer Ul Haq, University of Malakand, Pakistan
    Mohammad Faisal, University of Malakand, Pakistan
    Mahmood Ul Haq, University of Engineering and Technology,
        Peshawar, Pakistan

Three major advances are driving fundamental changes in the field of remote sensing. New satellite sensors will provide global imaging with excellent spatial and spectral resolution. Technological advancements have rendered earlier limitations on data scale, resolution, location, and availability obsolete. Economic restructuring in the remote sensing community will shift control and dissemination of imagery and related information from the government to the private sector. The internet and other digital infrastructures will speed up information distribution to a global user base. The combined results of these advancements may have serious legal and ethical ramifications for all remote sensing experts. Remote sensing technology may soon be able to provide detailed information, potentially violating privacy and leading to legal and ethical implications. This chapter discusses the legal history of remote sensing, recent innovations in satellite surveillance and information technology, and potential legal and ethical concerns for the remote sensing community. Self-regulation of the profession is essential for balancing individual rights with the economic objectives of the remote sensing community and nation.

Chapter 5

A Review of Capsule Network Limitations, Modifications, and Applications in Object Recognition 88

    Mahmood Ul Haq, University of Engineering and Technology,
        Peshawar, Pakistan
    Muhammad Athar Javed Sethi, University of Engineering and
        Technology, Peshawar, Pakistan
    Atiq Ur Rehman, Hamad Bin Khalifa University, Qatar

Modern computer vision and machine learning technologies have enabled numerous advances in a variety of domains, including pattern recognition and image classification. One of the most powerful machine learning methods is the capsule network, which encodes features based on their hierarchical relationships. A capsule network is a sort of neural network that uses inverted graphics to represent an item in distinct sections and see the existing link between these pieces, as opposed to CNNs, which lose most of the evidence relating to spatial placement and require a large amount of training data. As a result, the authors give a comparison of various capsule network designs utilized in diverse applications. The fundamental contribution of this study is that it summarizes and discusses the major current published capsule network topologies, including their advantages, limits, modifications, and applications.

Chapter 6

Advancements in Machine Learning and Deep Learning...................................113
Dina Darwish, Ahram Canadian University, Egypt

Among the most important methodologies in the field of modern intelligent technology is data-driven advanced machine learning methodology. In order to find rules, it makes use of data samples that have been observed, and it makes use of regular patterns in order to forecast unknown data in the future. In tandem with the development of artificial intelligence, the field of machine learning is making further strides forward. Due to this, there is a need for increased requirements for the training and applications of models, as well as the enhancement of the algorithm and the improvement of technological capabilities. This chapter discusses the recent technologies and trends in the artificial intelligence field, while giving examples and conclusions at the end of the chapter.

Chapter 7

COMSATS Face: An Image Database of Faces With Various Poses, Its
Structure, and Features..151
Mahmood Ul Haq, University of Engineering and Technology,
Peshawar, Pakistan
Muhammad Athar Javed Sethi, University of Engineering and
Technology, Peshawar, Pakistan
Aamir Shahzad, COMSATS University Islamabad, Abbottabad campus,
Pakistan
Muhammad Shahid Anwar, Gachon University, South Korea

The human face can appear different depending on the circumstances because of its flexibility and three-dimensional structure. Researchers are facing several obstacles relating to face poses, illumination, facial expressions, head direction, occlusion, hairdo, etc. in the process of developing dependable and efficient algorithms for face detection, face identification, and face expression analysis. To determine the algorithms' effectiveness, they need to be evaluated against a certain set of face image/database benchmarks. This work introduces a dataset of multiple-pose facial photographs. Eight hundred fifty photos from 50 people in 17 distinct stances are included in the collection ($0°, 5°, 10°, 15°, 20°, 25°, 30°, 35°, 55°, -5°, -10°, -15°, -20°, -25°, -30°, -35°, -55°$). Three distinct lighting conditions are also included in the dataset. Eight resolutions ($144 \times 256, 200 \times 200, 100 \times 100, 70 \times 70, 50 \times 50,$

$40 \times 40, 20 \times 20$ and 10×10 pixels) are available for the dataset. This dataset's facial image content can provide insight on the effectiveness and resilience of upcoming face detection and recognition systems. Additionally, based on the suggested face database, a comparison study of two face recognition methods, such as PAL and PCA, is performed.

Chapter 8
The Development, Applications, Challenges, and Analysis of a Cricket Player
Face Recognition Dataset..173

Mahmood Ul Haq, University of Engineering and Technology,
Peshawar, Pakistan
Muhammad Athar Javed Sethi, University of Engineering and
Technology, Peshawar, Pakistan
Subhan Ullah, University of Malakand, Pakistan
Abd Ullah, University of Malakand, Pakistan

Over the last decade, face recognition technology has played a critical role in various circumstances, such as airport boarding, security applications, biometric verification, and smart homes. Along with the major role of face recognition in the areas above, we must recognize the important role of face recognition in various sports (i.e., cricket and football). The importance of proper player surveillance and identification in sports, particularly cricket, cannot be overstated. Articles are saturated with many deep-face evaluation systems; however, they are not up to mark due to the lack of significant face posture data. To address the black box in facial expression datasets, this chapter presents a comprehensive cricket player facial recognition dataset. The authors have a wide selection of cricket player images from various teams, playing styles, and backgrounds. It includes images taken during games, practices, and official team photos, providing a diverse range of facial changes and challenges for facial recognition systems. Furthermore, they evaluate the efficacy of cutting-edge facial recognition algorithms on our dataset, providing insights into the effectiveness of current methodologies as well as potential areas for development. Eventually, the extensive experimental analyses demonstrate that the current work is significant in addressing the black box in facial expression datasets.

Chapter 9
Exploring COVID-19 Classification and Object Detection Strategies: X-Rays
Image Processing 198

Saifullah Jan, City University, Peshawar, Pakistan
Aiman, City University, Peshawar, Pakistan
Bilal Khan, City University, Peshawar, Pakistan
Muhammad Arshad, City University, Peshawar, Pakistan

The overlapping imaging characteristics of COVID-19 viral pneumonia and non-

COVID-19 viral pneumonia chest X-rays (CXRs) make differentiation difficult for radiologists. Machine learning (ML) has demonstrated promising outcomes in a range of medical sectors, enhancing diagnostic accuracy through its interaction with radiological tests. The potential contribution of ML models in assisting radiologists in discriminating COVID-19 from non-COVID-19 viral pneumonia from CXRs, on the other hand, deserves further examination and exploration. The goal of this study is to empirically assess ML models' capacity to classify X-ray images into COVID-19, pneumonia, and normal cases. The study evaluates the efficacy of K-nearest Neighbor (KNN), random forest (RF), AdaBoost (AB), and neural networks (NN) with various hidden neuron configurations using a wide range of performance measures. These metrics evaluate the area under the curve (AUC), classification accuracy (CA), F1 score (F1), precision, and recall, resulting in a comprehensive evaluation technique. ROC analysis is used to gain a thorough knowledge of the models' discriminating skills. The results show that NN models, particularly those with 100 and 150 hidden neurons, outperform in all criteria, proving their ability to reliably categorize medical disorders. Notably, the study emphasizes the difficulties in separating COVID-19 from pneumonia, emphasizing the importance of strong classification methods. While the study provides useful insights, its drawbacks include the use of a single dataset, the absence of more sophisticated deep learning architectures, and a lack of interpretability analyses. Nonetheless, the study adds to the developing area of medical picture categorization, directing future attempts to improve diagnosis accuracy and widen the use of machine learning in healthcare. The findings highlight the utility of NN models in medical diagnostics and pave the way for future study in this vital area of technology and healthcare.

Compilation of References 219

Related References 253

About the Contributors 278

Index 282

Preface

In recent years, the intersection of deep learning techniques and remote sensing technologies has ushered in a new era of possibilities in satellite image analysis. The demand for high-resolution imagery coupled with advancements in object recognition and deep learning methodologies has spurred innovation across various research domains. From multi-temporal scene classifications to semantic segmentation, the landscape of remote sensing image processing is rapidly evolving to meet the challenges posed by climate change, global security concerns, and the imperative for precise land surface analysis.

In response to these challenges, our book, *Deep Cognitive Modelling in Remote Sensing Image Processing*, endeavors to delve into the complexities of this burgeoning field. Led by esteemed editors Sadique Ahmad, Muhammad Shahid Anwar, Ala Saleh D. Alluhaidan, and Mohammed A. El-Affendi, this volume aims to unravel the intricacies of remote sensing data through the lens of cognitive modeling. By peering into the black box of image processing, we seek to illuminate pathways for optimized deep learning models, thereby enhancing our understanding and capabilities in analyzing high-resolution remote sensing imagery.

The edited reference book presents a diverse array of chapters encapsulating cutting-edge research and innovation across remote sensing, computer vision, and machine learning domains. Chapter contributors delve into intricate methodologies, ranging from object detection optimization in remote sensing imagery to ship classification techniques leveraging deep cognitive modeling. Legal and ethical implications of technological advancements in remote sensing are explored alongside transformative paradigms such as capsule networks and data-driven machine learning methodologies. Furthermore, chapters elucidate the practical applications of facial recognition algorithms in diverse contexts, from drone technology to sports surveillance in cricket. Through rigorous evaluations of classification models for medical disorders and comprehensive datasets aiding facial recognition advancements, this collection fosters interdisciplinary dialogue and propels the boundaries of knowledge in the field.

Chapter 1 delves into the intricate dynamics of synchronization among various technological advancements, focusing on the black box that influences object detection optimization. Through an extensive review spanning from 2019 to 2023, the study explores this black box in two distinct dimensions. Firstly, it elucidates diverse research methodologies and innovative features employed in literature to achieve object detection in remote sensing images. Secondly, it meticulously evaluates deep learning models' effectiveness, drawing attention to limitations and recent trends in object detection. By providing in-depth recommendations and future directions, this chapter endeavors to propel the trajectory of deep learning innovation in remote sensing.

Presenting a novel approach to ship classification in optical remote sensing imagery, chapter 2 underscores the significance of distinguishing between empty and full container ships. Leveraging deep cognitive modeling techniques and renowned pre-trained deep learning models, the study showcases the efficacy of InceptionV3 in particular, following strategic data augmentation to address class imbalance challenges. Rigorous analysis of evaluation metrics demonstrates the superiority of the proposed models, contributing significantly to the field of ship classification methodologies in ORS imagery. The findings hold implications for maritime applications and decision-making processes, emphasizing the pivotal role of deep cognitive modeling in addressing complex classification tasks.

Chapter 3 provides a comprehensive exploration of facial identification algorithms utilized in drones, addressing the burgeoning demand for face recognition technology across diverse domains. Through a comparative analysis of various techniques, including neural networks, PCA, LBPH, and others, the study elucidates the potential for hybrid algorithm design to meet real-time application requirements. Additionally, insights into face datasets collected via drones enrich the discourse, offering valuable guidance for facial recognition technologists seeking to enhance system efficacy and resilience in dynamic operational environments.

Technological advancements have ushered in an era of unprecedented data scale, resolution, and availability in remote sensing. Chapter 4 navigates the shifting landscape of control and dissemination of imagery from government to private sectors, facilitated by digital infrastructures. However, the confluence of these advancements raises significant legal and ethical concerns, particularly regarding privacy violations and individual rights. Through a historical lens, recent innovations, and discussions on potential ramifications, this chapter underscores the imperative for self-regulation within the remote sensing community to balance economic objectives with ethical considerations.

Chapter 5 elucidates the transformative potential of capsule networks in modern computer vision and machine learning domains. By encoding features based on hierarchical relationships, capsule networks offer a promising alternative to traditional

CNNs, particularly in contexts where spatial placement is crucial. Through a comparative analysis of diverse capsule network designs and their applications, the chapter contributes to understanding the advantages, limitations, and modifications of this innovative approach, paving the way for future advancements in pattern recognition and image classification.

Chapter 6 delves into the paradigm shift brought about by data-driven advanced machine learning methodologies in the realm of modern intelligent technology. By leveraging observed data samples to forecast future trends, these methodologies underscore the need for continual enhancements in algorithmic capabilities and technological infrastructure. Through discussions on recent technologies and trends, accompanied by illustrative examples, the chapter offers insights into the evolving landscape of artificial intelligence, setting the stage for further innovation and application in diverse domains.

Conducting a comparative analysis of current 2D face databases, chapter 7 introduces a dataset of multi-pose facial photographs, meticulously curated over five months in real-world scenarios. Comprising diverse stances and captured using sophisticated tools, this dataset offers invaluable insights into the effectiveness and resilience of upcoming face detection and recognition systems. By bridging the gap between theoretical concepts and practical applications, the chapter contributes to advancing facial recognition technologies, with implications for a wide range of domains.

Chapter 8 addresses the challenges in facial recognition systems within the realm of cricket, presenting a comprehensive dataset of cricket player images across various scenarios. By evaluating cutting-edge facial recognition algorithms on this dataset, the chapter offers critical insights into system efficacy and potential areas for development. Through extensive experimental analyses, the chapter underscores the significance of addressing the black box in facial expression datasets, thereby contributing to enhanced player surveillance and identification in sports, particularly cricket.

Chapter 9 evaluates the efficacy of classification models, including K-nearest Neighbor, Random Forest, AdaBoost, and Neural Networks, in categorizing medical disorders. Through comprehensive performance measures and ROC analysis, the study demonstrates the superiority of Neural Network models, particularly those with specific hidden neuron configurations. However, limitations such as dataset constraints and the absence of interpretability analyses highlight avenues for further research and refinement in classification methodologies, particularly in distinguishing COVID-19 from Pneumonia.

Our target audience comprises professionals and researchers immersed in the realms of computer science, cognitive modeling, image processing, remote sensing, and beyond. Through a curated selection of topics ranging from border monitoring to object localization, georeferencing, and mathematical modeling, this book endeavors to serve as a comprehensive resource for those navigating the complex landscape of remote sensing image analysis.

As editors, we are deeply grateful to the contributors who have shared their expertise and insights to make this endeavor possible. We hope that *Deep Cognitive Modelling in Remote Sensing Image Processing* will inspire further exploration, innovation, and collaboration in this dynamic field.

Sadique Ahmad
Prince Sultan University, Saudi Arabia

Muhammad Shahid Anwar
Gachon University, South Korea

Ala Saleh Alluhaidan
Princess Nourah bint Abdulrahman University, Saudi Arabia

Mohammed A. El-Affendi
Prince Sultan University, Saudi Arabia

Chapter 1

Deep Strategy of Object Detection in Remote Sensing Images:
A Systematic Review

Sadique Ahmad
https://orcid.org/0000-0001-6907-2318
Prince Sultan University, Saudi Arabia

Ala Saleh D. Alluhaidan
https://orcid.org/0000-0001-6829-9705
Princess Nourah bint Abdulrahman University, Saudi Arabia

Mohamed A. El Affendi
https://orcid.org/0000-0001-9349-1985
Prince Sultan University, Saudi Arabia

M. Shahid Anwar
https://orcid.org/0000-0001-8093-6690
Gachon University, South Korea

ABSTRACT

Recently, the demand for satellite image analysis increased with the recent advancement in various research areas (e.g., improvements in remote sensing image resolution, object recognition ideas, and deep learning techniques). Articles report new trends in object detection using remote sensing images, such as multi-temporal scene classifications, semantic segmentation, multiresolution, and large-scale optical image analysis. However, this study observed a black box of synchronization among the advancements in various technologies (i.e., new ideas in satellite image analysis) and advancements in deep learning techniques. Without investigating this black box, the optimization of object detection will be out of track. So, to keep the deep learning innovation on track, the primary goal of the current work is to explore the aforementioned black box to advance object detection in remote sensing images. In contrast, focusing on a specific technique, this study explores the black

DOI: 10.4018/979-8-3693-2913-9.ch001

box while reviewing 150 articles from 2019 to 2023. First, the study highlights research methodologies and novel features of literature to achieve object detection in remote sensing images. Second, it evaluates effective deep learning models and assesses various featured studies to draw a clear picture of limitations and recent object detection trends to provide in-depth recommendations and future directions.

1. INTRODUCTION

The rapid increase in global security issues, urbanization, and natural disasters have created numerous societal challenges. On the one hand, the community needs an autonomous land surface monitoring system for various problems, such as human rights violations, border conflict management, and terrorist attacks in different countries (Al-Bilbisi, 2019; Charrua et al., 2021; Vogels et al., 2019; Levin et al., 2020). On the other hand, urbanization and environmental changes lead to drought, ultimately increasing plant mortality, reducing agriculture production, and causing various diseases in crops (Wang et al., 2021; Mzid et al., 2021; Alhichri et al., 2021; Wen et al., 2021; Olson and Anderson, 2021; Yi et al., 2021; Osco et al., 2021). Articles produce new trends and deep strategies in remote sensing data analysis to address these issues. (Vivekananda et al., 2021; Ru et al., 2020; Du et al., 2019; Xu et al., 2019; Wang et al., 2019). However, extensive literature studies depict that some models are good in prediction accuracy, while few are significant in rapid image analysis with low accuracy. These studies show that deep learning models are insignificant during satellite image analysis, even though they produce good results while evaluating classical photography or aerial imagery. Satellite images have very high resolutions, while classical photography has a relatively small amount of pixels. Also, classical images have red, green, and blue colors, while satellite images can have multiple red, green, and blue channels. The object is very small in the satellite image, consisting of a few pixels (Signoroni et al., 2019; Gu et al., 2019; Zhang et al., 2019; Silver et al., 2019; Pan et al., 2020). Satellite image has specific data with one shot, such as dealing with rain, cloudy, sunny, and different satellite positions. Also, we have to assess the Geo-referencing of an image. It means that we must know the object's location in the picture. While evaluating remote sensing satellite images and various deep object detection strategies, we observed a black box of synchronization among the advancements in different technologies. The three main points of this black box are given below.

- Technologies for climate change monitoring.
- New ideas in satellite image analysis.
- Advancements in deep learning techniques.

The aforementioned extensive literature analysis shows that without investigating the black box, the optimization of object detection will be out of track. So, the current study proposes two basic questions to explore the black box and to keep the deep learning innovation on track. These questions are given below.

- **Q1: What are the trended in-depth strategies to achieve object detection? How the particular object in conventional photography and remote sensing image is further explored?**
- **Q2: What are the limitations of trended deep learning models while processing remote sensing images?**

In contrast, focusing on a specific technique, this study explores the proposed three black box questions in two-folds based on reviewing 150 articles from 2019 to 2023. The first fold examines the innovations and strategies of the selected featured studies. Second, the study highlights research methodologies and those features of the literature that played a key role in object detection using remote-sensing images. It also evaluates effective deep learning models, innovations, and targeted application areas of the selected studies. Finally, the study assesses various featured studies to draw a clear picture of limitations and recent object detection trends to provide in-depth recommendations and future directions. The rest of the work is organized as follows. Section two represents the literature review, section three discusses the review methodology, and section four demonstrates the review results and recommendations.

2. REVIEW METHODOLOGY

The review process is started from the initial screening within the scope of the selected research questions. We focused on recent and state-of-the-art contributions to remote sensing data analysis, object recognition, and deep learning techniques. In addition, Figure 1. demonstrates the current review methodology with the essential modules.

The current study comprehensively evaluates the prior contributions while following the procedures demonstrated by Petersen et al. (2008) and Keele et al. (2007). Also, the review methodology of Keele et al. (2007) is used to analyze the earlier work, and Petersen et al. (2008) method is used for study mapping. The review explores the black box in featured articles contributing to remote sensing object detection and scene classification either directly or indirectly.

Also, it put four research questions to illustrate the primary goal of a particular article. The research questions help us choose relevant work for screening and examining the primary challenges in remote sensing data analysis. In addition, each research question consists of a list of keywords to investigate deep learning models, interesting object recognition ideas, and applications in society (Ding et al., 2020; Sun et al., 2020; Xie et al., 2020; Li et al., 2021; Peng et al., 2019). We search for qualitative published work via these keywords, including peer-reviewed journal papers, conference proceedings, and qualitative book chapters. The procedure of articles' selection for screening and reviewing is described in the following section.

2.1. Research Questions

- Q1: What are the trended in-depth strategies to achieve object detection? How the particular object in conventional photography and remote sensing image is further explored?
- Q2: What are the limitations of trended deep learning models while processing remote sensing images?

2.2. Searching Keywords

- Recent advancement in AI towards Climate Change.
- Multi-temporal Scene Classification and Climate Change Applications.
- Recent Trend in Remote Sensing Technology for Climate Change.
- Multi-resolution and Large-scale Optical Image analysis.
- Deep Model role in Climate Change, Urbanization, and Agriculture Production.
- Objects Recognition in Remote Sensing Images.
- Different Features of Urbanization.
- Urbanization in Remote Sensing Images.
- Detecting Objects in Satellite Images. ● Detecting Objects in Remote Sensing. ● QGIS, Python, and Object Detection.
- Python API for Object Detection in Remote Sensing Images.
- Segmentation of Remote Sensing Images.
- Pixel Based Image Segmentation.

2.3. Screening

For intensive review, the study selects various studies for exploration based on the following terms and conditions.

- The first task of the team is to select relevant articles.
- Articles should have features as follows, comparatively good citations, good citations within a year, authors must critically evaluate it in their papers, and prioritize the article published from 2020 to 2024. • Eventually, an abstract-level review is performed to extract the target data. Targeting the particular paper for rapid review is a pilot project for further in-depth exploration and data extraction. In the in-depth review, we mainly focus on the challenges, motivations, and featured contributions.

3. INSIGHTFUL ANALYSIS

Based on the aforementioned three research questions, we achieved various findings and analyses that will definitely contribute to the optimization process of object recognition in remote-sensing images. In the Table 1, the proposal shows an In-depth Analysis of Findings, Research Methodologies, and Feature Extraction. These features can be used to achieve the goal of object recognition in remote sensing images. On the other hand, Table 2 represents the in-depth investigations and analysis of deep learning models, featured innovations of the particular article, and application areas (climate change). Furthermore, the best feature of this table is it highlights limitations in the prior studies and gives future directions. Figure 2 depicts different types of innovations and their intensity in digital libraries, and Figure 3 shows the intensity of effective deep-learning models in various application areas.

3.1. Question No. 1: What Are the Trended In-Depth Strategies to Achieve Object Detection? How Is the Particular Object in Conventional Photography and Remote Sensing Image Further Explored?

Under the investigation of question 1, the current work investigates the major contributions of different fields featured studies based on deep learning strategies. These strategies are shown in Table 1. The significant attributes reflect the core theme of a particular research article to make it easy for future researchers. The primary reason behind highlighting the significant contributions is to provide recent trends, the best track in research, and recent novelty in remote sensing image analysis. They contributed to different environments and application areas with excellent results. Each attribute has been investigated and used by many researchers. Also, every attribute is cited, and the future reader can find the particular article as well as related studies. Table 1 shows various attributes, i.e., strategies, innovation, frequency of publications, year of publication, deep learning model, and the particular publisher. The method section shows the acronym of a particular method. The objective section reflects the specific goal to achieve or a high-level challenge to address, such as addressing border monitoring issues, conflict management, and natural disaster monitoring. Furthermore, the methodology shows the approach and strategy adopted in research to achieve a particular goal or address specific challenges for the betterment of society. Also, figure 2, discusses the frequency and intensity of innovations in various digital libraries. The primary frame of references is selected to demonstrate the recent trend in remote sensing image analysis and object detection area of research. Their selection mainly depends upon the recently addressed challenging research gap and the effective object recognition ideas. Also, the major frame of reference includes various features of a particular study.

3.2. Question No. 2: What Are the Limitations of Trended Deep Learning Models While Processing Remote Sensing Images?

Deep learning plays a crucial role in analyzing the object in satellite images. Recent studies report many deep learning approaches that are claimed to be adequate for

Figure 1. Review framework

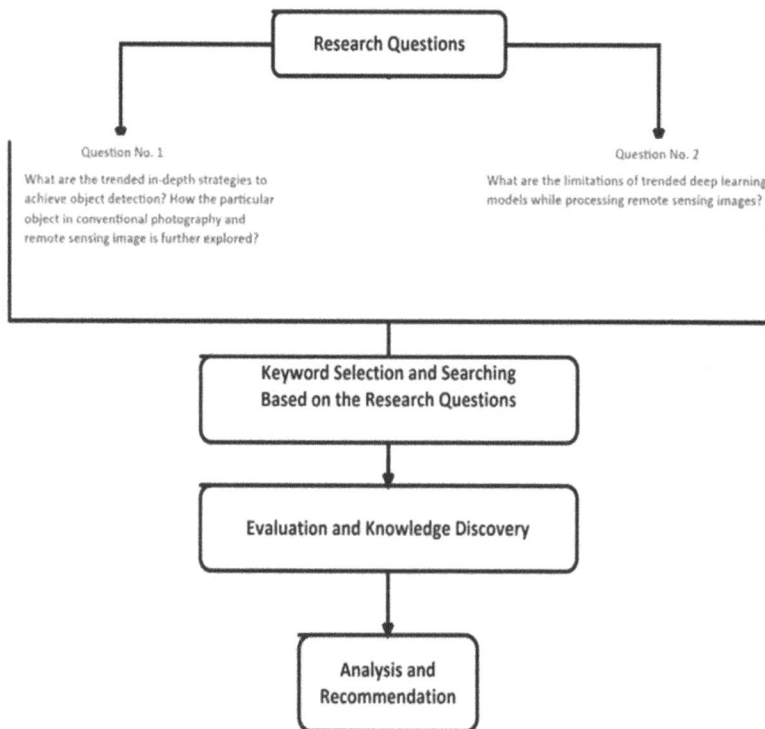

Table 1. Innovation analysis based on various deep learning strategies

Strategy	Innovation	Frequency of Publications	Year of Publication	Deep Learning Models	Publisher
Multitemporal scene classifications	1. Modification 2. New Method with existing dataset. 3. Iterative Framework 4. Innovation based on accuracy improvement.	25	2021-2023	1) VGGnet, 2) LSTM, 3) Deep Seasonal Network, 4) 4D U-Nets, 5) Single Tensor Network.	Elsevier IEEE Springer
Segmentation	1. Review work 2. Data analysis 3. Innovation based on processing time	20	2022-2023	1) CNN, 2) Transformer Complementary Network, 3) Autoencoder	Elsevier IEEE MDPI
Semantic segmentation	1. Modification 2. New Method with existing dataset. 3. Analysis of application 4. Accuracy and processing time	32	2018-2023	1) Autoencoder, 2)VGGnet, 3) Scaling-up Remote Sensing Segmentation	Elsevier IEEE MDPI Springer Wiley
Multiresolution	1. New Method with existing dataset.	24	2022-2023	1) YOLO with tiling on client/ server architectures 2)LSTM, 3) Multi-resolution CNN	Elsevier IEEE
Large-scale optical image analysis	1. Comparison 2. Review work 3. Psychological findings	38	2018-2023	1)CNN, 2) YOLOv3, and YOLOv4 3) VGGNet 4)LSTM	Elsevier IEEE MDPI Springer Wiley
Multi-scale object detection	1. Review work 2. Data analysis	12	2022-2023	1)YOLOv7, 2) EfficientNet, 3) DenseNet	IEEE MDPI
Scaling-up Remote Sensing Segmentation	1. Comparison 2. Modification 3. Analysis of various application	23	2019-2023	LSTM	Elsevier IEEE MDPI Wiley Hindawi
Global–local contextaware network	–global1. Modification 2. New Method 3. Review work	10	2023-2023	1) YOLO versions, 2) VGGNet.	IEEE Hindawi
MultiKernel Dilated Convolution and Transformer	1. Review work 2. Innovative Model	5	2022-2023	1) DenseNet, 2) Self Organizing Maps (SOMs), 3) VGGNet.	MDPI Hindawi
One-Stage Object Detection of Remote Sensing Images	1. New Methods 2. Comparative analysis 3. Innovation based on accuracy improvement and optimization.	5	2023-2023	1) InceptionV3, 2) YOLOv7, 3)CNN.	IEEE Elsevier MDPI

Figure 2. This figure shows different types of innovations and their intensity in digital libraries

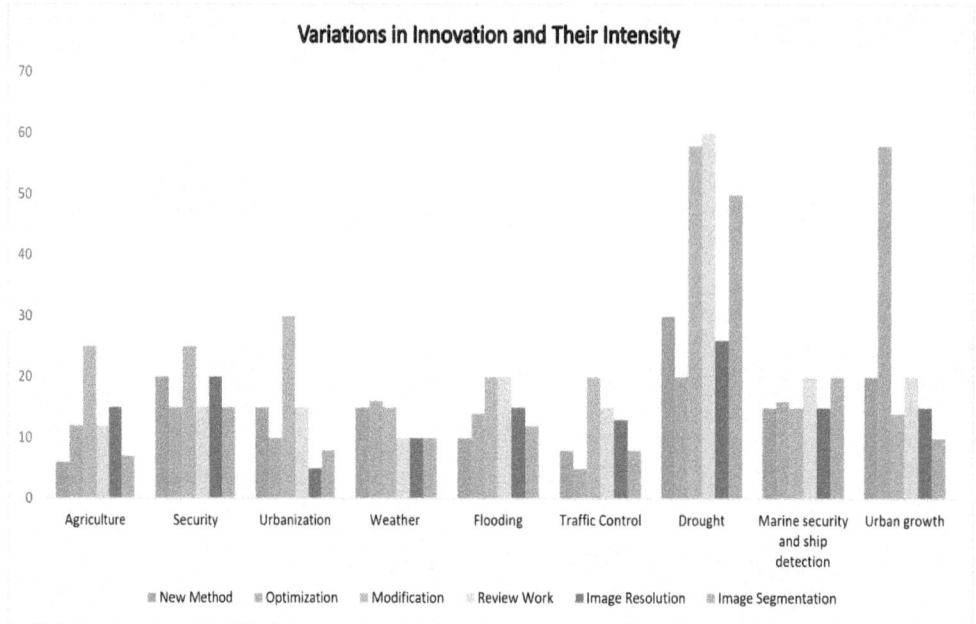

different scenarios and circumstances. Therefore, the basic deep learning techniques for evaluating such areas are critical to giving a clear pathway to future researchers in remote sensing image analysis. These basic techniques include but are not limited to various Yolo versions, R-CNN, Fast R-CNN, Faster R-CNN, Autoencoders, Deep Belief Networks, Recurrent Neural Networks, Deep Self-Organizing Maps, Random Forest, and last but not the least, Sparse Auto-encoders. These are the commonly used deep learning models in various research areas. Along with that, the studies show different ideas to achieve the goal of object recognition, e.g., single-shot detection, scene detection, multi-temporal scene classification, and semantic segmentation. Table 2 shows the contribution of deep learning in the form of new methods, modified models, optimization of existing methods, and the intensity of publications. The intensity shows how many and how frequently the research has been published. Along with that, the methodology attributes are also mentioned. It is an addition to the existing goal of this table 2. It is the best pathway for future researchers while addressing the methodology attributes.

Articles are saturated with machine learning and deep learning techniques to recognize objects in geospatial data. These techniques are applied to geospatial images in three broad categories, i.e., classification, clustering, and regression problems. The CNN or capsule network detects objects, but we must consider three things. The image size will be huge, so the deep learning techniques cannot process it simultaneously. Thus, it needs performance segmentation before the object detection and convolution process. Second, we have to deal with more color along with the RGB. Third, we need to evaluate the geo-position of an object on the earth's surface (Hossain and Chen, 2019; Fu et al., 2020; Zhang et al., 2020; Liu et al., 2020; Cui et al., 2020; Ma et al., 2019). They also used vector machine algorithms to create land-cover classification layers with classification objectives. In the clustering problem, they identify meaningful clusters within the geospatial image (Tokarev et al., 2020; Mangalraj et al., 2020; Du et al., 2019; Pires de Lima and Marfurt, 2019; Wang et al., 2019; Yin et al., 2020). Also, articles report geographically weighted regression to model spatially varying statistical associations. With the advancement in computer vision, the literature is evidenced by new models such as R CNN, Fast R CNN, and Faster R CNN, which outperformed the existing deep learning models. However, recent studies reveal that the YOLO (You Only Look Once) technique has outperformed all the existing techniques in computer vision. It splits the whole picture into multiple segments, resulting in multiple object detection in a single frame. Table 3 represents the frequency of publication in various digital libraries while targeting specific keywords. Finally, Table 5, gives the reflections of contribution frequencies using various deep-learning models. It also shows the contributions in different application areas.

4. LIMITATIONS AND RECOMMENDATIONS

In Ru et al. (2020), the authors propose the Correlation Based Fusion (CorrFusion) method to integrate and fuse those factors that are highly correlated with each other. Also, deep features have been extracted with CNN. Lastly softmax layers will finalize the fuses scene. Such contributions pave the way for an important concept of synchronization and transfer learning. Every model performs their best to achieve object detection and further explore based on Multi-temporal scene classification, segmentation, semantics segmentation, Multi-resolution, Large-scale optical image analysis, Multi-scale object detection, Global local aware network, Multi-Kernel Dilated Convolution and Transformer, and One-Stage Object Detection of Remote Sensing Images. In addition, in Ahmad et al. (2021), the authors propose Deep Frustration Severity Network (DFSN) to explore frustration severity in two 34 partitions. Bayesian inference methods were used to estimate the posterior probability of frustration severity. This model gives an important concept of iterative calculation of different segments of images. Such an iterative process can be applied to fuse highly correlated features of images or objects in the images. Finally, the current study highly recommends YOLOvs (versions) to boost the processing of image

Table 2. In-depth analysis of ML and DL models: Limitations and future directions

ML or DL Model	Innovation	Application Area	Limitations and Future Directions	Articles
VGG16	Innovation based on accuracy improvement and optimization.	Conventional image processing, such as face detection, and action detection.	It is hard to train because VGGNet is comparatively slow in training and testing. It is not recommended for the remote sensing image analysis.	5
ResNet	New Methods	Wildfire, Climate Change (agriculture, plant mortality)	Accuracy is while classifying images having features as follows, 1) spectral signatures, 2) temporal changes are frequent, 3) satellite images with complex texture, and 3) finally, the spatial relationships between objects. General mistakes such as misclassification of rivers, lakes, crops, urbanization and buildings, and forests (41).	13
YOLOv7	1. Comparative analysis 2. Comparison based on processing time.	Marine security and ship detection	A complete comparison of YOLOv3, YOLOv4, YOLOv5 and YOLOv6 can be found here (Laroca et al., 2018; Kasper-Eulaers et al., 2021; Zhang et al., 2022; Bochkovskiy et al., 2020; Redmon and Farhadi, 2018; Bohara et al., 2021; Jia et al., 2021; Lee and Kim, 2020; Zhu et la., 2021; Khan et al., 2020; Abdullah et al., 2022a, 2022b; Anwar et al., 2019; Ahmad et al., 2022; Ahmad et al., 2018; Rahim et al., 2023; Hosni et al., 2019, 2018)	3
Autoencoder	s Modification in the existing work	Climate Change(Urban growth)	Limitations in preserving latent space meaning during image analysis.	17
DenseNet	Comparison based on accuracy	Climate Change(Urban growth)	Accuracy is while classifying images having features as follows, 1) spectral signatures, 2) temporal changes are frequent, 3) satellite images with complex texture, and 4) finally, the spatial relationships between objects. General mistakes such as misclassification of rivers, lakes, crops, urbanization and buildings, and forests (41).	21
Convolutional Neural Networks (CNNs)	Modification in the existing models	Climate Change (Drought, Agriculture)	Traditional CNN is quite effective in processing conventional photography; however, it takes a huge amount of time to process satellite remote-sensing images. Need a large dataset to train against the particular area of interest in the images. Hard to optimize because of their black box feature. Can be overfitted due to memorizing noise in the data.	12
EfficientNet	Innovation based on accuracy improvement and optimization.	Climate Change(weather prediction)	Accuracy is while classifying images having features as follows, 1) spectral signatures, 2) temporal changes are frequent, 3) satellite images with complex texture, and 4) finally, the spatial relationships between objects. General mistakes such as misclassification of rivers, lakes, crops, urbanization and buildings, and forests (Adegun et al., 2023).	6

Figure 3. This figure shows the intensity of effective deep learning models in various application areas

Intensity of Effective Deep Learning Models in Various Application Areas

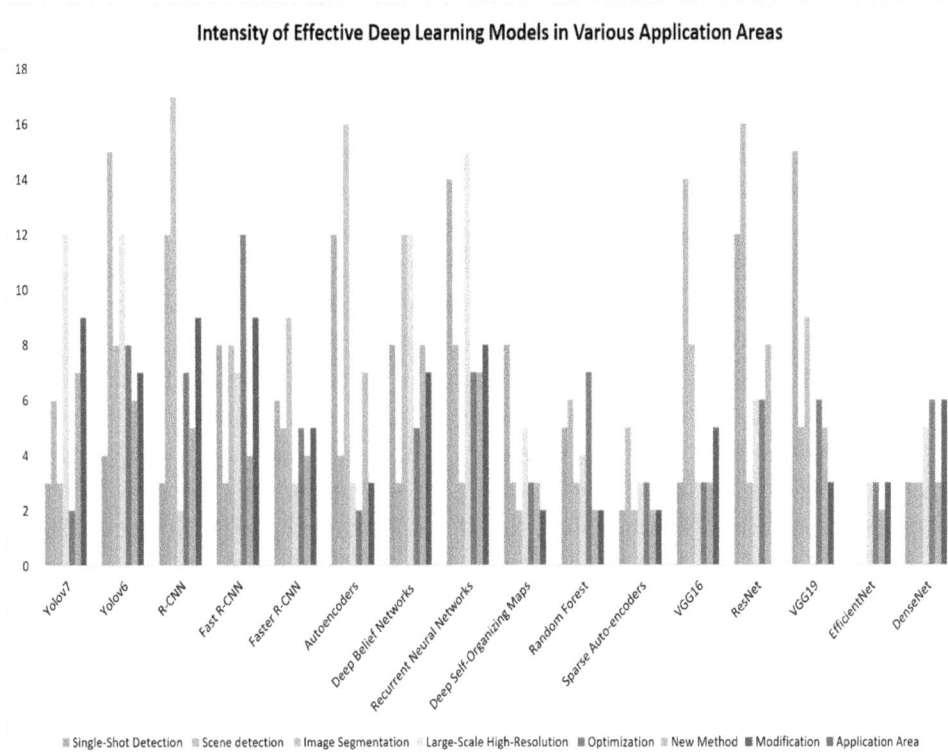

analysis because adding an iterative model like DFSN and transferring other features from aforementioned deep models will slow down the image processing, such as VGGNet, DenseNet, LSTM, CNN, autoencoder etc (Haq et al., 2019, 2022, 2023, 2024; Ullah et al., 2019, 2022; Munawar et al., 2019; Ahmad et al., 2018, 2020; Akhtar et al., 2022; Khan et al., 2023; Akhtar et al., 2022).

5. SUCCESS AND IMPACT OF THE EXPECTED RESULTS

The autonomous object recognition system using remote sensing images will revolutionize different domains in sustainability, such as monitoring climate change, enhancing agriculture productivity, preventing urbanization, focusing on natural resource management, and flooding prevention. It will also offer various advantages, such as providing researchers with a mindset focusing on optimizing the existing conventional systems to timely detection and mitigation of environmental challenges. Our review findings show that Object detection models will be critical in enhancing agricultural productivity. Such a system will enable us to monitor crop diseases and identify critical stages of spray to enhance crop health. Also, it will prevent our crops from any damage by optimizing irrigation and fertilizer usage. The farmers can use the autonomous crop monitoring system to make informed decisions regarding irrigation, spray, and fertilizer usage.

Literature is saturated with many findings showing climate change is a big threat to sustainable development nowadays (Memon et al., 2023; Rahim et al., 2024; Khan et al., 2023; Hosni et al., 2020; Anwar et al., n.d.; Gui et al., 2024; Sheng et al., 2024; Jiang et al., 2024; Lin et al., 2024; Zhang et al., 2024). It is a global concern that poses an influential threat to human well-being, animals, and plants. We need to explore climate dynamics while investigating and addressing this global issue. It demands a thorough understanding of the Earth's dynamics, accurate data collection, and compelling satellite monitoring systems. Such a system is only possible with an in-depth understanding of various loopholes in deep learning models, challenges in remote sensing image analysis, and targeting the aforementioned black box between existing research contributions (Fu et al., 2024; Zhang et al., 2024; Sun et al., 2024; Duan et al., 2024; Lian et al., 2024; Sagar et al., 2024; Li et al., 2024; Nan et al., 2024). Such remote sensing object detection systems will be powerful tools with massive edges in controlling climate change. Robust object detection system plays a critical role in urbanization management and control over urbanization, such as identifying buildings, roads, and infrastructure. We can use this information for a remarkable recommendation system. Government policymakers enhance their policies by tracking urban growth patterns. Furthermore, the government can optimize resource allocation and ensure sustainable administration and management of urban areas. Also the figure 4 show the intensity of contributions focusing the application areas which are shown by various keywords.

7. KEY SUMMARIES OF FEATURED RESEARCH CONTRIBUTIONS.

Table 5 illustrates key insights into recent trends in object detection using remote sensing images. Also, it highlights the contributions of individual features, shedding light on their relative importance in a specific context, which could be crucial for understanding trends in object detection using remote sensing image analyses. Meanwhile, Table 6 shows the acronyms and abbreviations used in the study.

Table 3. Frequency of publications in various digital libraries based on the proposed keywords

Keywords	ACM	IEEE	Springer	TSP	Elsevier	Wiley	MDPI	Frontiers
Recent advancement in AI towards Climate Change.	22	43	4	7	23	9	45	30
Multi-temporal Scene Classification and Climate Change Applications.	7	5	8	12	9	12	17	11
Recent Trend in Remote Sensing Technology for Climate Change.	27	33	13	15	19	10	59	53
Multi-resolution and Large-scale Optical Image analysis.	3	8	23	16	6	8	26	33
Deep Model Role in Climate Change, Urbanization, and Agriculture Production.	33	54	25	28	22	14	55	63
Objects Recognition in Remote Sensing Images.	5	34	21	55	32	17	25	28
Different Features of Urbanization.	12	4	32	12	54	37	31	35
Urbanization in Remote Sensing Images.	22	43	7	5	27	33	3	8
Buildings in Satellite Images.	34	21	55	4	32	32	12	5
Roads in Satellite Images.	4	65	23	5	9	23	32	29
Vehicles in Satellite Images.	8	23	54	25	28	22	14	24
People in Satellite Images.	27	33	55	4	25	32	12	8
Detecting Objects in Satellite Images.	34	21	55	32	17	43	0	1
Detecting Objects in Remote Sensing.	33	54	5	34	12	4	8	4
Deep Learning and Remote Sensing.	0	8	67	3	2	6	6	0
Deep Learning and Remote Sensing Object Recognition.	9	0	25	28	22	14	5	27
Remote Sensing Object Recognition Optimization.	33	3	23	5	9	10	4	32
QGIS, Python, and Object Detection.	32	12	54	25	28	8	5	9
Python API for Object Detection in Remote Sensing Images.	23	32	25	28	7	9	7	0
Segmentation of Remote Sensing Images.	5	0	21	55	32	17	4	6
Pixel Based Image Segmentation.	7	23	32	12	54	37	21	9
Object-Based Image Segmentation.	24	5	8	43	8	9	0	23
Geospatial Image Analysis	6	12	25	12	6	0	23	7

Figure 4. The above graph shows frequency of contributions focusing in various applications areas and keywords. Also, different digital libraries are given.

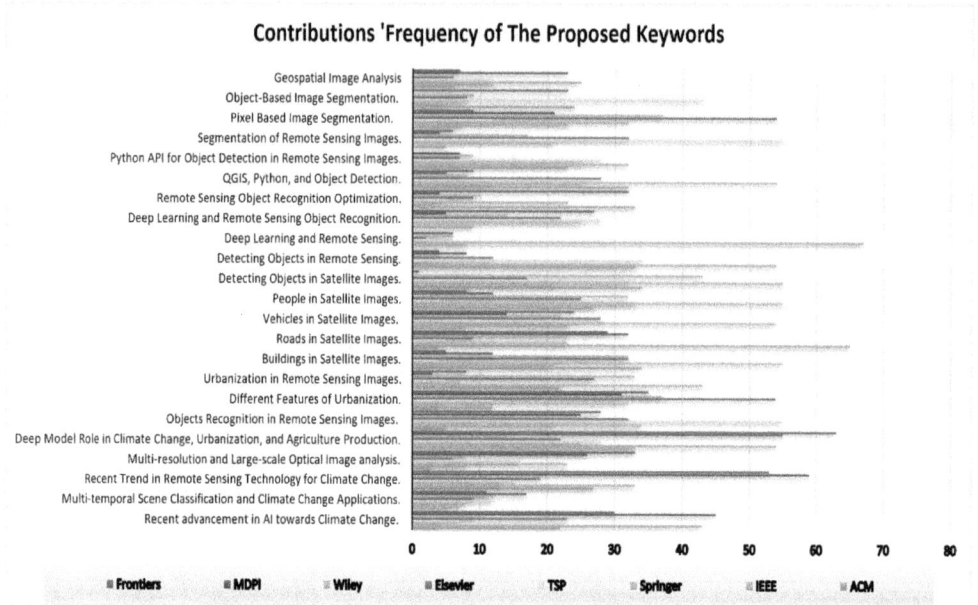

8. CONCLUSION

This study focused on the black box of synchronization between the three major technologies, i.e., 1) technologies for climate change monitoring, 2) new ideas in satellite image analysis, and 3) advancements in deep learning techniques. To keep our task on track, we selected three in-depth research questions. So, the black box was explored in four-folds while reviewing 750 articles from 2015 to 2023. In the first fold, we evaluated the innovations and strategies of the selected featured studies. The outcomes of this fold are shown in Table 1, such as an In-depth Analysis of Findings, Research Methodologies, and Feature Extraction. Second, it evaluates effective deep learning models, innovations, and targeted application areas of the selected studies. The results of this module can be found in Table 2 which shows the in-depth investigations and analysis of deep learning models, featured innovations of the particular article, and application areas (climate change). The limitations in the prior studies and future directions demonstrated by Table 2 play a critical role in influencing object detection in remote sensing images. Along with that, the studies show different ideas to achieve the goal of object recognition, e.g., single-shot detection, scene detection, multitemporal scene classification, and semantic segmentation. Table 3 shows the contribution of deep learning in the form of new

14

Table 4. Intensity of deep learning model in various application areas

	Single-Shot Detection	Scene Detection	Image Segmentation	Large-Scale High Resolution	Optimization	New Method	Modification	Application Area
Yolov7	3	6	3	12	2	7	9	Security, Urbanization, Agriculture
Yolov6	4	15	8	12	8	6	7	Security, Urbanization, Agriculture, traffic, Weather, flooding
R-CNN	3	12	17	2	7	5	9	Agriculture
Fast R-CNN	8	3	8	7	12	4	9	Traffic, Urbanization, Security
Faster R-CNN	6	5	9	3	5	4	5	Security, Urbanization, Agriculture, traffic, Weather, flooding
Autoencoders	12	4	16	3	2	7	3	Flooding, and weather
Deep Belief Networks	8	3	12	12	5	8	7	Security, Urbanization, Agriculture, traffic, Weather, flooding
Recurrent Neural Networks	14	8	3	15	7	7	8	General purpose
Deep Self-Organizing Maps	8	3	2	5	3	3	2	General purpose
Random Forest	5	6	3	4	7	2	2	General purpose, Security
Sparse Auto-encoders	2	5	2	3	3	2	2	Security, flooding and weather
VGG16	3	14	8	3	3	3	5	Marine security and ship detection
ResNet	12	16	3	6	6	8	0	General purpose
VGG19	15	5	9	0	6	5	3	Marine security and ship detection
EfficientNet	0	0		3	3	2	3	General purpose
DenseNet	3	3	3	5	6	3	6	Agriculture

Table 5. Key summaries of feature contributions.

Method	Objective/ Application Area	Methodology	Feature	Conf Journal Book Chapter
SM-Review	Climate Change	Review based on massive articles from 2015 to 2022. Various ML Models were for analysis and monitoring. The main focus points is Change and dynamic monitoring.	Shoreline Extraction and the coast line analysis.	Journal
ELST- GEE- RSD	Climate Change. Vegetative Land	Establishing association among LST, NDVI, and NDBI Estimation of the surface temperature Deploying Machine Learning Models to measure the NDVI and NDBI Satellite remote sensing data were used. Java script algorithm for analysis.	Estimating the surface temperature establishing association among LST, NDBI and NDVI.	Book Chapter
Spatiotempo variations	ral Climate Change	1. Extraction of phenological parameters from remote sensing data. 2. The authors have used time series data. 3. This is from 2000 to 2017 (MODIS-NDVI). 4. The authors analyzed 4 remote sensing phenological parameters were analyzed (SOS, EOS, MOS, and LOS).	1. Indicating climate change 2. various parameters to detect vegetation.	Journal
MDG- Surkhandary	Climate a Change	Analysis based on 2 stations data	1. climate-related hazards 2. High Drought	Conference
CGIS-RSA	Novel Method for climate change detection. Region types as well as soil types	In this research methodology, region types are as follow, 1) very long history 2) young soil with high diversity of	1. Monitoring and investigating landscape 2. Monitoring time-wise changes in land surface.	Journal
U-M- Climate change	AI for Climate Change	Analyzes the influence of vegetation over LST Second, the impact or influential nature of LST over the human health Landsat thermal Dataset was used for this research.	1. focusing on the temperature of land surface. 2. Climatology in Urban areas 3. Development in Urban areas	Journal
IA-RSGA	Climate change and food security	Here the main goal is food security. 1. Dataset is from 2002–2021. 2. Data belong to Landsat, MODIS remote sensing satellite images In technical methodology, 3. The authors used deep learning CNN for predictions. 4. DL-CNN based on the Google Earth Engine. 5. In the second stage, the authors analyzed the statistical association among various variable such, topographical variable, climatic and geospatial variables.	1. Influence of climate changes on food. 2. Food security 3. Diverse impact of climate change	Journal
RSRSMAGCC	Climate Change	The main focus is SRS. The authors explore their critical role in climate change research. The authors also investigated the advantages SRS in the aforementioned areas of research, such as climate change.	1. Climate change 2. Remote sensing data for climate change 3. Global climate change 4. adopt remote sensing to global climate change.	RSRSMAGCC
RSFS-Agri	Climate change and sustainable agriculture	Review work focusing on sustainable agriculture.	1. Sustainability of agriculture, i.e., crops and land 2. Precision farming 3. Extensive Review	A-Hyper- RSFS-Agri

continued on following page

Table 5. Continued

Method	Objective/ Application Area	Methodology	Feature	Conf Journal Book Chapter
NR-TMB	Border Monitoring	(1) It main focuses on multitemporal imagery.	Change in scene detection	NR-TMB
BorderSense	Border Monitoring	hybrid wireless sensor network architecture.	BorderSense utilizes the most effective sensor based methods and technologies, such WMSNs (wireless multimedia sensor networks) as well as WUSNs which is referred to as wireless underground sensor networks	BorderSense
RS-IPS	International Peace and Security	Here the author mainly focuses on reviewing and exploring the major role of remote sensing technologies in peace and controlling security issues	Review the research concepts and implementation based on the remote sensing ideas. Highlighting future challenges within the application area and remote sensing techniques	RS-IPS
DES	detection of excavated soil related to illegal tunnel activities	Deep CNN high resolution multispectral satellite images Spatial and spectral information incorporated to CNN	tunnel digging activities imbalance learning technique	Conference
RSGC	Validating the Remotely Sensed Geography of Crime	Reviewed 61 articles on crime consequences of misidentifying crimes	needless intrusion, intimidation, surveillance or violence. Potential hazards of the use of Google Earth to identify crimes and criminals.	Journal
PRSDT	Border security (Ship Detection Technique)	It detects pixels.	Ship detection using wake signature as a feature set.	Journal
FOLDOUT	Border surveillance	Review FOLDOUT system for border surveillance.	Reviewing multiple literatures.	Book Chapter
HBDL	Marine security (Ship detection)	Employs YOLOv3 Semantic segmentation and image segmentation Hashing (SHA-256) is applied	Inclusion of hashing. Ship count Location of bounding box	Journal
DLA	Climate Change Agricultural Field Extraction	multi-task segmentation model ResUNet-a linked UNet backbone	Detection of field borders Field mapping, Field borders, Individual fields based on Landsat-8 and Sentinel-2 images	Journal
HNM	DOTA: A Large-Scale Dataset for Object Detection in Aerial Images. An addition to Climate changes	It develop a hybrid network, such as TransConvNet,	adaptive feature fusion network	Journal
FSoD-Net	Climate change observation	MSE-Net	lower proportion of foreground target pixels and drastic differences in object scale	IEEE Transaction

continued on following page

17

Table 5. Continued

Method	Objective/ Application Area	Methodology	Feature	Conf Journal Book Chapter
FactSeg	Small object semantic segmentation: car and ship etc.	They used a FactSeg framework which is referred to as foreground activation driven small object semantic segmentation framework.	This model activates features of the small object	IEEE Transaction
ASSD	General remote sensing imagery analysis	Here they prposed single-shot detector is proposed which is also referred to as ASSD	Spatial misalignment	IEEE Transaction
RAEAN	General remote sensing imagery analysis	It works on the Relation-augmented embedded graph attention network (EGAT).	It Enables the full utilization and exploitation of the spatial as well as semantic association between objects whic mainly target is to improve the detection performance	IEEE Transaction
FMSSD	Single-Shot Detection for Multiscale Objects	It aggregates the context information. Also, it also work on multiple scales and the same scale feature maps.	Feature pyramid and multiple atrous rates	IEEE Transaction

Table 6. Abbreviations and acronym

Abbreviation	Acronym
SM-Review	Use of Machine Learning and Remote Sensing Techniques for Shoreline Monitoring: A Review of Recent Literature
ELST-GEE-RSD	Estimation of land surface temperature and urban heat island by using google earth engine and remote sensing data
Spatiotemporal variations	Spatiotemporal variations in remote sensing phenology of vegetation and its responses to temperature change of boreal forest in tundra-taiga transitional zone in the Eastern Siberia
MDG-Surkhandarya	Monitoring dynamics of green spaces in Surkhandarya region based on remote sensing data of climate change
CGIS-RSA	A combined GIS and remote sensing approach for monitoring climate change-related land degradation to support landscape preservation and planning tools: The Basilicata case study
U-M-Climate change	A geographic information systems and remote sensing–based approach to assess urban micro-climate change and its impact on human health in Bartin, Turkey
IA-RSGA	An integrated approach of remote sensing and geospatial analysis for modeling and predicting the impacts of climate change on food security
RSRSM-AGCC	The role of satellite remote sensing in mitigating and adapting to global climate change
A-Hyper-RSFS-Agri	Application of hyperspectral remote sensing role in precision farming and sustainable agriculture under climate change: A review
NR-TMB	Near real-time change detection for border monitoring
BorderSense	BorderSense: Border patrol through advanced wireless sensor networks
RS-IPS	Remote sensing for international peace and security: Its role and implications

continued on following page

Table 6. Continued

Abbreviation	Acronym
DES	Deep learning for effective detection of excavated soil related to illegal tunnel activities
RSGC	Validating the remotely sensed geography of crime: A review of emerging issues
PRSDT	Preliminary results of ship detection technique by wake pattern recognition in SAR images
FOLDOUT	Foldout: A through foliage surveillance system for border security
HBDL	Hash-based deep learning approach for remote sensing satellite imagery detection
DLM	Investigation of Deep Learning Methodologies in Satellite Image based Ship Detection
DLA	Agricultural Field Extraction with Deep Learning Algorithm and Satellite Imagery
HNM	Hybrid Network Model: TransConvNet for Oriented Object Detection in Remote Sensing Images
FSoD-Net	FSoD-Net: Full-scale object detection from optical remote sensing imagery
FactSeg	Factseg: Foreground activation-driven small object semantic segmentation in largescale remote sensing imagery
ORSIm	ORSIm detector: A novel object detection framework in optical remote sensing imagery using spatial-frequency channel features
ASSD	ASSD: Feature aligned single-shot detection for multiscale objects in aerial imagery
RAEAN	A Relation-Augmented Embedded Graph Attention Network for Remote Sensing Object Detection
FMSSD	FMSSD: Feature-merged single-shot detection for multiscale objects in large-scale remote sensing imagery

methods, modified models, optimization of existing methods, and the intensity of publications. In the third fold, the study highlights research methodologies and novel features of literature to achieve object detection in remote sensing images. Finally, the study drew a clear picture of limitations and recent object detection trends to provide in-depth recommendations and future directions.

Data Statement

Data can be provided upon a formal request.

ACKNOWLEDGMENT

The authors would like to acknowledge Prince Sultan University and the EIAS: Data Science and Blockchain Laboratory for their valuable support.

REFERENCES

Abdullah, F. B., Iqbal, R., Ahmad, S., El-Affendi, M. A., & Abdullah, M. (2022b). An empirical analysis of sustainable energy security for energy policy recommendations. *Sustainability (Basel)*, *14*(10), 6099. doi:10.3390/su14106099

Abdullah, F. B., Iqbal, R., Ahmad, S., El-Affendi, M. A., & Kumar, P. (2022a). Optimization of multidimensional energy security: An index based assessment. *Energies*, *15*(11), 3929. doi:10.3390/en15113929

Adegun, A. A., Viriri, S., & Tapamo, J.-R. (2023). Review of deep learning methods for remote sensing satellite images classification: Experimental survey and comparative analysis. *Journal of Big Data*, *10*(1), 93. doi:10.1186/s40537-023-00772-x

Ahmad, S., Anwar, M. S., Ebrahim, M., Khan, W., Raza, K., Adil, S. H., & Amin, A. (2020). Deep network for the iterative estimations of students' cognitive skills. *IEEE Access : Practical Innovations, Open Solutions*, *8*, 103100–103113. doi:10.1109/ACCESS.2020.2999064

Ahmad, S., Anwar, M. S., Khan, M. A., Shahzad, M., Ebrahim, M., & Memon, I. (2021). Deep frustration severity network for the prediction of declined students' cognitive skills. *2021 4th international conference on Computing & Information Sciences (ICCIS)*, 1–6.

Ahmad, S., El-Affendi, M. A., Anwar, M. S., & Iqbal, R. (2022). Potential future directions in optimization of students' performance prediction system. *Computational Intelligence and Neuroscience*, *2022*, 2022. doi:10.1155/2022/6864955 PMID:35619762

Ahmad, S., Li, K., Amin, A., & Khan, S. (2018). A novel technique for the evaluation of posterior probabilities of student cognitive skills. *IEEE Access : Practical Innovations, Open Solutions*, *6*, 53153–53167. doi:10.1109/ACCESS.2018.2870877

Ahmad, Li, Eddine, & Khan. (2018). A biologically inspired cognitive skills measurement approach. *Biologically Inspired Cognitive Architectures, 24*, 35–46.

Akhtar, S., Ali, A., Ahmad, S., Khan, M. I., Shah, S., & Hassan, F. (2022). The prevalence of foot ulcers in diabetic patients in Pakistan: A systematic review and meta-analysis. *Frontiers in Public Health*, *10*, 1017201. doi:10.3389/fpubh.2022.1017201 PMID:36388315

Akhtar, S., Ramzan, M., Shah, S., Ahmad, I., Khan, M. I., Ahmad, S., El-Affendi, M. A., & Qureshi, H. (2022). Forecasting exchange rate of Pakistan using time series analysis. *Mathematical Problems in Engineering*, *2022*, 2022. doi:10.1155/2022/9108580

Al-Bilbisi, H. (2019). Spatial monitoring of urban expansion using satellite remote sensing images: A case study of amman city, jordan. *Sustainability (Basel)*, *11*(8), 2260. doi:10.3390/su11082260

Alhichri, H., Alswayed, A. S., Bazi, Y., Ammour, N., & Alajlan, N. A. (2021). Classification of remote sensing images using efficientnet-b3 cnn model with attention. *IEEE Access : Practical Innovations, Open Solutions*, *9*, 14078–14094. doi:10.1109/ACCESS.2021.3051085

Anwar, M. S., Ullah, I., Ahmad, S., Choi, A., Ahmad, S., Wang, J., & Aurangzeb, K. (n.d.). Immersive Learning and AR/VR-Based Education: Cybersecurity Measures and Risk Management. In Cybersecurity Management in Education Technologies (pp. 1-22). CRC Press.

Anwar, M. S., Wang, J., Ullah, A., Khan, W., Ahmad, S., & Li, Z. (2019). Impact of stalling on qoe for 360-degree virtual reality videos. *2019 IEEE International Conference on Signal, Information and Data Processing (ICSIDP)*, 1–6. 10.1109/ ICSIDP47821.2019.9173042

Bochkovskiy, A., Wang, C.-Y., & Liao, H.-Y. M. (2020). *Yolov4: Optimal speed and accuracy of object detection.* arXiv preprint arXiv:2004.10934.

Bohara, M., Patel, K., Patel, B., & Desai, J. (2021). An ai based web portal for cotton price analysis and prediction. *3rd International Conference on Integrated Intelligent Computing Communication & Security (ICIIC 2021)*, 33–39. 10.2991/ ahis.k.210913.005

Charrua, A. B., Padmanaban, R., Cabral, P., Bandeira, S., & Romeiras, M. M. (2021). Impacts of the tropical cyclone idai in mozambique: A multitemporal landsat satellite imagery analysis. *Remote Sensing (Basel)*, *13*(2), 201. doi:10.3390/rs13020201

Cui, B., Chen, X., & Lu, Y. (2020). Semantic segmentation of remote sensing images using transfer learning and deep convolutional neural network with dense connection. *IEEE Access : Practical Innovations, Open Solutions*, *8*, 116744–116755. doi:10.1109/ACCESS.2020.3003914

Ding, L., Zhang, J., & Bruzzone, L. (2020). Semantic segmentation of large-size vhr remote sensing images using a two-stage multiscale training architecture. *IEEE Transactions on Geoscience and Remote Sensing*, *58*(8), 5367–5376. doi:10.1109/ TGRS.2020.2964675

Du, B., Ru, L., Wu, C., & Zhang, L. (2019). Unsupervised deep slow feature analysis for change detection in multi-temporal remote sensing images. *IEEE Transactions on Geoscience and Remote Sensing, 57*(12), 9976–9992. doi:10.1109/TGRS.2019.2930682

Du, B., Ru, L., Wu, C., & Zhang, L. (2019). Unsupervised deep slow feature analysis for change detection in multi-temporal remote sensing images. *IEEE Transactions on Geoscience and Remote Sensing, 57*(12), 9976–9992. doi:10.1109/TGRS.2019.2930682

Duan, S., Cheng, P., Wang, Z., Wang, Z., Chen, K., Sun, X., & Fu, K. (2024). MDCNet: A Multi-platform Distributed Collaborative Network for Object Detection in Remote Sensing Imagery. *IEEE Transactions on Geoscience and Remote Sensing, 62*, 1–15. doi:10.1109/TGRS.2024.3353192

Fatima, R., Samad Shaikh, N., Riaz, A., Ahmad, S., El-Affendi, M. A., Alyamani, K. A., Nabeel, M., Ali Khan, J., Yasin, A., & Latif, R. M. A. (2022). A natural language processing (nlp) evaluation on covid-19 rumour dataset using deep learning techniques. *Computational Intelligence and Neuroscience, 2022*, 2022. doi:10.1155/2022/6561622 PMID:36156967

Fu, K., Chang, Z., Zhang, Y., Xu, G., Zhang, K., & Sun, X. (2020). Rotation-aware and multi-scale convolutional neural network for object detection in remote sensing images. *ISPRS Journal of Photogrammetry and Remote Sensing, 161*, 294–308. doi:10.1016/j.isprsjprs.2020.01.025

Fu, R., Chen, C., Yan, S., Zhang, R., Wang, X., & Chen, H. (2024). FADL-Net: Frequency-Assisted Dynamic Learning Network for Oriented Object Detection in Remote Sensing Images. *IEEE Transactions on Industrial Informatics*, 1–13. doi:10.1109/TII.2024.3378841

Gu, Y., Wang, Y., & Li, Y. (2019). A survey on deep learning-driven remote sensing image scene understanding: Scene classification, scene retrieval and scene-guided object detection. *Applied Sciences (Basel, Switzerland), 9*(10), 2110. doi:10.3390/app9102110

Gui, S., Song, S., Qin, R., & Tang, Y. (2024). Remote Sensing Object Detection in the Deep Learning Era—A Review. *Remote Sensing (Basel), 16*(2), 327. doi:10.3390/rs16020327

Haq, M. U., Sethi, M. A. J., Ahmad, S., ELAffendi, M. A., & Asim, M. (2024). Automatic Player Face Detection and Recognition for Players in Cricket Games. *IEEE Access: Practical Innovations, Open Solutions, 12*, 41219–41233. doi:10.1109/ACCESS.2024.3377564

Haq, M. U., Sethi, M. A. J., & Rehman, A. U. (2023). Capsule Network with Its Limitation, Modification, and Applications—A Survey. *Machine Learning and Knowledge Extraction*, *5*(3), 891–921. doi:10.3390/make5030047

Haq, M. U., Sethi, M. A. J., Ullah, R., Shazhad, A., Hasan, L., & Karami, G. M. (2022). COMSATS face: A dataset of face images with pose variations, its design, and aspects. *Mathematical Problems in Engineering*, *2022*, 2022. doi:10.1155/2022/4589057

Haq, M. U., Shahzad, A., Mahmood, Z., Shah, A. A., Muhammad, N., & Akram, T. (2019). Boosting the face recognition performance of ensemble based LDA for pose, non-uniform illuminations, and low-resolution images. [TIIS]. *KSII Transactions on Internet and Information Systems*, *13*(6), 3144–3164.

Hosni, A. I. E., Li, K., & Ahmad, S. (2019). Darim: Dynamic approach for rumor influence minimization in online social networks. *International Conference on Neural Information Processing*, 619–630. 10.1007/978-3-030-36711-4_52

Hosni, A. I. E., Li, K., & Ahmad, S. (2020). Minimizing rumor influence in multiplex online social networks based on human individual and social behaviors. *Information Sciences*, *512*, 1458–1480. doi:10.1016/j.ins.2019.10.063

Hosni, A. I. E., Li, K., Ding, C., & Ahmed, S. (2018). Least cost rumor influence minimization in multiplex social networks. *International Conference on Neural Information Processing*, 93–105. 10.1007/978-3-030-04224-0_9

Hossain, M. D., & Chen, D. (2019). Segmentation for object-based image analysis (obia): A review of algorithms and challenges from remote sensing perspective. *ISPRS Journal of Photogrammetry and Remote Sensing*, *150*, 115–134. doi:10.1016/j.isprsjprs.2019.02.009

Jia, W., Xu, S., Liang, Z., Zhao, Y., Min, H., Li, S., & Yu, Y. (2021). Real-time automatic helmet detection of motorcyclists in urban traffic using improved yolov5 detector. *IET Image Processing*, *15*(14), 3623–3637. doi:10.1049/ipr2.12295

Jiang, H., Luo, T., Peng, H., & Zhang, G. (2024). MFCANet: Multiscale Feature Context Aggregation Network for Oriented Object Detection in Remote-Sensing Images. *IEEE Access*.

Kasper-Eulaers, M., Hahn, N., Berger, S., Sebulonsen, T., Myrland, Ø., & Kummervold, P. E. (2021). Detecting heavy goods vehicles in rest areas in winter conditions using yolov5. *Algorithms*, *14*(4), 114. doi:10.3390/a14040114

Khan, M., Rahim, M., Alanzi, A. M., Ahmad, S., Fatlane, J. M., Aphane, M., & Khalifa, H. A. E. W. (2023). Dombi Aggregation Operators For p, q, r–Spherical Fuzzy Sets: Application in the Stability Assessment of Cryptocurrencies. *IEEE Access*.

Khan, M. I., Qureshi, H., Bae, S. J., Khattak, A. A., Anwar, M. S., Ahmad, S., ... Ahmad, S. (2023). *Malaria prevalence in Pakistan: A systematic review and meta-analysis (2006–2021)*. Heliyon.

Khan, S., Zhang, Z., Zhu, L., Rahim, M. A., Ahmad, S., & Chen, R. (2020). Scm: Secure and accountable tls certificate management. *International Journal of Communication Systems*, *33*(15), e4503. doi:10.1002/dac.4503

Laroca, R., Severo, E., Zanlorensi, L. A., Oliveira, L. S., Gon͵calves, G. R., Schwartz, W. R., & Menotti, D. (2018). *A robust real-time automatic license plate recognition based on the yolo detector. In 2018 international joint conference on neural networks (ijcnn)*. IEEE. doi:10.1109/IJCNN.2018.8489629

Levin, N., Kyba, C. C., Zhang, Q., de Miguel, A. S., Roma'n, M. O., Li, X., Portnov, B. A., Molthan, A. L., Jechow, A., & Miller, S. D. (2020). Remote sensing of night lights: A review and an outlook for the future. *Remote Sensing of Environment*, *237*, 111443. doi:10.1016/j.rse.2019.111443

Li, J., Tian, P., Song, R., Xu, H., Li, Y., & Du, Q. (2024). PCViT: A Pyramid Convolutional Vision Transformer Detector for Object Detection in Remote Sensing Imagery. *IEEE Transactions on Geoscience and Remote Sensing*, *62*, 1–15. doi:10.1109/TGRS.2024.3360456

Li, R., Zheng, S., Duan, C., Su, J., & Zhang, C. (2021). Multistage attention resunet for semantic segmentation of fine-resolution remote sensing images. *IEEE Geoscience and Remote Sensing Letters*, *19*, 1–5.

Lian, Y., Shi, X., Shen, S., & Hua, J. (2024). Multitask learning for image translation and salient object detection from multimodal remote sensing images. *The Visual Computer*, *40*(3), 1395–1414. doi:10.1007/s00371-023-02857-3

Lin, J., Zhao, Y., Wang, S., & Tang, Y. (2024). A robust training method for object detectors in remote sensing image. *Displays*, *81*, 102618. doi:10.1016/j.displa.2023.102618

Liu, X., Zhai, H., Shen, Y., Lou, B., Jiang, C., Li, T., Hussain, S. B., & Shen, G. (2020). Large-scale crop mapping from multisource remote sensing images in google earth engine. *IEEE Journal of Selected Topics in Applied Earth Observations and Remote Sensing*, *13*, 414–427. doi:10.1109/JSTARS.2019.2963539

Ma, L., Liu, Y., Zhang, X., Ye, Y., Yin, G., & Johnson, B. A. (2019). Deep learning in remote sensing applications: A meta-analysis and review. *ISPRS Journal of Photogrammetry and Remote Sensing*, *152*, 166–177. doi:10.1016/j. isprsjprs.2019.04.015

Mangalraj, P., Sivakumar, V., Karthick, S., Haribaabu, V., Ramraj, S., & Samuel, D. J. (2020). A review of multi-resolution analysis (mra) and multigeometric analysis (mga) tools used in the fusion of remote sensing images. *Circuits, Systems, and Signal Processing*, *39*(6), 3145–3172. doi:10.1007/s00034-019-01316-6

Memon, F. S., Abdullah, F. B., Iqbal, R., Ahmad, S., Hussain, I., & Abdullah, M. (2023). Addressing women's climate change awareness in Sindh, Pakistan: An empirical study of rural and urban women. *Climate and Development*, *15*(7), 565–577. doi:10.1080/17565529.2022.2125784

Munawar, F., Khan, U., Shahzad, A., Haq, M. U., Mahmood, Z., Khattak, S., & Khan, G. Z. (2019, January). An empirical study of image resolution and pose on automatic face recognition. In *2019 16th International Bhurban Conference on Applied Sciences and Technology (IBCAST)* (pp. 558-563). IEEE. 10.1109/IBCAST.2019.8667233

Mzid, N., Pignatti, S., Huang, W., & Casa, R. (2021). An analysis of bare soil occurrence in arable croplands for remote sensing topsoil applications. *Remote Sensing (Basel)*, *13*(3), 474. doi:10.3390/rs13030474

Nan, G., Zhao, Y., Fu, L., & Ye, Q. (2024). Object Detection by Channel and Spatial Exchange for Multimodal Remote Sensing Imagery. *IEEE Journal of Selected Topics in Applied Earth Observations and Remote Sensing*, *17*, 8581–8593. doi:10.1109/JSTARS.2024.3388013

Olson, D., & Anderson, J. (2021). Review on unmanned aerial vehicles, remote sensors, imagery processing, and their applications in agriculture. *Agronomy Journal*, *113*(2), 971–992. doi:10.1002/agj2.20595

Osco, L. P., Junior, J. M., Ramos, A. P. M., de Castro Jorge, L. A., Fatholahi, S. N., de Andrade Silva, J., Matsubara, E. T., Pistori, H., Gonçalves, W. N., & Li, J. (2021). A review on deep learning in uav remote sensing. *International Journal of Applied Earth Observation and Geoinformation*, *102*, 102456. doi:10.1016/j. jag.2021.102456

Pan, Z., Xu, J., Guo, Y., Hu, Y., & Wang, G. (2020). Deep learning segmentation and classification for urban village using a worldview satellite image based on u-net. *Remote Sensing (Basel)*, *12*(10), 1574. doi:10.3390/rs12101574

Peng, C., Li, Y., Jiao, L., Chen, Y., & Shang, R. (2019). Densely based multiscale and multi-modal fully convolutional networks for high-resolution remote-sensing image semantic segmentation. *IEEE Journal of Selected Topics in Applied Earth Observations and Remote Sensing, 12*(8), 2612–2626. doi:10.1109/JSTARS.2019.2906387

Petersen, K., Feldt, R., Mujtaba, S., & Mattsson, M. (2008). Systematic mapping studies in software engineering. *12th International Conference on Evaluation and Assessment in Software Engineering (EASE), 12*, 1–10. 10.14236/ewic/EASE2008.8

Pires de Lima, R., & Marfurt, K. (2019). Convolutional neural network for remotesensing scene classification: Transfer learning analysis. *Remote Sensing (Basel), 12*(1), 86. doi:10.3390/rs12010086

Rahim, A., Zhong, Y., Ahmad, T., Ahmad, S., Pławiak, P., & Hammad, M. (2023). P. P lawiak, M. Hammad, Enhancing smart home security: Anomaly detection and face recognition in smart home iot devices using logit-boosted cnn models. *Sensors (Basel), 23*(15), 6979. doi:10.3390/s23156979 PMID:37571762

Rahim, M., Amin, F., Tag Eldin, E. M., Khalifa, A. E. W., & Ahmad, S. (2024). p, q-Spherical fuzzy sets and their aggregation operators with application to third-party logistic provider selection. *Journal of Intelligent & Fuzzy Systems*, (Preprint), 1-24.

Redmon, J., & Farhadi, A. (2018). *Yolov3: An incremental improvement.* arXiv preprint arXiv:1804.02767.

Ru, L., Du, B., & Wu, C. (2020). Multi-temporal scene classification and scene change detection with correlation based fusion. *IEEE Transactions on Image Processing, 30*, 1382–1394. doi:10.1109/TIP.2020.3039328 PMID:33237858

Sagar, A. S., Chen, Y., Xie, Y., & Kim, H. S. (2024). MSA R-CNN: A comprehensive approach to remote sensing object detection and scene understanding. *Expert Systems with Applications, 241*, 122788. doi:10.1016/j.eswa.2023.122788

Sheng, Z. H. A. N. G., Shanshan, L. I., Guofang, W. E. I., Xinnai, Z. H. A. N. G., & Jianwei, G. A. O. (2024). Refined multi-scale feature-oriented object detection of the remote sensing images. *National Remote Sensing Bulletin, 26*(12), 2616–2628.

Signoroni, A., Savardi, M., Baronio, A., & Benini, S. (2019). Deep learning meets hyperspectral image analysis: A multidisciplinary review. *Journal of Imaging, 5*(5), 52. doi:10.3390/jimaging5050052 PMID:34460490

Silver, M., Tiwari, A., & Karnieli, A. (2019). Identifying vegetation in arid regions using object-based image analysis with rgb-only aerial imagery. *Remote Sensing (Basel), 11*(19), 2308. doi:10.3390/rs11192308

Sun, H., Fu, L., Li, J., Guo, Q., Meng, Z., Zhang, T., ... Yu, H. (2024). Defense against Adversarial Cloud Attack on Remote Sensing Salient Object Detection. In *Proceedings of the IEEE/CVF Winter Conference on Applications of Computer Vision* (pp. 8345-8354). 10.1109/WACV57701.2024.00816

Sun, X., Shi, A., Huang, H., & Mayer, H. (2020). Bas ^{4} net: Boundary-aware semi-supervised semantic segmentation network for very high resolution remote sensing images. *IEEE Journal of Selected Topics in Applied Earth Observations and Remote Sensing*, *13*, 5398–5413. doi:10.1109/JSTARS.2020.3021098

Tokarev, K., Orlova, Y. A., Rogachev, A., Chernyavsky, A., & Tokareva, Y. M. (2020). The intelligent analysis system and remote sensing images segmentation engineering by using methods of advanced machine learning and neural network modeling. *IOP Conference Series: Materials Science and Engineering*, 734, 012124. 10.1088/1757-899X/734/1/012124

Ullah, A., Jami, A., Aziz, M. W., Naeem, F., Ahmad, S., Anwar, M. S., & Jing, W. (2019, December). Deep Facial Expression Recognition of facial variations using fusion of feature extraction with classification in end to end model. In *2019 4th International Conference on Emerging Trends in Engineering, Sciences and Technology (ICEEST)* (pp. 1-6). IEEE. 10.1109/ICEEST48626.2019.8981687

Ullah, A., Ullah, A., Ahmad, S., Haq, M. U., Shah, K., & Mlaiki, N. (2022). Series type solution of fuzzy fractional order Swift–Hohenberg equation by fuzzy hybrid Sumudu transform. *Mathematical Problems in Engineering*, *2022*, 2022. doi:10.1155/2022/3864053

Ullah, H., Haq, M. U., Khattak, S., Khan, G. Z., & Mahmood, Z. (2019, August). A robust face recognition method for occluded and low-resolution images. In *2019 International Conference on Applied and Engineering Mathematics (ICAEM)* (pp. 86-91). IEEE. 10.1109/ICAEM.2019.8853753

Vogels, M. F., de Jong, S. M., Sterk, G., & Addink, E. A. (2019). Mapping irrigated agriculture in complex landscapes using spot6 imagery and object-based image analysis–a case study in the central rift valley, ethiopia–. *International Journal of Applied Earth Observation and Geoinformation*, *75*, 118–129. doi:10.1016/j.jag.2018.07.019

Wang, C., Liu, B., Liu, L., Zhu, Y., Hou, J., Liu, P., & Li, X. (2021). A review of deep learning used in the hyperspectral image analysis for agriculture. *Artificial Intelligence Review*, *54*(7), 5205–5253. doi:10.1007/s10462-021-10018-y

Wang, P., Sun, X., Diao, W., & Fu, K. (2019). Fmssd: Feature-merged single-shot detection for multiscale objects in large-scale remote sensing imagery. *IEEE Transactions on Geoscience and Remote Sensing, 58*(5), 3377–3390. doi:10.1109/TGRS.2019.2954328

Wang, W., Liu, K., Tang, R., & Wang, S. (2019). Remote sensing image-based analysis of the urban heat island effect in shenzhen, china. *Physics and Chemistry of the Earth Parts A/B/C, 110*, 168–175. doi:10.1016/j.pce.2019.01.002

Wen, D., Huang, X., Bovolo, F., Li, J., Ke, X., Zhang, A., & Benediktsson, J. A. (2021). Change detection from very-high-spatial-resolution optical remote sensing images: Methods, applications, and future directions. *IEEE Geoscience and Remote Sensing Magazine, 9*(4), 68–101. doi:10.1109/MGRS.2021.3063465

Xie, Y., Tian, J., & Zhu, X. X. (2020). Linking points with labels in 3d: A review of point cloud semantic segmentation. *IEEE Geoscience and Remote Sensing Magazine, 8*(4), 38–59. doi:10.1109/MGRS.2019.2937630

Xu, Y., Du, B., Zhang, L., Cerra, D., Pato, M., Carmona, E., Prasad, S., Yokoya, N., Ha̎nsch, R., & Le Saux, B. (2019). Advanced multi-sensor optical remote sensing for urban land use and land cover classification: Outcome of the 2018 ieee grss data fusion contest. *IEEE Journal of Selected Topics in Applied Earth Observations and Remote Sensing, 12*(6), 1709–1724. doi:10.1109/JSTARS.2019.2911113

Yi, D., Su, J., & Chen, W.-H. (2021). Probabilistic faster r-cnn with stochastic region proposing: Towards object detection and recognition in remote sensing imagery. *Neurocomputing, 459*, 290–301. doi:10.1016/j.neucom.2021.06.072

Yin, S., Li, H., Teng, L., Jiang, M., & Karim, S. (2020). An optimised multi-scale fusion method for airport detection in large-scale optical remote sensing images. *International Journal of Image and Data Fusion, 11*(2), 201–214. doi:10.1080/19479832.2020.1727573

Zhang, C., Yue, P., Tapete, D., Jiang, L., Shangguan, B., Huang, L., & Liu, G. (2020). A deeply supervised image fusion network for change detection in high resolution bi-temporal remote sensing images. *ISPRS Journal of Photogrammetry and Remote Sensing, 166*, 183–200. doi:10.1016/j.isprsjprs.2020.06.003

Zhang, G., Yu, W., & Hou, R. (2024). MFIL-FCOS: A Multi-Scale Fusion and Interactive Learning Method for 2D Object Detection and Remote Sensing Image Detection. *Remote Sensing (Basel), 16*(6), 936. doi:10.3390/rs16060936

Zhang, H., Tian, M., Shao, G., Cheng, J., & Liu, J. (2022). Target detection of forward-looking sonar image based on improved yolov5. *IEEE Access : Practical Innovations, Open Solutions, 10*, 18023–18034. doi:10.1109/ACCESS.2022.3150339

Zhang, S., Wu, R., Xu, K., Wang, J., & Sun, W. (2019). R-cnn-based ship detection from high resolution remote sensing imagery. *Remote Sensing (Basel), 11*(6), 631. doi:10.3390/rs11060631

Zhang, T., Zhang, X., Zhu, X., Wang, G., Han, X., Tang, X., & Jiao, L. (2024). Multistage Enhancement Network for Tiny Object Detection in Remote Sensing Images. *IEEE Transactions on Geoscience and Remote Sensing, 62*, 1–12. doi:10.1109/TGRS.2024.3396874

Zhang, Y., Liu, T., Yu, P., Wang, S., & Tao, R. (2024). SFSANet: Multi-scale Object Detection In Remote Sensing Image Based on Semantic Fusion and Scale Adaptability. *IEEE Transactions on Geoscience and Remote Sensing*.

Zhu, X., Lyu, S., Wang, X., & Zhao, Q. (2021). Tph-yolov5: Improved yolov5 based on transformer prediction head for object detection on drone-captured scenarios. *Proceedings of the IEEE/CVF international conference on computer vision*, 2778–2788.

Chapter 2
Deep Learning Classification of Empty and Full Container Ships in Satellite Imagery:
A Cognitive Modeling Approach for Maritime Applications

Serif Ali Sadik

https://orcid.org/0000-0003-2883-1431
Faculty of Engineering, Kutahya Dumlupinar University, Turkey

ABSTRACT

This chapter presents a novel approach to ship classification in optical remote sensing (ORS) imagery, focusing on the distinction between empty and full container ships. Leveraging deep cognitive modeling techniques, the study employs renowned pre-trained deep learning models, including VGG-16, VGG-19, and InceptionV3, with fine-tuning for enhanced performance. The investigation addresses the challenges posed by class imbalance through strategic data augmentation. Results demonstrate the efficacy of the proposed models, with InceptionV3 exhibiting superior performance. Evaluation metrics encompassing accuracy, precision, recall, F1-score, AUC-ROC, and AUC-PR are meticulously analyzed. These findings contribute to the advancement of ship classification methodologies in ORS imagery, with implications for maritime applications and decision-making processes. The work underscores the importance of deep cognitive modeling in addressing complex classification tasks and paves the way for future enhancements and applications in the field.

DOI: 10.4018/979-8-3693-2913-9.ch002

INTRODUCTION

Satellite imagery has become an essential tool in the field of remote sensing, offering a unique perspective for observing and understanding our living planet. The integration of advanced computational techniques, particularly those rooted in deep learning, has significantly enhanced our ability to extract meaningful information from these vast datasets (Sowmya et al., 2017). In the domain of maritime applications, where the identification and classification of vessels from satellite imagery are critical, innovative approaches are continuously sought to address the challenges posed by the ever-changing maritime landscape (Gallego et al., 2018; L. Huang et al., 2018; Lang & Wu, 2017).

In the area of remote sensing, the advent of optical remote sensing (ORS) technology has improved our ability to observe and analyze maritime activities with clarity. ORS, operating within the visible and near-infrared spectrum, offers a powerful means to extract intricate visual details from the Earth's surface. In the maritime context, this technology becomes instrumental in detecting and classifying vessels based on their distinct visual signatures. The high spatial resolution and multispectral capabilities inherent in ORS imagery provide a unique vantage point for discerning fine details such as ship dimensions, superstructure features, and color patterns (Cheng & Han, 2016; Li et al., 2020).

The precise identification and classification of ships from optical imagery hold leading significance across various domains, including maritime security, environmental monitoring, and traffic management. Understanding vessel types, tracking movements, and discerning anomalous activities contribute to enhanced situational awareness, aiding in both commercial and security-related applications. The scientific framework employed in ship classification involves the integration of machine learning, computer vision, and spectral analysis techniques (Dong et al., 2018, 2019). Supervised learning algorithms, such as Support Vector Machines (SVM) (Dong et al., 2018) and Convolutional Neural Networks (CNN) (Xiao et al., 2022), leverage labeled datasets to discern intricate patterns in ship characteristics. Supervised learning thrives on labeled datasets, where each instance of ship imagery is meticulously annotated. These algorithms adeptly learn and discern intricate patterns within the labeled data, enabling them to generalize and classify unseen ship images accurately. SVM, with its capacity to identify complex patterns in high-dimensional spaces, and CNN, which excels in hierarchical feature extraction, stand out as powerful tools for ship classification in optical remote sensing imagery (Caruana & Niculescu-Mizil, 2006).

Complementary to machine learning, computer vision techniques play a critical role in ship classification. Computer vision algorithms analyze the visual content of images, extracting meaningful features such as ship superstructures, hull shapes, and spatial relationships. This process involves image segmentation, object recognition, and feature extraction, collectively enhancing the algorithm's ability to characterize and classify diverse ship types based on visual cues. The synergy of machine learning and computer vision ensures a comprehensive and adaptive approach to ship classification challenges (Wang et al., 2021; Y. Zhang et al., 2020).

Spectral analysis forms another critical component of the ship classification framework, providing insights into the unique spectral signatures of various ship materials and conditions. By examining the reflectance or emission properties of different wavelengths in the optical spectrum, spectral analysis aids in discriminating between materials with distinct reflective characteristics. This proves invaluable in differentiating ship features from the surrounding environment, especially in scenarios where visual cues alone may be insufficient. The fusion of spectral information with spatial features enhances the overall robustness of ship classification models, making them more resilient to variations in environmental conditions and ship appearances (Zhuang et al., 2018). These methodologies create a robust and adaptive scientific framework for ship classification, effectively leveraging the wealth of information embedded in ORS imagery. The fusion of machine learning, computer vision, and spectral analysis not only facilitates the automatic identification of ships but also contributes to the development of sophisticated models capable of discerning intricate details and adapting to the diverse challenges posed by maritime environments.

Various other methods are employed for ship classification, each with its distinct scientific framework and advantages. Synthetic Aperture Radar (SAR) utilizes microwave frequencies to capture detailed Earth surface images. Notably, SAR excels in ship detection owing to the pronounced radar reflections from metallic ship surfaces. This quality facilitates the identification of vessels amidst adverse weather conditions and low-light scenarios. Beyond mere detection, ship classification through SAR involves the categorization of vessels based on inherent characteristics such as size, shape, and radar cross-section. The fine resolution of SAR imagery is pivotal, allowing for the discernment of intricate structural features and facilitating the categorization of vessels into specific classes (Lang et al., 2016; Moreira et al., 2013).

The Automatic Identification System (AIS) constitutes a crucial component in contemporary maritime technologies, playing a lead role in ship classification and tracking. Essentially, AIS is a transponder system installed on vessels that continuously broadcasts key navigational and identification information. This includes the ship's name, Maritime Mobile Service Identity (MMSI), position, course, speed, and other relevant details. While AIS was primarily designed for collision avoidance and enhancing navigational safety, its data has become instrumental in ship classification (I.-L. Huang et al., 2024a).

AIS contributes real-time information directly from ships, allowing for accurate and timely vessel tracking. The system relies on VHF radio frequencies, enabling communication between nearby vessels and shore-based stations. This broadcasted information forms a rich dataset that can be leveraged for ship classification purposes. Analysts and researchers utilize AIS data to discern patterns in vessel movements, identify traffic routes, and classify ships based on their AIS-transmitted attributes. The information is particularly valuable for classifying vessels engaged in commercial activities, understanding their behavior, and enhancing overall maritime domain awareness. However, it's crucial to note that AIS has limitations, especially in scenarios where vessels may intentionally disable or manipulate their AIS transponders for security or privacy reasons. Despite these constraints, AIS remains a fundamental tool in ship classification, providing valuable insights into the real-time dynamics of maritime traffic and contributing to comprehensive monitoring and management of vessel activities (Chen et al., 2020; Zhou et al., 2019).

ORS stands out among these methods with its various advantages. ORS imagery offers high spatial resolution, facilitating detailed visualization of ships and their features. It captures visual appearance, color, shape, and superstructure details crucial for ship classification. The accessibility of ORS data from various sources, coupled with its versatility in capturing imagery across different spectral bands, makes it a valuable and widely applicable tool in maritime surveillance and monitoring. In summary, while each ship classification method has its merits, ORS, with its high-resolution visual information, proves instrumental in accurately classifying ships based on their visual characteristics.

RECENT ADVANCES FROM LITERATURE

In 2016, the researchers introduced a method known as joint feature and classifier selection (JFCS), which integrates a classifier selection strategy into a wrapper feature selection framework. This approach employs an enhanced Sequential Forward Floating Search (SFFS) algorithm to efficiently identify the optimal combination of feature subsets, scaling methods, and classifiers. The proposed method demonstrated increased robustness across various datasets and noise levels, yielding the highest classification accuracy at 94.62% for the initial dataset they examined (Lang et al., 2016).

Another work introduces a ship classification approach leveraging superstructure scattering features in RadarSat-2 SAR images. The novel ratio of dimensions (RoD) feature, defined as a fusion of 2-D and 3-D features, is presented by the researchers. They estimate a parametric vector F to describe ship characteristics and utilize SVM for distinguishing various ship types. Experimental results affirm the effectiveness and reliability of the proposed feature in SAR image ship classification, showcasing its superiority over the FL method with an overall classification accuracy exceeding 80% (Jiang et al., 2016).

In a 2017 study, the researchers introduced the BDA-KELM model, a fusion of the Binary Differential Algorithm (BDA) and the Kernel Extreme Learning Machine (KELM), designed for automatic feature selection and classifier parameter optimization in ship classification from high-resolution SAR images. The authors conducted tests on TerraSAR-X images featuring diverse ship types, comparing the performance of BDA-KELM with other well-known classification algorithms. Notably, BDA-KELM demonstrated higher performance across metrics such as overall accuracy, precision, recall, and F1-score, underscoring its efficacy in high-resolution SAR image classification (Wu et al., 2017).

In 2019, Dong et al. introduced a ship classification framework utilizing a deep residual network designed for high-resolution SAR images. Their approach involved the development of a deep learning framework incorporating data augmentation and ResNet. The outcomes of their study revealed that the proposed framework exhibited better performance compared to other CNN based networks. Specifically, it achieved an overall accuracy of 99% on the augmented dataset, demonstrating its efficacy under the optimal fine-tuning strategy (Yingbo Dong et al., 2019).

In 2020, researchers presented an approach to enhance ship classification accuracy in SAR images with moderate resolution. Addressing a notable challenge associated with medium-resolution images, which often lack the intricate details required for conventional classification methods, they introduced the Densely Connected Triplet CNN. This novel framework comprises three interconnected branches, each handling an image within a meticulously constructed "triplet." The triplets consist of an anchor image (representing the ship to be classified), a positive image of the same class, and a negative image of a different class. This architecture prompts the network to learn discriminative features capable of distinguishing between various ship types, even in the presence of limited image detail. The proposed Tri-DenseNet exhibited the highest performance, even under the lowest simulated resolution, the accuracy approached 87.97% (He et al., 2021).

In a more recent study, the researchers introduced a Multisource Heterogeneous Transfer Learning (MS-HeTL) method that combines information from multiple sources to enhance classification accuracy. MS-HeTL employs three main stages. Firstly, with Multisource Heterogeneous Feature Augmentation (MS-HFA), features were extracted from each source (SAR, AIS, ORS) individually. Then MS-HFA augmented the features from each source, enriching them with complementary information extracted from the other sources. Finally, the augmented features are fed into an SVM classifier trained on the target SAR data. With this novel method, the researchers achieved highest classification accuracy of 86.88% with the HR-SAR dataset (Lang et al., 2022).

Utilizing novel data analyzing techniques, AIS data emerges as a valuable source for analyzing ship behavior. Operating on VHF radio frequencies, the AIS system broadcasts information, creating a rich dataset that holds significant potential for ship classification purposes. In a study conducted in 2019, a novel approach to ship classification within ports was devised, focusing on behavior clustering. This methodology involved the analysis of AIS data obtained from the port of Rotterdam. The researchers utilized clustering algorithms to analyze AIS datasets, identifying groups of ships that exhibited similar movement patterns. Subsequently, distinct features were extracted from each cluster, encapsulating the essence of these movement patterns. The research demonstrated that ship length and beam could serve as explanatory variables for classifying ships into corresponding behavior clusters. Classifiers were developed utilizing both unsupervised discretization (equal width binning) and supervised discretization (Chi2). The results showed that classifications based on Chi2 discretization surpassed those using equal width discretization in terms of performance (Zhou et al., 2019).

In another study, an architecture for trajectory-based fishing ship classification has presented using AIS data. It focused on the dynamics of ship movement rather than static features, aiming to improve accuracy and capture the specific behavior patterns of fishing vessels. AIS data containing ship positions and timestamps were processed to create smooth and representative trajectories for each vessel and a set of features was extracted from the trajectories. The extracted features were fed into a classification model, such as SVM or Decision Tree (DT), to categorize ships as either "fishing" or "non-fishing". The proposed approach achieves classification results that are close to those from the complete process experiment using minimal features (Sánchez Pedroche et al., 2020).

In 2022, Yan et al. presented a method that incorporated the extraction and analysis of ship behavior characteristics alongside conventional geometric features. This method demonstrated proficiency in ship classification and anomaly detection using Spaceborne AIS Data. The outcomes indicated a noteworthy classification accuracy of 92.70% for five distinct ship types, achieved through SVM and Random Forest (RF) algorithms (Yan et al., 2022).

In a recent investigation, a novel classification approach was introduced, focusing on AIS data for cargo ships. The study introduced a classifier specifically designed for bulk carriers, containers, general cargo, and vehicle carriers. To evaluate the model's effectiveness, eight classification algorithms spanning tree-structure-based, proximity-based, and regression-based categories were employed. The findings revealed that tree-structure-based algorithms, specifically XGBoost and RF, exhibited commendable performance, achieving a highest classification accuracy of 96.9% (I.-L. Huang et al., 2024b).

ORS imagery stands as an additional data source for addressing ship classification and detection challenges. Possessing higher resolution, ORS images offer valuable target texture and color features that enhance their efficacy in detection and classification tasks. In 2018, researchers proposed a method for ship classification from ORS images, consisting of two stages. In the initial prescreening stage, a novel visual saliency detection method was employed based on the distinction in statistical characteristics between highly non-uniform regions and homogeneous backgrounds. For the discrimination stage, the method utilized the radial gradient histogram to capture rotational-invariant features of the candidate regions, combining these features with Local Binary Patterns (LBP) for detailed texture information. Subsequently, an SVM classifier was applied to detect ships based on the extracted visual features, achieving the optimal detection performance of 94% (Dong et al., 2018).

In a study based on deep learning, researchers utilized a pre-trained CNN to generate candidate regions that potentially contain ships within high-resolution ORS images. Subsequently, a dedicated Regional Convolutional Neural Network (R-CNN) model, specifically trained on high-resolution ship images, further refined these candidate regions and classified them as either ships or non-ships and achieved a classification accuracy over 95% (S. Zhang et al., 2019).

In 2019, Dong et al. introduced an innovative approach that systematically analyzes images at different scales, effectively capturing ships of diverse sizes. During the target identification phase, the study explored a rotation-invariant descriptor, combining the histogram of oriented gradients (HOG) cells and the Fourier basis. This descriptor proved instrumental in detecting ships. The detection process employed an SVM classifier, achieving an overall accuracy of 93%.

In a separate study, the researchers acquired high-resolution land-look images from United States Geological Survey (USGS) datasets. To enrich variability and optimize training efficiency, data augmentation techniques, including color and geometric transformations, were employed. The implemented CNN, featuring multiple pooling and convolutional layers followed by a softmax classifier, demonstrated remarkable performance. The trained CNN achieved an impressive accuracy of 99.78% for ship detection and 99.24% for non-ship detection (Aminuddin et al., 2023).

A recent paper introduced a novel method known as fused feature and rebuilt (FFR) YOLOv3, an enhancement of the popular YOLOv3 network designed to elevate ship detection in optical remote-sensing images. This network strategically combined features from various levels within the YOLOv3 architecture, capturing both fine-grained and high-level information about the ships and their surroundings. Additionally, the authors incorporated a squeeze-and-excitation (SE) structure into

Table 1. Summary of recent studies on ship classification and detection

Data	Method	Brief Result	Reference
SAR	JFCS	The JFCS method, proposed to enhance ship classification performance in SAR images, considers features, scaling, and classifiers as an integrated whole. It achieved a peak classification accuracy of 94.62% across three distinct classes.	(Lang et al., 2016)
	SVM	With a novel feature called ratio of dimensions (RoD), the authors achieved overall classification accuracy exceeding 80% with SVM algorithm for three ship classes from SAR images.	(Jiang et al., 2016)
	BDA-KELM	The proposed model can achieve a better classification performance than conventional four widely used algorithms with a classification accuracy as high as 97% for three ship classes.	(Wu et al., 2017)
	ResNet	A ResNet based framework was proposed and its performance tested with augmented SAR imagery dataset. The proposed method achieved an overall accuracy of 99% for three classes.	(Yingbo Dong et al., 2019)
	SAR- Tri-DenseNet	The introduced the Densely Connected Triplet CNN (Sar-Tri-DenseNet) achieved ship classification accuracy as 87.97% in SAR images even under the lowest simulated resolution.	(He et al., 2021)
	MS-HeTL	With the proposed method, features were extracted from different sources and the augmented features are fed into an SVM classifier trained on the target SAR data. Highest classification accuracy of 86.88% was achieved for four ship classes.	(Lang et al., 2022)
AIS	K-means clustering and Chi2	The researchers utilized K-means clustering algorithms to analyze AIS datasets and extracted features from each cluster. The results showed that Chi2 discretization-based classifier has better performance for both inbound and outbound ships.	(Zhou et al., 2019)
	SVM and DT	Trajectory-based fishing ship classification has presented for AIS data. With minimal features, the proposed approach has classification results similar to more complex classification processes.	(Sánchez Pedroche et al., 2020)
	SVM and RF	A method that incorporated the extraction and analysis of ship behavior characteristics alongside conventional geometric features using Spaceborne AIS Data was presented. For five ship classes, classification accuracy of 92.70% was achieved with RF.	(Yan et al., 2022)
	XGBoost	With AIS data for four type of cargo ships, the study introduced a comparative study among eight classifier algorithms. With tree-based classifier algorithms the researchers achieved a classification accuracy of %96.9.	(I.-L. Huang et al., 2024b)
ORS	SVM	A novel visual saliency detection method was employed to extract region of interests (RoIs). With the extracted visual features from RoIs, detection performance of 94% was achieved with SVM classifier.	(Dong et al., 2018)
	R-CNN	An R-CNN model refined candidate regions extracted from ORS images and classified them as either ships or non-ships and achieved a classification accuracy over 95%.	(S. Zhang et al., 2019)
	SVM	A descriptor, combining the histogram of oriented gradients (HOG) cells and the Fourier basis was utilized and with SVM classifier, classification accuracy of 93% was achieved.	(Dong et al., 2019)
	CNN	High-resolution images obtained from USGS datasets were augmented and used in training the proposed CNN network. The CNN network provided 99% accuracy for ship detection during the testing phase.	(Aminuddin et al., 2023)
	FFR-YOLOv3	A novel deep learning method was introduced as FFR-YOLOv3. The proposed network demonstrated an impressive detection accuracy of 95% for ship detection from ORS images.	(Wang et al., 2021)

the backbone network to fortify its ability to extract features. The proposed network demonstrated an impressive detection accuracy of 95% (Wang et al., 2021)

In the literature, ship classification and detection studies typically revolve around three primary types of data (Table 1). The SAR imaging system stands out for its advantage in day-night discrimination and adverse weather conditions. However, it lacks the ability to extract visual distinctive features of ships. AIS, a widely used maritime system, enables ships to broadcast crucial information, such as identity, position, course, and speed, providing real-time and accurate data for vessel tracking and monitoring. ORS imagery, with its high spatial resolution, facilitates detailed visualization of ships and their features. Optical imagery, on the other hand, captures the visual appearance, including color, shape, and superstructure details, making it valuable for classification.

This study specifically aims to classify whether container ships are empty or full, a classification that, to the author's knowledge, has not been explored in the literature. The ability to identify and distinguish between empty and full container ships holds significant importance, influencing decision-making processes in shipping, logistics, and port management. Traditional classification methods often struggle with the complexities and scale of satellite imagery. Our approach leverages deep cognitive modeling, integrating the power of Convolutional Neural Networks (CNNs).

MATERIALS AND METHODS

The flow diagram illustrating the proposed method in this study is presented in Figure 1. Full and empty container images within the ORS dataset exhibited variations in sizes and resolutions. To standardize the image data dimensions, a preprocessing step was employed, involving either cropping or padding. Addressing class imbalance within the dataset, the data augmentation step utilized the rotation method to duplicate empty container images, effectively rectifying the imbalance. The dataset was then split into training and testing sets in an 80:20 ratio.

Three CNN-based networks were chosen for implementing the transfer learning method in classification. Transfer learning, particularly using pre-trained models, proves beneficial in deep learning tasks for various reasons. When datasets are limited, pre-trained models offer a resource-efficient solution, having learned valuable features from extensive training on diverse data. This is especially advantageous when computational resources are constrained. Transfer learning adapts knowledge from a source domain to a related target domain, enhancing model performance. Models pre-trained on diverse tasks provide generalized features, contributing to robustness and adaptability for new, unseen tasks. Leveraging pre-trained models as feature extractors, retaining lower-level features relevant to specific tasks, is another advantage. Therefore, transfer learning facilitates improved convergence, faster training, and enhanced performance in various applications.

A fine-tuning step was subsequently applied to ensure these networks were tailored to address the specific challenges posed by the classification problem. The obtained classification results were interpreted based on predefined metrics.

Dataset and Data Preparation

The study utilized a publicly available Optical Remote Sensing (ORS) dataset (Lang & Yang, 2022), comprising eight ship classes and totaling 8678 images. All images, collected through Google Earth with sub-meter resolution, were matched with corresponding class information from the official website (http://www. marinetraffic.com). Focusing specifically on container ships within the dataset, a subclass of interest, there were 431 images of empty container ships and 826 images of full container ships.

Each image, stored in a TIFF extension file, represented a single ship, but the images varied in size. To standardize inputs for deep networks, all images were resized to 512x512 pixels with a resolution of 96 dpi, applying clipping or padding as needed. To address class imbalance, the number of empty container ship images was doubled using the rotation method, effectively rectifying the imbalance. The data preparation step concluded with the division of the processed data into training and test sets at an 80:20 ratio. Example images from the dataset were given in Figure 2. Figure 2(a) shows an image of a empty container ship from the dataset. However, the size of the image was not standardized. Thus, padding method was utilized to resize the image to 512x512 pixels (Figure 2(b)). Further, an image of a empty container ship was given in Figure 2(c). To balance the classes in the dataset, empty container ship images were augmented with rotation method (Figure 2(d)).

Figure 1. Block diagram of the proposed classification method

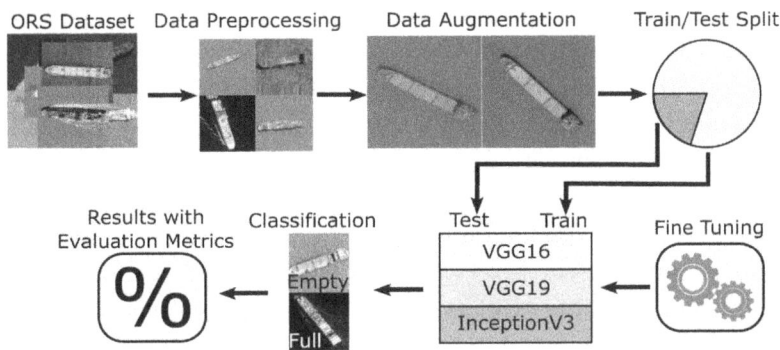

Figure 2. Example images from dataset: (a) Raw image of a full container ship, (b) processed image of a full container ship resized with padding, (c) raw image of an empty container ship, (d) augmented image with rotation technique

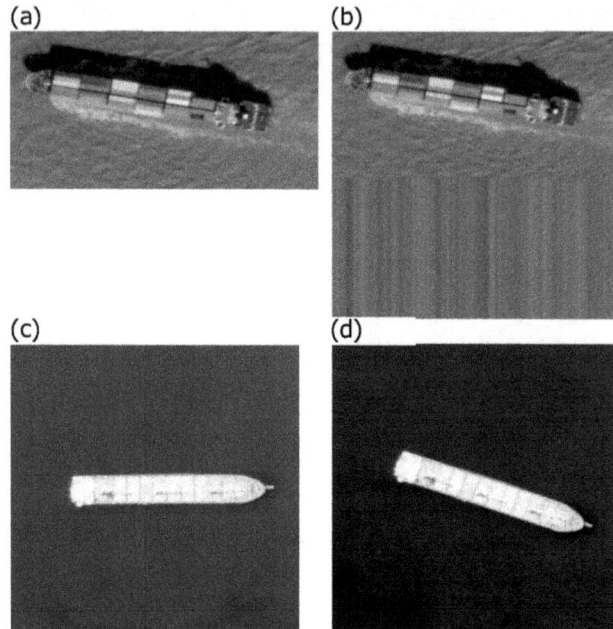

VGG16

The VGG16 (Visual Geometry Group 16) architecture is a deep convolutional neural network renowned for its simplicity and effectiveness in image classification tasks. Developed by the Visual Geometry Group at the University of Oxford, VGG16 is characterized by its distinctive structure, consisting of 16 weight layers, including 13 convolutional layers and three fully connected layers. Each convolutional block comprises multiple 3x3 convolutional layers, followed by max-pooling layers, allowing the network to learn hierarchical features of increasing complexity. VGG16 gained prominence during the ImageNet Large Scale Visual Recognition Challenge (ILSVRC) in 2014, where it demonstrated remarkable performance in object classification and localization tasks (Simonyan & Zisserman, 2014). The architecture's uniformity and straightforward design make it a valuable benchmark in the field of deep learning, aiding researchers in understanding the impact of network depth and convolutional filter sizes on model performance. Despite its depth, VGG16's architecture remains accessible, serving as an educational resource and a foundation for more advanced neural network designs (Theckedath & Sedamkar, 2020).

VGG19

The VGG19 (Visual Geometry Group 19) architecture stands as a formidable extension of the VGG16 model, both conceived and developed by the Visual Geometry Group at the University of Oxford (Simonyan & Zisserman, 2014). As its nomenclature implies, VGG19 incorporates 19 weight layers, further enhancing the network's capacity to learn intricate hierarchical features from input images. This architecture maintains the fundamental design principles of VGG16, featuring convolutional blocks comprised of 3x3 filters, interspersed with max-pooling layers. The additional layers in VGG19 contribute to an increased model depth, empowering it to capture and represent more nuanced and abstract features. Similar to its predecessor, VGG19 has demonstrated exceptional efficacy in image classification tasks, notably during the ImageNet Large Scale Visual Recognition Challenge. The architecture's accessibility and straightforward design have rendered it a valuable benchmark for researchers and practitioners alike, fostering a deeper understanding of the impact of network depth on performance. VGG19's legacy lies not only in its superior capabilities but also in its role as an educational asset and a foundation for subsequent advancements in deep learning architectures (Bansal et al., 2023).

InceptionV3

The InceptionV3 architecture, an evolution of its predecessors, represents a paradigm shift in CNN design, offering a novel approach to feature extraction. Developed by researchers at Google, InceptionV3 is characterized by its utilization of inception modules, a set of carefully crafted convolutional filters of varying sizes within the same layer (Szegedy et al., 2016). This innovative architectural choice enables the network to capture features at multiple scales simultaneously, enhancing its ability to recognize patterns of diverse complexities. Notably, InceptionV3 incorporates factorized convolutions, employing 1x1 convolutions for dimensionality reduction before applying larger filters. This design strategy not only reduces computational complexity but also facilitates the efficient learning of spatial hierarchies. InceptionV3 has demonstrated outstanding performance in image classification and object detection tasks, earning acclaim for its robustness and efficiency. The architecture's incorporation of auxiliary classifiers at intermediate layers aids in mitigating the vanishing gradient problem during training. InceptionV3's impact extends beyond its achievements, serving as a blueprint for subsequent Inception models and influencing the broader landscape of deep learning architectures (Liu et al., 2020).

Evaluation Metrics

The choice of evaluation metrics holds significant importance in assessing the performance of deep learning models for binary classification tasks. These metrics serve as practical benchmarks, offering clear insights into how well a model is performing across different aspects. While accuracy provides an overall measure of correctness, precision and recall become crucial, especially in scenarios with imbalanced datasets. The F1 Score strikes a balance between precision and recall, offering a comprehensive performance indicator. Receiver Operating Characteristic (ROC) and Precision and Recall (PR) curves, along with Area Under the Curve (AUC) values, provide nuanced perspectives on the ability of the model to distinguish between classes at different decision thresholds. Specificity and Matthews Correlation Coefficient (MCC) contribute additional insights, particularly in situations with imbalanced class distributions. A thoughtful consideration of these metrics helps practitioners make informed decisions, ensuring that the model's strengths and limitations align with the specific needs of the binary classification problem at hand.

The confusion matrix (Figure 3) is a straightforward and practical tool used in evaluating the performance of a classification model. It provides a clear summary of how well the model is making predictions by breaking down the results into four categories: true positives (correctly predicted positives - TP), true negatives (correctly predicted negatives - TN), false positives (incorrectly predicted positives - FP), and false negatives (incorrectly predicted negatives - FN). Essentially, it gives a snapshot of the ability of the model to correctly classify instances and where it might be making mistakes. This simple tabulation is particularly useful in assessing the performance of binary classification models in a transparent and easily interpretable manner. The confusion matrix facilitates the computation of various performance metrics, including accuracy, precision, recall, and the F1 score. These metrics play a crucial role in assessing the classification model's performance, particularly in scenarios with unbalanced class distributions (Luque et al., 2019).

Figure 3. An example of confusion matrix

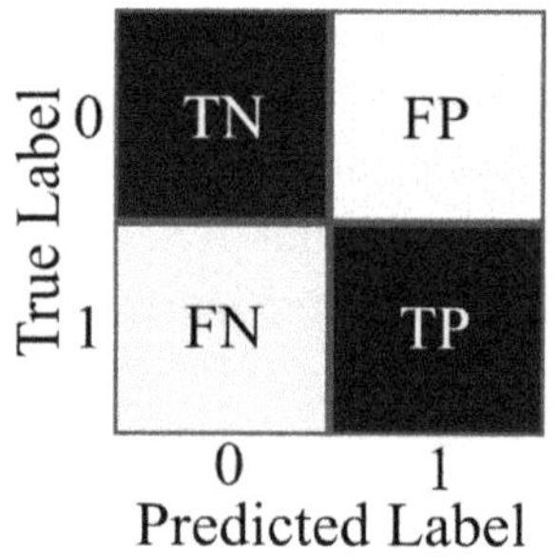

a) Accuracy: Accuracy measures how often the model's predictions are correct across all classes. It represents the ratio of correctly predicted instances (both true positives and true negatives) to the total number of instances in the dataset. While accuracy is a commonly used metric, it may not be suitable for imbalanced datasets where one class significantly outweighs the other.

$$Accuracy = \frac{TP+TN}{TP+TN+FP+FN} \tag{1}$$

b) Precision: Precision measures the ratio of correctly predicted positive instances to the total instances predicted as positive. A high precision indicates that when the model predicts a positive instance, it is likely to be correct.

$$Precision = \frac{TP}{TP+FP} \tag{2}$$

c) Recall: Recall, also known as sensitivity or true positive rate, is another important metric in classification that focuses on the model's ability to capture all relevant positive instances. Recall measures the ratio of correctly predicted positive instances to the total actual positive instances. A high recall indicates that the model is effective at identifying most of the positive instances in the dataset.

$$Recall = \frac{TP}{TP+FN} \tag{3}$$

d) F1-Score: The F1 score is a combined metric that balances precision and recall, providing a single value that reflects both the ability of the model to make accurate positive predictions and to capture all relevant positive instances. It is particularly useful when there is an uneven distribution between positive and negative instances in the dataset.

$$F1\,Score = 2 \times \frac{Precision \times Recall}{Precision + Recall} \tag{4}$$

e) Specificity: Specificity, also known as the true negative rate, is a performance metric in classification that focuses on the model's ability to correctly identify negative instances. It is particularly relevant when the cost of false positives (incorrectly predicting a negative instance) is high. specificity measures the ratio of correctly predicted negative instances to the total instances that are actually negative. A high specificity indicates that the model is effective at correctly identifying instances that truly belong to the negative class.

$$Specificity = \frac{TN}{TN + FP} \tag{5}$$

f)   Matthews Correlation Coefficient – MCC: The Matthews Correlation Coefficient (MCC) is a metric used to assess the performance of a binary classification model, taking into account all four values in the confusion matrix (true positives, true negatives, false positives, and false negatives). It is particularly useful when dealing with imbalanced datasets. MCC ranges from -1 to 1, where 1 indicates perfect prediction, 0 denotes no better than random prediction, and -1 indicates total disagreement between predictions and actual outcomes.

$$MCC = \frac{TP \times TN - FP \times FN}{\sqrt{(TP + FP)(TP + FN)(TN + FP)(TN + FN)}} \tag{6}$$

In addition to the numerical metrics listed above, the performance of classifier algorithms can also be interpreted with graphically displayed metrics. The Receiver Operating Characteristic (ROC) curve and the Area Under the ROC Curve (AUC-ROC) are graphical and numeric metrics, respectively, used to evaluate the performance of binary classification models. The ROC curve is a graphical representation of a model's ability to distinguish between the positive and negative classes across different threshold settings. It plots the true positive rate (sensitivity) against the false positive rate at various classification thresholds. The curve helps visualize the trade-off between sensitivity and specificity. A diagonal line (the line of no-discrimination) represents random guessing, and a curve above it indicates better-than-random performance. An example representation of ROC curve is given in Figure 4.

The Area Under the ROC Curve (AUC-ROC) quantifies the overall performance of a model by calculating the area under the ROC curve. AUC-ROC ranges from 0 to 1, where a higher value indicates better discrimination ability. An AUC-ROC of 0.5 suggests no discrimination (equivalent to random guessing), while an AUC-ROC of 1.0 represents perfect discrimination (Bradley, 1997).

Another useful visual metric is Precision-Recall (PR) curve. The PR curve is a graphical representation of a model's performance in binary classification, particularly when there is an imbalance between positive and negative instances (Figure 5).

Figure 4. An example to ROC curve

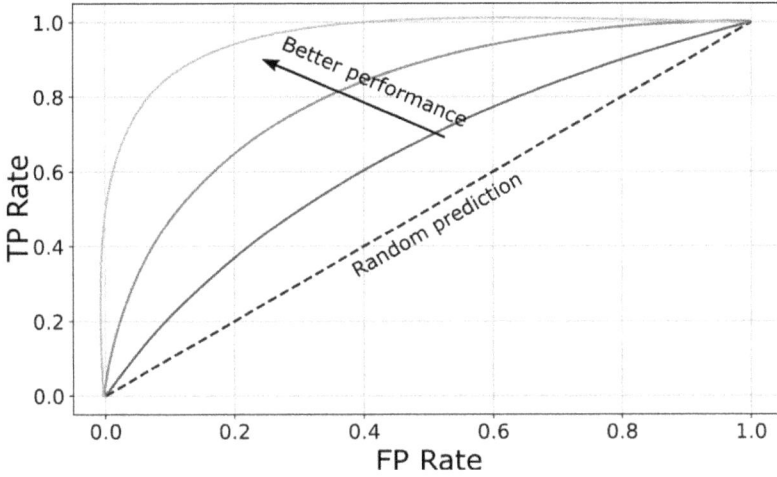

It plots precision (positive predictive value) against recall (sensitivity) at various probability thresholds. The PR curve helps visualize the trade-off between precision and recall at different classification thresholds. The curve typically starts from the point (0,1) representing perfect precision and recall but might curve downward as the threshold decreases, reflecting the compromise between precision and recall.

The Area Under the PR Curve (AUC-PR) quantifies the overall performance of a model, similar to AUC-ROC. A higher AUC-PR indicates better model performance, especially in situations with imbalanced datasets where precision and recall are crucial metrics (Miao & Zhu, 2022).

RESULTS

In this study, binary classification was performed between empty and full container ship classes using the Optical Remote Sensing (ORS) dataset. Pre-trained deep learning models, namely VGG-16, VGG-19 and InceptionV3, were utilized with a fine-tuning process. The models underwent fine-tuning by initializing them with weights pre-trained on the 'imagenet' dataset and constructing a new model on top of them. This process involved excluding the top classification layer, adding a Flatten layer to convert the output into a one-dimensional array, and introducing two Dense layers for further feature extraction and classification.

To prevent overfitting on the relatively small dataset and leverage the knowledge captured by the pre-trained models, a fine-tuning technique was applied by freezing the weights of the pre-trained layers. This meant that, during training, only the

Figure 5. An example to PR curve

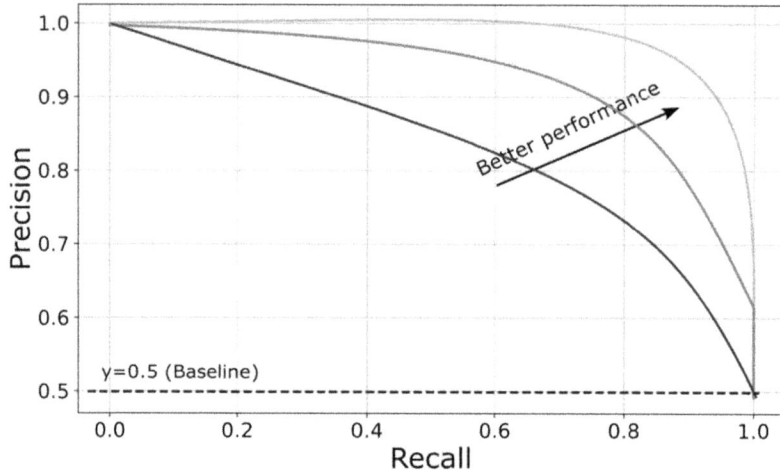

weights of the newly added layers (i.e., the Flatten and Dense layers) were updated, while the pre-trained layers remained fixed.

The models were compiled using the Adaptive Moment Estimation (Adam) optimizer and a binary crossentropy loss function (suitable for binary classification). The learning rate for the Adam optimizer was set to 0.001, carefully chosen to contribute to stable and effective convergence during training.

The ORS dataset containing images of empty and full container ships underwent pre-processing to generate inputs for the aforementioned models. The dataset comprised 431 images of empty container ships and 826 images of full container ships. To address the imbalanced class issue, the images of empty container ships were augmented to a total of 862 images. Subsequently, the dataset was divided into training and testing sets, following an 80:20 ratio. The models were then trained using the prepared training data and evaluated on unseen test data, with the results assessed using the aforementioned metrics.

The VGG16, VGG19, and InceptionV3 models underwent training for 25 epochs, and the evolution of accuracy and loss across these epochs is depicted in Figure 6. Notably, the accuracy curves for all three models surpassed 98% after 10 epochs. Additionally, the loss curves exhibited a significant decline, approaching near-zero values after 10 epochs and maintaining stability thereafter.

The assessment of all three trained networks involved the evaluation of previously unseen test data, allowing for a comparative analysis of their classification performances. Focusing on the VGG16 network, the Figure 7 illustrates key metrics obtained during the testing phase, including the confusion matrix, ROC curve, and PR curve.

Figure 6. Model training evaluation over 25 epochs: (a) accuracy and (b) loss

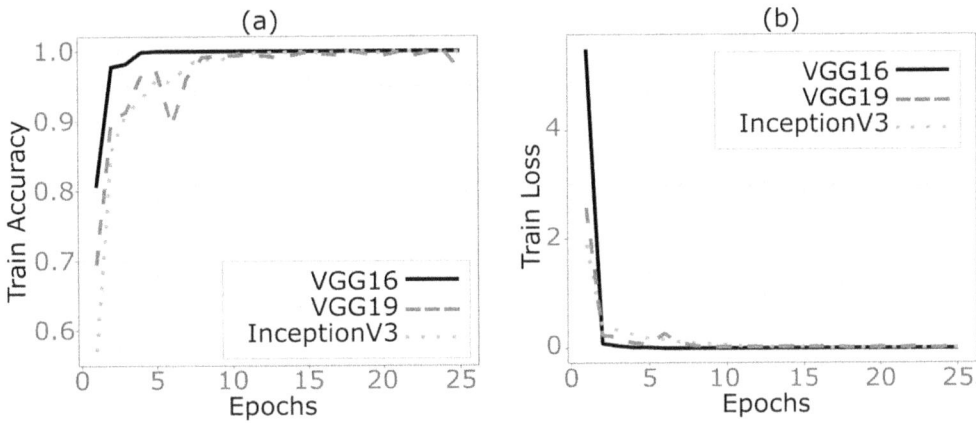

Figure 7. (a) Confusion matrix, (b) ROC curve, and (c) PR curve illustrating the classification results of the VGG16 network

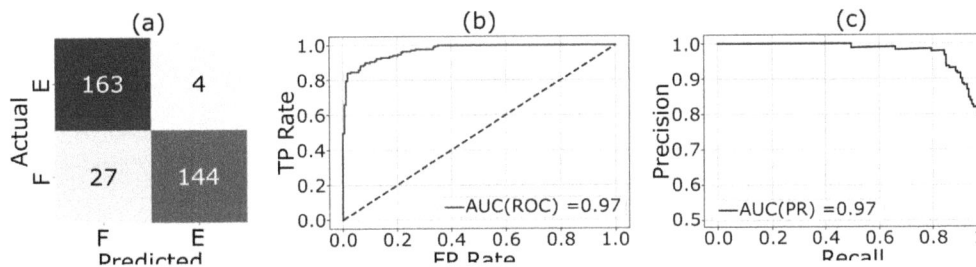

Specifically, the VGG16 network demonstrated commendable performance by accurately identifying 163 instances of empty container ships and 144 instances of full container ships. The misclassification rate was minimal, with only 4 false positives (FPs) and 27 false negatives (FNs).

The achieved AUC ROC (0.97) and AUC PR (0.97) underscore the robustness of the model. These scores, approaching the ideal value of 1.0, signify the network's exceptional discriminatory power and precision in the task of classifying empty and full container ships. Overall, these results validate the effectiveness of the VGG16 network in this classification task.

Figure 8 shows the evaluation metrics obtained during the testing phase of VGG19 network which are the confusion matrix, ROC curve, and PR curve. Comparing the two networks, VGG19 has a lower TP count but also fewer FPs. This suggests that while VGG19 may be slightly less sensitive in identifying full container ships, it is also more conservative in avoiding FPs. The AUC ROC and AUC PR values

Figure 8. (a) Confusion matrix, (b) ROC curve, and (c) PR curve illustrating the classification results of the VGG19 network

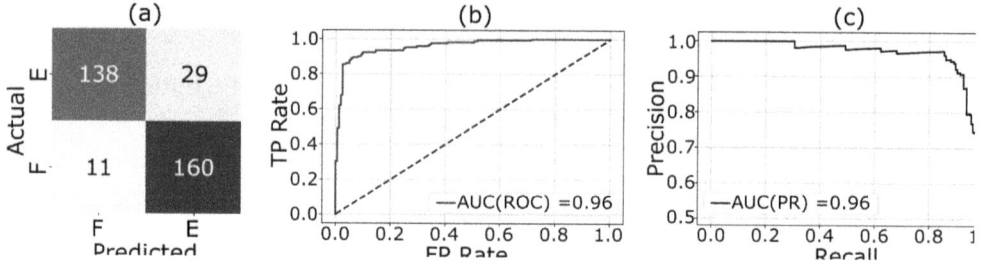

Figure 9. (a) Confusion matrix, (b) ROC curve, and (c) PR curve illustrating the classification results of the InceptionV3 network

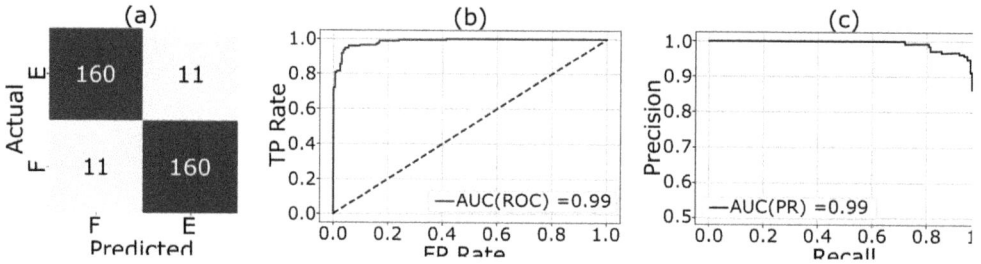

of 0.96 for VGG19 are still quite high, indicating strong discriminatory power and precision in classifying empty and full container ships.

Lastly, Figure 9 illustrates the confusion matrix, ROC curve and PR curve of the InceptionV3 network. The InceptionV3 network achieved the highest true positive count while maintaining a low false positive rate. This suggests that it is more sensitive in identifying full container ships while still effectively minimizing false positives. The AUC ROC and AUC PR values of 0.99 for InceptionV3 are highest among three networks, indicating superior discriminatory power and precision compared to both VGG19 (AUC ROC: 0.96) and VGG16 (AUC ROC: 0.97).

Table 2 presents the comparative results of the classification task of this study. InceptionV3 stands out as the top-performing model in terms of precision, recall, F1-score, MCC, and accuracy. VGG16 also performs admirably, particularly in terms of specificity. VGG19, while still performing well, shows a slightly higher false positive rate and lower specificity compared to the other two models.

Table 2. Comparative performances of fine-tuned deep learning models

Model	Precision	Recall	F1-Score	Specificity	MCC	Accuracy
VGG16	0.92	0.91	0.91	**0.98**	0.82	0.91
VGG19	0.89	0.88	0.88	0.83	0.77	0.88
InceptionV3	**0.95**	**0.95**	**0.95**	0.96	**0.89**	**0.95**

DISCUSSION

This study focused on the problem of classification of empty and full container ships from ORS images using fine-tuned deep learning models. For this purpose, a ORS dataset containing images of empty and full container ships was used. To obtain the input images for the fine-tuned VGG16, VGG19 and the InceptionV3 networks, firstly the images were preprocessed to a standard image size of 512x512 pixels. Then, to overcome the imbalance problem among the classes, the empty container ship images were augmented and a balanced class distribution was obtained. The top three layers of the networks abovementioned were fine-tuned for this specific classification problem. Then the networks were trained with training data and their performance were evaluated with unseen test data.

The comparative results obtained show that all three models obtained remarkable performance on this classification task. Further, the InceptionV3 shows the best classification performance in terms of precision (95%), recall (95%), F1-score (95%), MCC (89%), and accuracy (95%).

The performance disparity between VGG16 and VGG19, despite the latter being a deeper variant, can be comprehended through several considerations in their architectural differences. Firstly, the notion that deeper neural networks invariably result in improved performance is nuanced, as increasing depth introduces complexities such as vanishing gradients or overfitting, particularly when confronted with limited training data. The added depth of VGG19 contributes more parameters, potentially leading to overfitting if not adequately regularized. Moreover, the computational complexity of deeper models demands substantial resources, and the anticipated benefits may not be proportional, necessitating careful consideration of the trade-offs involved. In practice, the efficacy of a deeper model like VGG19 often hinges on meticulous hyperparameter tuning, adapting the network's intricacies to the specific characteristics of the dataset and task complexity.

The superior performance of InceptionV3 compared to the VGG models, VGG16 and VGG19, can be attributed to the distinctive architectural features of InceptionV3. InceptionV3 leverages the concept of inception modules, which utilize multiple convolutional filters of different sizes in parallel, enabling the model to capture information at various scales. This design promotes more efficient feature extraction, enabling the model to discern intricate patterns and representations within the data.

Moreover, InceptionV3 incorporates advanced architectural elements like batch normalization and factorized convolutions, contributing to faster convergence during training. The inception modules, with their parallelized operations, facilitate the extraction of diverse features, enhancing the model's ability to capture complex relationships within the dataset.

Additionally, the InceptionV3 architecture includes global average pooling, reducing the risk of overfitting by decreasing the number of parameters in the fully connected layers. This regularization technique aids in creating a more generalized and robust model, especially beneficial when dealing with limited training data.

Limitations of the Present Study

In acknowledging the study's limitations, several aspects warrant consideration. The relatively modest dataset size and potential class imbalance may impact the generalization capacity of the employed deep learning models, especially given the intricacies of VGG16, VGG19, and InceptionV3 architectures. The data augmentation technique applied to address imbalance introduces a need for cautious interpretation due to the potential memorization of duplicated instances. Further, hyperparameter choices, model interpretability, and the need for comprehensive evaluation across diverse datasets underscore areas for improvement. Lastly, recognizing the study's focus on a specific dataset emphasizes the necessity for caution when generalizing results to other scenarios. These limitations prompt consideration for future research directions and refinement strategies to enhance the study's applicability and robustness.

ACKNOWLEDGMENT

The numerical calculations reported in this chapter were fully performed at TUBITAK ULAKBIM, High Performance and Grid Computing Center (TRUBA resources).

REFERENCES

Aminuddin, J., Abdullatif, R. F., Anggraini, E. I., Gumelar, S. F., & Rahmawati, A. (2023). Development of convolutional neural network algorithm on ships detection in Natuna Islands-Indonesia using land look satellite imagery. *Remote Sensing Applications: Society and Environment, 32*, 101025. https://doi.org/https://doi.org/10.1016/j.rsase.2023.101025

Bansal, M., Kumar, M., Sachdeva, M., & Mittal, A. (2023). Transfer learning for image classification using VGG19: Caltech-101 image data set. *Journal of Ambient Intelligence and Humanized Computing*, *14*(4), 3609–3620. doi:10.1007/s12652-021-03488-z PMID:34548886

Bradley, A. P. (1997). The use of the area under the ROC curve in the evaluation of machine learning algorithms. *Pattern Recognition*, *30*(7), 1145–1159. doi:10.1016/S0031-3203(96)00142-2

Caruana, R., & Niculescu-Mizil, A. (2006). An Empirical Comparison of Supervised Learning Algorithms. *Proceedings of the 23rd International Conference on Machine Learning*, 161–168. 10.1145/1143844.1143865

Chen, X., Liu, Y., Achuthan, K., & Zhang, X. (2020). A ship movement classification based on Automatic Identification System (AIS) data using Convolutional Neural Network. *Ocean Engineering*, *218*, 108182. doi:10.1016/j.oceaneng.2020.108182

Cheng, G., & Han, J. (2016). A survey on object detection in optical remote sensing images. *ISPRS Journal of Photogrammetry and Remote Sensing*, *117*, 11–28. doi:10.1016/j.isprsjprs.2016.03.014

Dong, C., Liu, J., & Xu, F. (2018). Ship Detection in Optical Remote Sensing Images Based on Saliency and a Rotation-Invariant Descriptor. *Remote Sensing (Basel)*, *10*(3), 400. Advance online publication. doi:10.3390/rs10030400

Dong, C., Liu, J., Xu, F., & Liu, C. (2019). Ship Detection from Optical Remote Sensing Images Using Multi-Scale Analysis and Fourier HOG Descriptor. *Remote Sensing (Basel)*, *11*(13), 1529. Advance online publication. doi:10.3390/rs11131529

Dong, Y., Zhang, H., Wang, C., & Wang, Y. (2019). Fine-grained ship classification based on deep residual learning for high-resolution SAR images. *Remote Sensing Letters*, *10*(11), 1095–1104. doi:10.1080/2150704X.2019.1650982

Gallego, A.-J., Pertusa, A., & Gil, P. (2018). Automatic ship classification from optical aerial images with convolutional neural networks. *Remote Sensing (Basel)*, *10*(4), 511. doi:10.3390/rs10040511

He, J., Wang, Y., & Liu, H. (2021). Ship Classification in Medium-Resolution SAR Images via Densely Connected Triplet CNNs Integrating Fisher Discrimination Regularized Metric Learning. *IEEE Transactions on Geoscience and Remote Sensing*, *59*(4), 3022–3039. doi:10.1109/TGRS.2020.3009284

Huang, I.-L., Lee, M.-C., Nieh, C.-Y., & Huang, J.-C. (2024). Ship Classification Based on AIS Data and Machine Learning Methods. *Electronics (Basel)*, *13*(1), 98. Advance online publication. doi:10.3390/electronics13010098

Huang, L., Li, W., Chen, C., Zhang, F., & Lang, H. (2018). Multiple features learning for ship classification in optical imagery. *Multimedia Tools and Applications*, *77*(11), 13363–13389. doi:10.1007/s11042-017-4952-y

Jiang, M., Yang, X., Dong, Z., Fang, S., & Meng, J. (2016). Ship Classification Based on Superstructure Scattering Features in SAR Images. *IEEE Geoscience and Remote Sensing Letters*, *13*(5), 616–620. doi:10.1109/LGRS.2016.2514482

Lang, H., Li, C., & Xu, J. (2022). Multisource heterogeneous transfer learning via feature augmentation for ship classification in SAR imagery. *IEEE Transactions on Geoscience and Remote Sensing*, *60*, 1–14. doi:10.1109/TGRS.2022.3178703

Lang, H., & Wu, S. (2017). Ship classification in moderate-resolution SAR image by naive geometric features-combined multiple kernel learning. *IEEE Geoscience and Remote Sensing Letters*, *14*(10), 1765–1769. doi:10.1109/LGRS.2017.2734889

LangH.YangG. (2022). *ORS Dataset*. IEEE Dataport. doi:10.21227/y4f3-wh22

Lang, H., Zhang, J., Zhang, X., & Meng, J. (2016). Ship classification in SAR image by joint feature and classifier selection. *IEEE Geoscience and Remote Sensing Letters*, *13*(2), 212–216. doi:10.1109/LGRS.2015.2506570

Li, K., Wan, G., Cheng, G., Meng, L., & Han, J. (2020). Object detection in optical remote sensing images: A survey and a new benchmark. *ISPRS Journal of Photogrammetry and Remote Sensing*, *159*, 296–307. doi:10.1016/j.isprsjprs.2019.11.023

Liu, K., Yu, S., & Liu, S. (2020). An improved InceptionV3 network for obscured ship classification in remote sensing images. *IEEE Journal of Selected Topics in Applied Earth Observations and Remote Sensing*, *13*, 4738–4747. doi:10.1109/JSTARS.2020.3017676

Luque, A., Carrasco, A., Martín, A., & de las Heras, A. (2019). The impact of class imbalance in classification performance metrics based on the binary confusion matrix. *Pattern Recognition*, *91*, 216–231. doi:10.1016/j.patcog.2019.02.023

Miao, J., & Zhu, W. (2022). Precision–recall curve (PRC) classification trees. *Evolutionary Intelligence*, *15*(3), 1545–1569. doi:10.1007/s12065-021-00565-2

Moreira, A., Prats-Iraola, P., Younis, M., Krieger, G., Hajnsek, I., & Papathanassiou, K. P. (2013). A tutorial on synthetic aperture radar. *IEEE Geoscience and Remote Sensing Magazine*, *1*(1), 6–43. doi:10.1109/MGRS.2013.2248301

Sánchez Pedroche, D., Amigo, D., García, J., & Molina, J. M. (2020). Architecture for Trajectory-Based Fishing Ship Classification with AIS Data. *Sensors (Basel)*, *20*(13), 3782. Advance online publication. doi:10.3390/s20133782 PMID:32640561

Simonyan, K., & Zisserman, A. (2014). Very deep convolutional networks for large-scale image recognition. *ArXiv Preprint ArXiv:1409.1556*.

Sowmya, D. R., Deepa Shenoy, P., & Venugopal, K. R. (2017). Remote sensing satellite image processing techniques for image classification: A comprehensive survey. *International Journal of Computer Applications*, *161*(11), 24–37. doi:10.5120/ijca2017913306

Szegedy, C., Vanhoucke, V., Ioffe, S., Shlens, J., & Wojna, Z. (2016). Rethinking the inception architecture for computer vision. *Proceedings of the IEEE Conference on Computer Vision and Pattern Recognition*, 2818–2826. 10.1109/CVPR.2016.308

Theckedath, D., & Sedamkar, R. R. (2020). Detecting affect states using VGG16, ResNet50 and SE-ResNet50 networks. *SN Computer Science*, *1*(2), 1–7. doi:10.1007/s42979-020-0114-9

Wang, Q. W., Shen, F., Cheng, L., Jiang, J., He, G., Sheng, W., Jing, N., & Zhigang, M. (2021). Ship detection based on fused features and rebuilt YOLOv3 networks in optical remote-sensing images. *International Journal of Remote Sensing*, *42*(2), 520–536. doi:10.1080/01431161.2020.1811422

Wu, J., Zhu, Y., Wang, Z., Song, Z., Liu, X., Wang, W., Zhang, Z., Yu, Y., Xu, Z., Zhang, T., & Zhou, J. (2017). A novel ship classification approach for high resolution SAR images based on the BDA-KELM classification model. *International Journal of Remote Sensing*, *38*(23), 6457–6476. doi:10.1080/01431161.2017.1356487

Xiao, S., Zhang, Y., & Chang, X. (2022). Ship Detection Based on Compressive Sensing Measurements of Optical Remote Sensing Scenes. *IEEE Journal of Selected Topics in Applied Earth Observations and Remote Sensing*, *15*, 8632–8649. doi:10.1109/JSTARS.2022.3209024

Yan, Z., Song, X., Zhong, H., Yang, L., & Wang, Y. (2022). Ship Classification and Anomaly Detection Based on Spaceborne AIS Data Considering Behavior Characteristics. *Sensors (Basel)*, *22*(20), 7713. Advance online publication. doi:10.3390/s22207713 PMID:36298063

Zhang, S., Wu, R., Xu, K., Wang, J., & Sun, W. (2019). R-CNN-Based Ship Detection from High Resolution Remote Sensing Imagery. *Remote Sensing (Basel)*, *11*(6), 631. Advance online publication. doi:10.3390/rs11060631

Zhang, Y., Sheng, W., Jiang, J., Jing, N., Wang, Q., & Mao, Z. (2020). Priority Branches for Ship Detection in Optical Remote Sensing Images. *Remote Sensing (Basel)*, *12*(7), 1196. Advance online publication. doi:10.3390/rs12071196

Zhou, Y., Daamen, W., Vellinga, T., & Hoogendoorn, S. P. (2019). Ship classification based on ship behavior clustering from AIS data. *Ocean Engineering*, *175*, 176–187. doi:10.1016/j.oceaneng.2019.02.005

Zhuang, Y., Qi, B., Chen, H., Bi, F., Li, L., & Xie, Y. (2018). Locally Oriented Scene Complexity Analysis Real-Time Ocean Ship Detection from Optical Remote Sensing Images. *Sensors (Basel)*, *18*(11), 3799. Advance online publication. doi:10.3390/s18113799 PMID:30404224

Chapter 3
Mapping Faces From Above:
Exploring Face Recognition Algorithms and Datasets for Aerial Drone Images

Sadique Ahmad
ⓘ https://orcid.org/0000-0001-6907-2318
College of Computer and Information Sciences, Prince Sultan University, Saudi Arabia

Mohammed A. El Affendi
ⓘ https://orcid.org/0000-0001-9349-1985
College of Computer and Information Sciences, Prince Sultan University, Saudi Arabia

Mahmood Ul Haq
ⓘ https://orcid.org/0000-0002-1514-0300
University of Engineering and Technology, Peshawar, Pakistan

Zahid Farid
Abasyn University, Pakistan

Alaa Sal. Al Luhaidan
ⓘ https://orcid.org/0000-0001-6829-9705
College of Computer and Information Sciences, Princess Nourah bint Abdulrahman University, Pakistan

Muhammad Athar Javed Sethi
ⓘ https://orcid.org/0000-0001-7847-831X
University of Engineering and Technology, Peshawar, Pakistan

ABSTRACT

This chapter explains various facial identification algorithms used in drones. The burgeoning field of face recognition technology makes use of image processing to identify faces in people. Face recognition is becoming increasingly popular for a variety of reasons, such as the growing population, which necessitates higher security and surveillance systems, identity verification in the digital age, combat in rural regions, disaster relief, and so forth. This research compares and contrasts various face identification techniques, including neural networks, PCA, LBPH (local binary pattern histogram), PAL, capsule network, and LDA (linear discriminant analysis).
DOI: 10.4018/979-8-3693-2913-9.ch003

Additionally, a comparison of face datasets collected with drones is included in this chapter. The results of this study will help facial recognition technologists design a hybrid algorithm that meets the requirements of real-time applications.

INTRODUCTION

Drones, or unmanned aerial vehicles (UAVs), are a crucial component of military and public security scenarios in a world where video surveillance has become a crucial aspect. Nowadays, everyone can afford and operate these vehicles due to their ease of use (Hosni et al., 2019). The fact that they are nearly imperceptible to radar and are able to transport a payload such as an optical zoom high-resolution camera is a major advantage. Facial recognition is one of the most popular aerial surveillance applications, partly due to the significant advancements in deep learning that have fueled this field of study (Ullah et al., 2019).

The face is the essential component of an individual that is mostly utilized to identify them. Even if people are able to identify familiar faces, it gets harder and harder to identify unfamiliar faces. This is the point at which an automated system that can recognize people just as well as humans did was created. Nowadays, face recognition technology is widely employed in practically every aspect of life (Anwar et al., 2020; Rahim et al., 2023). Face recognition is being used in real-time in a variety of industries, including law enforcement, immigration checks, forensic investigations, attendance tracking, disease diagnosis, and more (Hosni et al., 2018). The demand for items with additional features has increased due to these rising technologies. As a result, technological fusion developed (Haq et al., 2022). But there are different benefits and drawbacks associated with each algorithm. To make the most of face recognition technology, this research has integrated it with drone technology (Haq et al., 2024). Different machine learning and deep learning algorithms can be used to create face recognition drones, but each approach has pros and cons of its own.

There are three main steps in the facial recognition system's identification process (Ahmad et al., 2022). As seen in Fig. 1, they are Acquisition, Extraction, and Recognition. Acquisition is the process of gathering and organizing photographs of people taken from various viewpoints, lights, and expressions into a database. The process of extraction involves gathering distinct facial traits, such as cheekbones, jaw line length, and eye spacing, in order to compare and identify each one (Siddiqui et al., 2023). By comparing the individual face with the faces in the database, the match is discovered (Fatima et al., 2022).

This research aims to examine various algorithms and datasets that have nearly resolved the predetermined parameters influencing recognition accuracy and weigh the advantages and disadvantages of each approach.

Figure 1. Face identification

ALGORITHMS FOR DRONE FACE RECOGNITION

Face identification for photos taken by drones poses special difficulties because of differences in resolutions, positions, and illumination. Drone face photos have led to the adaptation or creation of numerous face recognition algorithms.

Eigen Faces

It involves calculating the principal components and applying them to alter the data; occasionally, only the first few principal components are used, with the remaining principal components being ignored (Kshirsagar et al., 2011). It is a dimensionality reduction approach that minimizes the amount of variance in the dataset while reducing it from a larger set to a smaller set of variables. In machine learning, this algorithm is an unsupervised learning algorithm. This algorithm's main constituents are Eigen values, Eigen vectors, variance, and covariance. Obtaining the dataset and splitting it into a training set and a validation set is the first stage. Then, a two-dimensional matrix with rows representing data items and columns representing features should be used to represent these datasets. Eigen values and Eigen vectors are calculated using the covariance of the matrix, which is discovered after the data has been standardized. These eigenvectors are ordered from large to small, and the irrelevant features are eliminated from the dataset.

Linear Discriminant Analysis

Projecting higher-dimensional features into lower-dimensional regions is the method of dimensionality reduction known as LDA, also referred to as fisher's discriminant analysis (Tharwat et al., 2017). This algorithm is used to segregate images from different classes and to group images from the same class together. As a supervised machine learning algorithm, LDA uses two classes: in-class scatter and between-class dispersion. LDA finds the set of projecting vectors by increasing the ratio of the determinant of the in-class disperse matrix to the determinant of the between-class disperse matrix. The first step in LDA is the calculation of variance within and between classes. Next, by increasing the between-class variance and lowering

57

the within-class variation, a lower dimensional space is produced, as shown in Fig. 2. Similar to PCA, LDA searches for linear combinations of variables that provide the most consistent explanation for the data . A lot of work goes into modeling the difference between knowledge classes in LDA.

LDA is inexpensive, simple to use, and highly efficient in using memory. LDA's drawbacks include its sensitivity to overfitting and its limited sample size problem. The engineers looked into more advanced algorithms as a result of LDA's limitations.

Local Binary Pattern (LBP/LBPH)

LBPH is one of the texture descriptors that is frequently used in face identification. In LBPH, the image is divided into many regions, and the surrounding pixels are used to mine the features (Ahonen et al., 2004). Once the neighborhood of each pixel is thresholded, the result is regarded as a binary number. One is assigned to the neighbor pixel if the threshold pixel (central pixel) has a lower gray value than the neighbor pixel; if not, zero is assigned.

As illustrated in Fig. 3, the four key stages of LBPH are dataset generation, face acquisition, face extraction, and classification. The LBPH has little trouble converting grayscale to monotonic. The Euclidean distance is computed by comparing the attributes of the test image with those of the dataset. The smallest distance between the test and the original image determines the matching rate. The process of LBPH face detection involves gray-scaling and cropping the face.

Figure 2. LDA

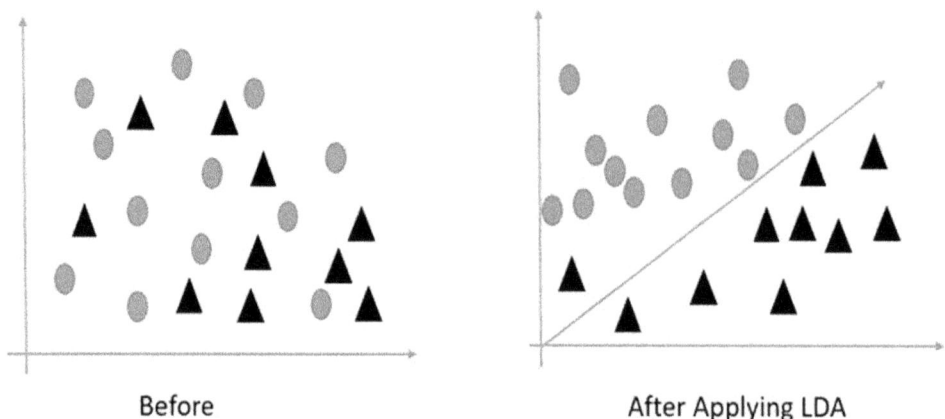

Before   After Applying LDA

Figure 3. Local binary pattern

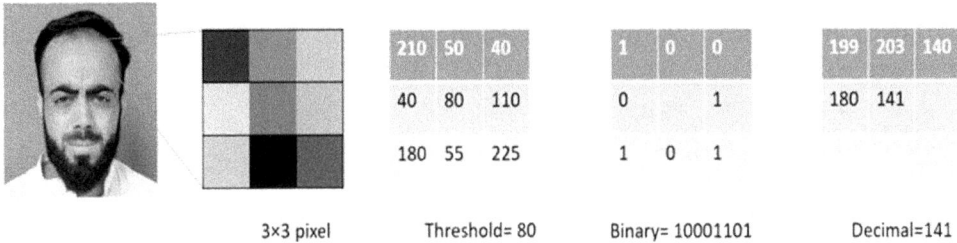

		210	50	40		1	0	0		199	203	140	
		40	80	110		0		1		180	141		
		180	55	225		1	0	1					

3×3 pixel	Threshold= 80	Binary= 10001101	Decimal=141

Neural Networks (NN)

This is a powerful facial recognition approach that aids in the identification of both known and unknown faces. Nodes—also referred to as artificial neurons—make up neural networks. Each and every node is connected to every other node. Based on the strength of the nodes, values are assigned. A higher value indicates that the nodes are more strongly connected. Neural networks' several layers are the output, while their hidden input is the input. Deep learning is the foundation of neural networks. A perspective called a neural network is utilized to teach computers through programs or to analyze data for certain objectives (Kasar et al., 2016). Neural networks are used in several ways for face identification. Radial Basis Function Networks (RBF), Convolutional Neural Networks (CNN), Back Propagation Networks (BPN), and Artificial Neural Networks (ANN) are frequently employed.

An ANN is thought of as a directed weighted graph. There are weighted directed edges connecting the inputs and outputs of the neurons. The input signal for an ANN is obtained as a pattern, and the images are obtained as vectors (Agarwal et al., 2010). CNN uses image identification and classification to identify objects and distinguish faces.

CNN is made up of neurons that have dynamic weights and biases. After receiving a high volume of inputs, each neuron applies a weighted aggregate to them. After passing the weighted sum through an activation function, it produces an output in response. The convolutional layer, activation layer, and pooling layer are the stages of the CNN network (Hangaragi et al., 2023).

Back propagation can be used to train multilayer feed-forward networks with differentiable transfer functions so they can be used for pattern classification, function approximation, and pattern connection. The BPN is composed of one input layer, one hidden layer, and one output layer. This network receives feature vectors as input from the feature extraction techniques.

In order to locate faces effectively, neural networks have a number of advantages. These days, many people employ these algorithms. Researchers are still working toward full accuracy even though these algorithms have achieved 98% accuracy.

PAL-Based FR Algorithm

PAL based face recognition technique, Haq et al. (2019) first locates a face in the input image using 68 points, and then it utilizes PCA to extract the mean image in part. Face feature extraction is done in the interim using the AdaBoost and LDA algorithms. Face categorization is done in the last step using the traditional closest center classifier. This technique have surpasses current cutting-edge face recognition algorithms in high recognition rate and significantly lower error rate under extremely difficult conditions. PAL based face recognition model has been tested on several FR dataset such as the LFW and CMU Multi-PIE databases.

Capsule Network

In essence, a capsule network is a kind of neural network that can produce reversed visuals. For instance, in object detection, the item is divided into smaller components (Sreekala et al., 2022). To represent that object, a hierarchical relationship is created between each subpart. The implementation of CapsNet is divided into three main parts. The hidden layer, the output layer, and the input layer are these. Figure presents the structural network of capsule network.

Figure 4. Capsule network

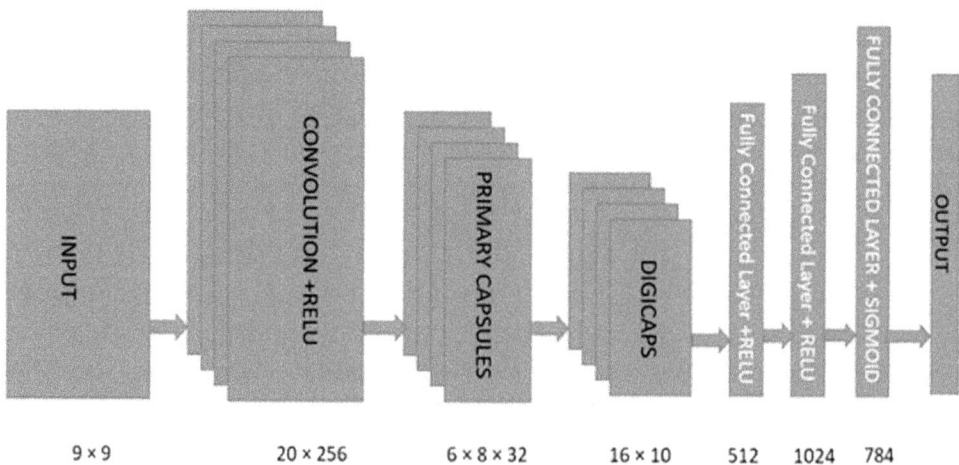

Using CapsNet for face recognition, the authors of Chui et al. (2019) extended their research after obtaining better results on the MNIST dataset. A three-layer capsule network including two convolutional layers and one fully connected layer was employed in the study. They outperformed conventional CNNs with an accuracy of 93.7% using the LFW dataset. A comparative examination of CapsNet with Fisherfaces, LeNet, and ResNet was reported by Mukhometzianov and Carrillo (2018). Four datasets containing faces, traffic signs, and other items were used to evaluate these techniques. According to the authors, the CapsNet method performs well on a limited dataset. For each class, CapsNet needs additional example photos in order to achieve a lower mistake rate.

CapsNet has been extensively used by researchers in a variety of computer vision applications such as face recognition, human activity recognition (HAR), human position estimation, object detection, object segmentation, image classification, and so on. CapsNets have shown promising results in a variety of computer vision applications. Fault identification (Wang et al., 2020; Chen et al., 2020; Chen et al., 2019; Wang et al., 2019), human action recognition (Jaysundara et al., 2021; Duarte et al., 2019), medical imaging (Wang et al., 2020; Mobiny and Van Ngyuen, 2018; Quan et al., 2021), forged images (Afchar et al., 2018; Sridhar and Sanagavarapu, 2021), agricultural image classification and segmentation (Waweru et al., 2021; Cai et al., 2021), monitoring evaluation (Tampubolon et al., 2019; Janakiramaiah et al., 2023), multi-label emotion classification (Fei et al., 2020; Liu et al., 2023), and other real-world problems (Marchisio et al., 2023; Kanth and Jacob, 2023) have been solved using these networks.

Table 1 provides a clear illustration of the benefits and drawbacks of facial recognition algorithms, allowing the researcher to choose the optimal method for their needs.

DATASET FOR DRONE FACE RECOGNITION

The process of building a drone facial recognition dataset requires a thorough collection and annotation of pictures or videos taken by drones in various situations. This procedure makes sure that people's consent is acquired for the use of their facial data, which necessitates careful consideration of consent, privacy protection, and adherence to legal and ethical requirements. Age, gender, ethnicity, and facial expressions should all be included in the dataset, along with difficult circumstances like changing lighting, unfavorable weather, and diverse backgrounds. A comprehensive dataset benefits from proper structure, documentation, and augmentation approaches. While keeping an eye on building a balanced dataset to reduce biases, researchers may also investigate already-existing face recognition

Table 1. Face recognition algorithms comparison

Algorithm	Merits	Demerits
Eigen Face	- Efficient for dimensionality reduction, reducing computational complexity. - Simple and easy to implement. - Works well with small datasets.	- Assumes linearity in data distribution, which may limit its effectiveness in complex scenarios. - May not handle non-linear variations well.
LDA (Linear Discriminant Analysis)	- Maximizes class separability, leading to enhanced discrimination. - Effective when classes are well-separated. - Reduces dimensionality while preserving discriminative information.	- Sensitive to outliers. - Requires the assumption of normality and homoscedasticity. - May not perform well in the presence of small sample sizes.
LBP (Local Binary Pattern)	- Robust to changes in illumination and facial expressions. - Computationally efficient. - Texture-based approach that captures facial details effectively.	- May not handle significant variations in pose. - Sensitive to noise. - Global LBP may not capture fine-grained details in local facial regions.
Neural Network	- Highly adaptable to complex, non-linear data distributions. - Effective feature learning capabilities. - Can handle large datasets.	- Requires substantial computational resources for training. - May suffer from overfitting, especially with limited data. - Prone to adversarial attacks.
PAL	- Robust to changes in illumination and facial expressions. - Works well with small datasets. Maximizes class separability, leading to enhanced discrimination. - Effective when classes are well-separated. - Reduces dimensionality while preserving discriminative information	-Sensitive to outliers. - Requires the assumption of normality and homoscedasticity.
Capsule Network	- Inherently designed to handle hierarchical and spatial relationships in data. - Potential to address pose variations better than traditional CNNs. - Robust to small changes in input.	- Computationally intensive, making training slower. - Limited research compared to traditional CNNs. - Limited availability of pre-trained models and datasets.

datasets and modify them to fit drone-based scenarios. Overall, the efficacy of the resulting drone facial recognition model is greatly influenced by the caliber and variety of the dataset.

There are additional difficulties associated with facial identification from drones, including motion, stance, lighting, background, and height. Due to these difficulties, as well as the comparatively low resolution of the faces that are collected and the fluctuating distance between the drone and the subjects, the problem of drone-based face identification becomes even more difficult. To the best of our knowledge, not much work has been done on automating face recognition using drones. Recently, Bindemann et al. (2017) examined how well people performed when trying to identify people in drone-captured films. The difficulty of the task and the low face identification performance were established by the authors. The DroneFace dataset was proposed by Hsu and Chen (2019), who also assessed the effectiveness of commercial systems and current methods. It's vital to remember that the DroneFace dataset uses a stationary GoPro camera to replicate data taken by a drone. To capture simply the camera's height effect, the subjects are instructed to look in a single direction while maintaining neutral emotions and staying motionless. They are also asked to look without glasses. Due to the very limited conditions, the study does not give a true image of drone-based facial recognition, but it does give an overview of the difficult nature of the problem. The IJB-S dataset was recently proposed by Kalka et al. (2018) and includes ten UAV-based movies for face recognition. In addition to the aforementioned study, several law enforcement and military groups use unmanned aerial vehicles (UAVs) to monitor borders and are investigating the possibility of utilizing drone technology for rescue operations. Table 2 presents comparison of these drone face datasets

CONCLUSION

In conclusion, the landscape of facial recognition algorithms applied to drone technology is examined in this chapter. Facial recognition has become a key technology as the demand for strong security measures and identity verification has increased. This research offers insights essential for creating hybrid algorithms suitable for real-time applications in a diversity of contexts, from security to disaster assistance, by contrasting and comparing different methodologies and datasets.

Table 2. Drone face datasets

Ref	Dataset	Images/ Videos	Subjects	Conditions	Drone Mobility
(Kalra et al., 2019)	DroneSURF	200 videos	58 subjects	changes in resolution, altitude, background, illumination, and attitude	Yes
(Hsu and Chen, 2017)	DroneFace	2057 images	11	Pose with background variations	No
(Luo et al., 2020)	Dronelams	261 drone videos	10k	Variability in facial expression and situation	yes
(Kalka et al., 2018)	IJB-S	5656 images and 404 videos	202	changes in resolution, altitude, background, illumination, and attitude	Yes

REFERENCES

Afchar, D., Nozick, V., Yamagishi, J., & Echizen, I. (2018, December). Mesonet: a compact facial video forgery detection network. In *2018 IEEE international workshop on information forensics and security (WIFS)* (pp. 1-7). IEEE. doi:10.1109/WIFS.2018.8630761

Agarwal, M., Jain, N., Kumar, M. M., & Agrawal, H. (2010). Face recognition using eigen faces and artificial neural network. *International Journal of Computer Theory and Engineering*, 2(4), 624–629. doi:10.7763/IJCTE.2010.V2.213

Ahmad, S., El-Affendi, M. A., Anwar, M. S., & Iqbal, R. (2022). Potential future directions in optimization of students' performance prediction system. *Computational Intelligence and Neuroscience*, 2022, 2022. doi:10.1155/2022/6864955 PMID:35619762

Ahonen, T., Hadid, A., & Pietikäinen, M. (2004). Face recognition with local binary patterns. In *Computer Vision-ECCV 2004: 8th European Conference on Computer Vision, Prague, Czech Republic, May 11-14, 2004.* [Springer Berlin Heidelberg.]. *Proceedings*, 8(Part I), 469–481.

Anwar, M. S., Wang, J., Ahmad, S., Khan, W., Ullah, A., Shah, M., & Fei, Z. (2020). Impact of the impairment in 360-degree videos on users VR involvement and machine learning-based QoE predictions. *IEEE Access : Practical Innovations, Open Solutions*, 8, 204585–204596. doi:10.1109/ACCESS.2020.3037253

Bindemann, M., Fysh, M. C., Sage, S. S., Douglas, K., & Tummon, H. M. (2017). Person identification from aerial footage by a remote-controlled drone. *Scientific Reports*, 7(1), 13629. doi:10.1038/s41598-017-14026-3 PMID:29051619

Cai, W., Wei, Z., Song, Y., Li, M., & Yang, X. (2021). Residual-capsule networks with threshold convolution for segmentation of wheat plantation rows in UAV images. *Multimedia Tools and Applications*, 80(21-23), 32131–32147. doi:10.1007/s11042-021-11203-5

Chen, L., Qin, N., Dai, X., & Huang, D. (2020). Fault diagnosis of high-speed train bogie based on capsule network. *IEEE Transactions on Instrumentation and Measurement*, 69(9), 6203–6211. doi:10.1109/TIM.2020.2968161

Chen, T., Wang, Z., Yang, X., & Jiang, K. (2019). A deep capsule neural network with stochastic delta rule for bearing fault diagnosis on raw vibration signals. *Measurement*, 148, 106857. doi:10.1016/j.measurement.2019.106857

Chui, A., Patnaik, A., Ramesh, K. & Wang, L. (2019). *Capsule Networks and Face Recognition*. Lindawangg. github. io.

Duarte, K., Rawat, Y. S., & Shah, M. (2019). Capsulevos: Semi-supervised video object segmentation using capsule routing. In *Proceedings of the IEEE/CVF international conference on computer vision* (pp. 8480-8489). 10.1109/ICCV.2019.00857

Fatima, R., Samad Shaikh, N., Riaz, A., Ahmad, S., El-Affendi, M. A., Alyamani, K. A., Nabeel, M., Ali Khan, J., Yasin, A., & Latif, R. M. A. (2022). A natural language processing (NLP) evaluation on COVID-19 rumour dataset using deep learning techniques. *Computational Intelligence and Neuroscience, 2022*, 2022. doi:10.1155/2022/6561622 PMID:36156967

Fei, H., Ji, D., Zhang, Y., & Ren, Y. (2020). Topic-enhanced capsule network for multi-label emotion classification. *IEEE/ACM Transactions on Audio, Speech, and Language Processing, 28*, 1839–1848. doi:10.1109/TASLP.2020.3001390

Hangaragi, S., Singh, T., & Neelima, N. (2023). Face detection and Recognition using Face Mesh and deep neural network. *Procedia Computer Science, 218*, 741–749. doi:10.1016/j.procs.2023.01.054

Haq, M. U., Sethi, M. A. J., Ahmad, S., ELAffendi, M. A., & Asim, M. (2024). Automatic Player Face Detection and Recognition for Players in Cricket Games. *IEEE Access: Practical Innovations, Open Solutions, 12*, 41219–41233. doi:10.1109/ACCESS.2024.3377564

Haq, M. U., Sethi, M. A. J., Ullah, R., Shazhad, A., Hasan, L., & Karami, G. M. (2022). COMSATS Face: A Dataset of Face Images with Pose Variations, Its Design, and Aspects. *Mathematical Problems in Engineering, 2022*, 2022. doi:10.1155/2022/4589057

Haq, M. U., Shahzad, A., Mahmood, Z., Shah, A. A., Muhammad, N., & Akram, T. (2019). Boosting the face recognition performance of ensemble based LDA for pose, non-uniform illuminations, and low-resolution images. *KSII Transactions on Internet and Information Systems, 13*(6), 3144–3164.

Hosni, A. I. E., Li, K., & Ahmad, S. (2019, December). DARIM: Dynamic approach for rumor influence minimization in online social networks. In *International Conference on Neural Information Processing* (pp. 619-630). Cham: Springer International Publishing. 10.1007/978-3-030-36711-4_52

Hosni, A. I. E., Li, K., & Ahmed, S. (2018). HISBmodel: A rumor diffusion model based on human individual and social behaviors in online social networks. *Neural Information Processing: 25th International Conference, ICONIP 2018, Siem Reap, Cambodia, December 13–16, 2018 Proceedings, 25*(Part II), 14–27.

Hsu, H. J., & Chen, K. T. (2017, June). DroneFace: an open dataset for drone research. In *Proceedings of the 8th ACM on multimedia systems conference* (pp. 187-192). 10.1145/3083187.3083214

Janakiramaiah, B., Kalyani, G., Karuna, A., Prasad, L. N., & Krishna, M. (2023). Military object detection in defense using multi-level capsule networks. *Soft Computing, 27*(2), 1045–1059. doi:10.1007/s00500-021-05912-0

Jayasundara, V., Roy, D., & Fernando, B. (2021). Flowcaps: Optical flow estimation with capsule networks for action recognition. In *Proceedings of the IEEE/CVF winter conference on applications of computer vision* (pp. 3409-3418). 10.1109/WACV48630.2021.00345

Kalka, N. D., Maze, B., Duncan, J. A., O'Connor, K., Elliott, S., Hebert, K., . . . Jain, A. K. (2018, October). Ijb–s: Iarpa janus surveillance video benchmark. In *2018 IEEE 9th international conference on biometrics theory, applications and systems (BTAS)* (pp. 1-9). IEEE.

Kalra, I., Singh, M., Nagpal, S., Singh, R., Vatsa, M., & Sujit, P. B. (2019, May). Dronesurf: Benchmark dataset for drone-based face recognition. In *2019 14th IEEE International Conference on Automatic Face & Gesture Recognition (FG 2019)* (pp. 1-7). IEEE.

Kanth, R. R., & Jacob, T. P. (2023, April). Enhanced capsule generative adversarial network with Blockchain fostered Intrusion Detection System for Enhancing Cyber security in Cloud. In *2023 2nd International Conference on Smart Technologies and Systems for Next Generation Computing (ICSTSN)* (pp. 1-6). IEEE. 10.1109/ICSTSN57873.2023.10151609

Kasar, M. M., Bhattacharyya, D., & Kim, T. H. (2016). Face recognition using neural network: A review. *International Journal of Security and Its Applications, 10*(3), 81–100. doi:10.14257/ijsia.2016.10.3.08

Kshirsagar, V. P., Baviskar, M. R., & Gaikwad, M. E. (2011, March). Face recognition using Eigenfaces. In *2011 3rd International Conference on Computer Research and Development* (Vol. 2, pp. 302-306). IEEE. 10.1109/ICCRD.2011.5764137

Liu, S., Wang, Z., An, Y., Zhao, J., Zhao, Y., & Zhang, Y. D. (2023). EEG emotion recognition based on the attention mechanism and pre-trained convolution capsule network. *Knowledge-Based Systems*, *265*, 110372. doi:10.1016/j.knosys.2023.110372

Luo, Y., Chen, S., & Ma, X. G. (2020). Drone lams: A drone-based face detection dataset with large angles and many scenarios. *arXiv preprint arXiv:2011.07689*.

Marchisio, A., Nanfa, G., Khalid, F., Hanif, M. A., Martina, M., & Shafique, M. (2023). SeVuc: A study on the Security Vulnerabilities of Capsule Networks against adversarial attacks. *Microprocessors and Microsystems*, *96*, 104738. doi:10.1016/j.micpro.2022.104738

Mobiny, A., & Van Nguyen, H. (2018, September). Fast capsnet for lung cancer screening. In *International conference on medical image computing and computer-assisted intervention* (pp. 741-749). Cham: Springer International Publishing.

Mukhometzianov, R., & Carrillo, J. (2018). *CapsNet comparative performance evaluation for image classification.* arXiv preprint arXiv:1805.11195.

Quan, H., Xu, X., Zheng, T., Li, Z., Zhao, M., & Cui, X. (2021). DenseCapsNet: Detection of COVID-19 from X-ray images using a capsule neural network. *Computers in Biology and Medicine*, *133*, 104399. doi:10.1016/j.compbiomed.2021.104399 PMID:33892307

Rahim, A., Zhong, Y., Ahmad, T., Ahmad, S., Pławiak, P., & Hammad, M. (2023). Enhancing smart home security: Anomaly detection and face recognition in smart home iot devices using logit-boosted cnn models. *Sensors (Basel)*, *23*(15), 6979. doi:10.3390/s23156979 PMID:37571762

Siddiqui, H. U. R., Younas, F., Rustam, F., Flores, E. S., Ballester, J. B., Diez, I. D. L. T., Dudley, S., & Ashraf, I. (2023). Enhancing Cricket Performance Analysis with Human Pose Estimation and Machine Learning. *Sensors (Basel)*, *23*(15), 6839. doi:10.3390/s23156839 PMID:37571624

Sreekala, K., Cyril, C. P. D., Neelakandan, S., Chandrasekaran, S., Walia, R., & Martinson, E. O. (2022). Capsule network-based deep transfer learning model for face recognition. *Wireless Communications and Mobile Computing*, *2022*, 1–12. doi:10.1155/2022/2086613

Sridhar, S., & Sanagavarapu, S. (2021, January). Fake news detection and analysis using multitask learning with BiLSTM CapsNet model. In *2021 11th International Conference on Cloud Computing, Data Science & Engineering (Confluence)* (pp. 905-911). IEEE. 10.1109/Confluence51648.2021.9377080

Tampubolon, H., Yang, C. L., Chan, A. S., Sutrisno, H., & Hua, K. L. (2019). Optimized capsnet for traffic jam speed prediction using mobile sensor data under urban swarming transportation. *Sensors (Basel)*, *19*(23), 5277. doi:10.3390/s19235277 PMID:31795519

Tharwat, A., Gaber, T., Ibrahim, A., & Hassanien, A. E. (2017). Linear discriminant analysis: A detailed tutorial. *AI Communications*, *30*(2), 169–190. doi:10.3233/AIC-170729

Ullah, H., Haq, M. U., Khattak, S., Khan, G. Z., & Mahmood, Z. (2019, August). A robust face recognition method for occluded and low-resolution images. In *2019 International Conference on Applied and Engineering Mathematics (ICAEM)* (pp. 86-91). IEEE. 10.1109/ICAEM.2019.8853753

Wang, Y., Ning, D., & Feng, S. (2020). A novel capsule network based on wide convolution and multi-scale convolution for fault diagnosis. *Applied Sciences (Basel, Switzerland)*, *10*(10), 3659. doi:10.3390/app10103659

Wang, Z., Zheng, L., Du, W., Cai, W., Zhou, J., Wang, J., Han, X., & He, G. (2019). A novel method for intelligent fault diagnosis of bearing based on capsule neural network. *Complexity*, *2019*, 2019. doi:10.1155/2019/6943234

Waweru, L. W., Kipyego, B. T., & Muchangi, D. M. (2021). Classification of plant leaf diseases based on capsule network-support vector machine model. *Int J Electr Eng Technol*, *12*, 188–199.

ADDITIONAL READING

Abdullah, F. B., Iqbal, R., Ahmad, S., El-Affendi, M. A., & Abdullah, M. (2022). An empirical analysis of sustainable energy security for energy policy recommendations. *Sustainability (Basel)*, *14*(10), 6099. doi:10.3390/su14106099

Akhtar, S., Ali, A., Ahmad, S., Khan, M. I., Shah, S., & Hassan, F. (2022). The prevalence of foot ulcers in diabetic patients in Pakistan: A systematic review and meta-analysis. *Frontiers in Public Health*, *10*, 1017201. doi:10.3389/fpubh.2022.1017201 PMID:36388315

Amiri, Z., Heidari, A., Navimipour, N. J., Unal, M., & Mousavi, A. (2024). Adventures in data analysis: A systematic review of Deep Learning techniques for pattern recognition in cyber-physical-social systems. *Multimedia Tools and Applications*, *83*(8), 22909–22973. doi:10.1007/s11042-023-16382-x

Anwar, M. S., Wang, J., Ullah, A., Khan, W., Ahmad, S., & Li, Z. (2019, December). Impact of stalling on QoE for 360-degree virtual reality videos. In *2019 IEEE International Conference on Signal, Information and Data Processing (ICSIDP)* (pp. 1-6). IEEE. 10.1109/ICSIDP47821.2019.9173042

Heidari, A., Jafari Navimipour, N., Dag, H., & Unal, M. (2024). Deepfake detection using deep learning methods: A systematic and comprehensive review. *Wiley Interdisciplinary Reviews. Data Mining and Knowledge Discovery*, *14*(2), e1520. doi:10.1002/widm.1520

Katzman, B. D., Alabousi, M., Islam, N., Zha, N., & Patlas, M. N. (2024). Deep learning for pneumothorax detection on chest radiograph: A diagnostic test accuracy systematic review and meta analysis. *Canadian Association of Radiologists Journal*. doi:10.1177/08465371231220885 PMID:38189265

Kaur, J., & Singh, W. (2024). A systematic review of object detection from images using deep learning. *Multimedia Tools and Applications*, *83*(4), 12253–12338. doi:10.1007/s11042-023-15981-y

Nguyen, H. S., Ho, D. K. N., Nguyen, N. N., Tran, H. M., Tam, K. W., & Le, N. Q. K. (2024). Predicting EGFR mutation status in non–small cell lung cancer using artificial intelligence: A systematic review and meta-analysis. *Academic Radiology*, *31*(2), 660–683. doi:10.1016/j.acra.2023.03.040 PMID:37120403

Prabhat, P., Gupta, H., & Vishwakarma, A. K. (2024). Face Detection: Present State and Research Directions. arXiv preprint arXiv:2402.03796.

Rostami, M., Farajollahi, A., & Parvin, H. (2024). Deep learning-based face detection and recognition on drones. *Journal of Ambient Intelligence and Humanized Computing*, *15*(1), 373–387. doi:10.1007/s12652-022-03897-8

Shaheed, K., Szczuko, P., Kumar, M., Qureshi, I., Abbas, Q., & Ullah, I. (2024). Deep learning techniques for biometric security: A systematic review of presentation attack detection systems. *Engineering Applications of Artificial Intelligence*, *129*, 107569. doi:10.1016/j.engappai.2023.107569

Theodore Armand, T. P., Nfor, K. A., Kim, J. I., & Kim, H. C. (2024). Applications of Artificial Intelligence, Machine Learning, and Deep Learning in Nutrition: A Systematic Review. *Nutrients*, *16*(7), 1073. doi:10.3390/nu16071073 PMID:38613106

Chapter 4
Ethical Considerations in Remote Sensing and Cognitive Modeling

Baseer Ul Haq
University of Malakand, Pakistan

Mohammad Faisal
(iD) https://orcid.org/0000-0001-6823-5592
University of Malakand, Pakistan

Mahmood Ul Haq
(iD) https://orcid.org/0000-0002-1514-0300
University of Engineering and Technology, Peshawar, Pakistan

ABSTRACT

Three major advances are driving fundamental changes in the field of remote sensing. New satellite sensors will provide global imaging with excellent spatial and spectral resolution. Technological advancements have rendered earlier limitations on data scale, resolution, location, and availability obsolete. Economic restructuring in the remote sensing community will shift control and dissemination of imagery and related information from the government to the private sector. The internet and other digital infrastructures will speed up information distribution to a global user base. The combined results of these advancements may have serious legal and ethical ramifications for all remote sensing experts. Remote sensing technology may soon be able to provide detailed information, potentially violating privacy and leading to legal and ethical implications. This chapter discusses the legal history of remote sensing, recent innovations in satellite surveillance and information technology, and potential legal and ethical concerns for the remote sensing community. Self-regulation of the profession is essential for balancing individual rights with the economic objectives of the remote sensing community and nation.

DOI: 10.4018/979-8-3693-2913-9.ch004

1. INTRODUCTION

Ethics are crucial to responsible research and innovation in remote sensing and cognitive modeling. With the development of advanced sensing technology and AI-powered algorithms, our ability to extract insights from remote data has reached new heights. However, this astonishing accomplishment raises a number of ethical concerns that require our undivided attention (Slonecker et al., 1998). The ethical landscape of remote sensing and cognitive modeling is broad and dynamic, with problems ranging from privacy and data protection to prejudice and fairness in algorithmic decision-making. In this setting, the ideals of human dignity, fairness, transparency, and accountability serve as the foundation for ethical inquiry. As researchers, practitioners, and stewards of technological innovation, we must respect these principles and ensure that our actions are consistent with the overarching aims of social well-being and environmental sustainability (Fernandez-Diaz et al., 2018). As we go through the ethical elements of remote sensing and cognitive modeling, we must acknowledge the interdependence of our decisions and their far-reaching consequences. Our decisions about data gathering, algorithm design, and application distribution have far-reaching social, cultural, and ecological implications. By engaging in meaningful thinking and debate, we may create an ethical framework that guides our efforts and promotes trust, equity, and inclusivity. In the upcoming sections, we will look into specific ethical considerations that apply to distant sensing and cognitive modeling. From protecting individual privacy to fostering environmental stewardship, each topic sheds light on the intricate relationship between technology and society (Davis and Sanger, 2021). Through critical investigation and collaborative interaction, we hope to traverse these ethical issues with integrity and foresight, harnessing the revolutionary potential of remote sensing and cognitive modeling for the greater good.

Remote sensing involves using electromagnetic energy to detect and measure features of a target, such as the Earth's surface (Cole, 1987). Remote sensing is the process of collecting and analyzing data about the electromagnetic radiation reflected or emitted from an item in order to gain meaningful information about the object (Lillesand et al., 2015). Aerial photography and satellite imaging are conventional remote sensing techniques used for weather forecasting, mapping, intelligence gathering, global process research, land-use planning, conservation, and drug interdiction. Furthermore, a new generation of advanced remote sensing techniques is expected to play an increasingly important role in the future of an information-rich society. This study explores how remote sensing technology affects personal privacy, constitutional guarantees against excessive searches, and law enforcement. Remote sensing techniques improve monitoring efficiency by providing an area perspective, temporal definition, change detection, and accurate measurement. Aerial photographs from the 1930s and satellite images from the 1970s and 1980s are widely available and have played an important part in public programming and policy formulation. Aerial pictures and satellite data have long been used in litigation [6].

Remote sensing is currently undergoing a tremendous shift in terms of technical monitoring capabilities. Advances in spectral and spatial resolutions, new sensors, new platforms, and ever-improving digital analysis and transmission tools are transforming and expanding the level and types of detail that can be retrieved from raw imagery (Mehmood et al., 2020). Previously fundamental imaging constraints on scale, resolution, availability, location, and cost may become entirely irrelevant. Furthermore, the expanding number of orbital and airborne sensors, and the resulting volume of available imaging data, is radically changing the overall worldwide capability for overhead monitoring.

Remote sensing is revolutionizing information management, control, and communication. Historically, the remote sensing community had a close relationship with the US government. This included designing and launching sensors and orbiting vehicles, selling and distributing data, and receiving grants for application development. Foreign governments and multinational firms are increasingly entering the remote sensing sector as the industry undergoes economic transformation. Diversification and worldwide information infrastructure have led to a new way of distributing and analyzing high-resolution spatial and spectral data (Ullah et al., 2021)

Advancements in spatial and spectral monitoring, combined with global information management systems, raise the risk of remote sensing data misuse. The digital information revolution has raised concerns about protecting privacy and quality of life (Brennan and Macauley, 1995). Remote sensing technology may also have the potential to violate legal and ethical boundaries regarding privacy for individuals and corporations.

Emerging technology typically outpaces society's ability to establish laws, policies, and ethics to govern its use. New scientific advancements may have unintended consequences, including the possibility of misuse, which society may not fully acknowledge (Faisal et al., 2020).

This chapter examines the legal basis of remote sensing technology and highlights potential regulatory and ethical concerns for the remote sensing community.

2. THE CHANGING LANDSCAPE

Remote sensing specialists are facing three major developments. The first, and most evident, are technological advancements in monitoring systems. While aerial photography remains the most popular kind of remote sensing, satellite imaging sensors have grown significantly during the last 20 years. For almost 25 years, the US Landsat satellite series has given global multispectral photography. Other governments, including Russia, France, Japan, India, and Canada, also market and sell imagery from their orbiting systems. The intelligence community is now considering commercializing high-resolution imaging technology, which was previously only used for gathering intelligence (Slonecker et al., 1998). Over 20 polar-orbiting remote sensing satellites are scheduled to be launched over the next decade. Most are multispectral, with spatial resolutions of fewer than 5 metres, and some as low as 1 metre (ASPRS, 1996a; Brennan and Macauley, 1995; Steele, 1991). Morain and Budge (1995) also include over 20 available and proposed hyperspectral sensors with several spectral channels and ground sample lengths of less than 1 metre. In addition to advances in spatial and spectral resolution made by a new generation of satellites, new processing techniques and algorithms, such as the use of neural networks and/or sub-pixel land-cover classification, are increasing the amount of information that can be confidently extracted from imagery (Foody, 1996).

The second change pertains to the commercial and global restructuring of remote sensing infrastructure. Remote sensing infrastructure is evolving from government to commercial, domestic to worldwide. Remote sensing technologies and data are often linked to government operations, particularly in the United States. This ensures oversight and control over disseminated information, preventing data misuse. Currently, countries such as Canada's RADARSAT, France's SPOT, and Russia's SOVINFORMSPUTNIK openly sell global imagery from their own orbiting sensors, unlike the US. Data rights, pricing, and distribution rules are determined by the nations that own the spacecraft (United Nations, 1987).Recently, numerous countries, including the United States, have turned to quasi-public entities to market and sell imagery captured by their orbiting satellites. Further, the remote sensing world is already experiencing the totally private creation, launch, operation and commercialization of remote sensing satellites, some with spatial resolutions at the 1- to 3-metre level (Miller and Small, 2003).

The third issue concerns the rapidly changing nature of information distribution and access in modern society. The growing global information infrastructure opens many new surveillance opportunities. Many vendors of remotely sensed data market and distribute digital imagery directly via the Internet (Faisal et al., 2020a). The next generation of remotely sensed information is likely to come from foreign satellites and multinational corporations dealing with information technology in a global marketplace, which may operate without oversight from the US or international governments. The globalization of remote sensing information systems will cause the flow of data and information products to transcend traditional jurisdictional and national boundaries, as well as existing ways of effective legal control (Faisal et al., 2020b).

Remote sensing has expanded beyond traditional aerial photography to include diverse uses. Lidars identify chemical compounds, radars can see beneath tree canopies or overhanging structures, and thermal infrared wavelengths provide detailed information on occupancy or discharge. Multispectral instruments with sub-meter pixel resolutions offer opportunities for research into bandwidth combinations, logical associations, and "fusion" or combined data analysis, especially with hyper-spectral data. In brief, future technical capabilities will not be restricted to the use of traditional remote sensing analysis processes, but will also include a new generation of high resolution sensors and analytical procedures.

The general remote sensing community should understand that these fundamental changes in technology and infrastructure could have far-reaching consequences. There is a significant difficulty with inconsistent remote sensing law, creating information policy, and the interface between the two (Gabrynowicz, 1993). We are about to enter an era where we will have the technical ability to determine and distribute extremely fine details about the individual's home and life in society, and that this ability, at least in the interim, may be largely uncontrolled because of underdeveloped policy, complex and sometimes incoherent remote sensing laws, and a general inability to enforce at the international level.

3. LEGAL BACKGROUND

The legal community has long been interested in remote sensing technology due to its usefulness and intrusiveness. Although constitutional issues have persisted, technological limits have not produced significant problems (Latin et al., 1976). Previously, the level of information was considered acceptable for purposes such as map making, land-use planning, and environmental protection.

Latin et al. (1976) and Uhlir (1990) identified three types of remote sensing applications in the legal field: (1) for public policy creation, (2) for investigative purposes, and (3) for producing admissible evidence. Aerial images and maps have been used successfully and extensively as evidence in judicial trials (Gillen, 1986; Quinn, 1979).Satellite imagery has also been effectively submitted as evidence in pollution control trials such as U.S. v. Reserve Mining and State v. Inland Steel Company (Latin et al., 1976).

The Fourth Amendment of the US Constitution addresses the basic issue of improving remote sensing technology due to its historical association with government:

"The right to be secure in one's person, house, papers, and effects against unreasonable searches and seizures is protected. Warrants must be issued based on probable cause, supported by oath or affirmation, and specifically describing the location and items to be searched."

Overhead remote sensing technology for monitoring and law enforcement raise legal concerns related to the history and interpretation of the Fourth Amendment of the United States Constitution. According to Koplow (1992), the legal guarantee against unreasonable searches is a core component of U.S. constitutional law, despite its ambiguity and open-ended nature. Since Weeks v. United States in 1914, the Supreme Court has prohibited the use of evidence obtained through an improper search (without a warrant based on reasonable cause) in federal criminal prosecutions. Since then, the Supreme Court has expanded the definition of unreasonable searches to encompass non-physical trespassing acts like wiretapping and electronic eavesdropping (Kennicott, 1974). Repeated Supreme Court judgments have reaffirmed the Fourth Amendment's vitality and importance, even in situations involving critical national security concern (Koplow, 1992). In Katz v. United States, the foundational case in search and seizure law, the Supreme Court established two lines of inquiry to identify searches that can be conducted without a warrant. First, has there been evidence of a genuine, legitimate expectation of privacy? Second, is this expectation one that society is willing to recognize as reasonable? These two conditions comprise the present concept of an unwanted but legal search. If the individual has a reasonable expectation of privacy, the search cannot be done without a warrant. The 1986 Supreme Court judgment, Dow Chemical Company v. The United States (hereinafter referred to as Dow), remains a landmark in the field of remote sensing and law enforcement. The EPA attempted to enforce Clean Air Act requirements by seeking entry to the Dow Chemical factory in Midland, Michigan. After a follow-up visit was denied, the EPA conducted an aerial photo overflight with a standard mapping camera to ensure correct equipment installation and detect any illegal discharges. Dow filed a lawsuit against the EPA, alleging violations of trade-secrets legislation, acting outside its authority under the Clean Air Act, and conducting an unconstitutional search under the Fourth Amendment. After a passionate dissent, the Supreme Court concluded 5-4 that the EPA acted legally in acquiring aerial pictures. The Dow ruling considered trade secrets law, the EPA's power under the Clean Air Act, and Fourth Amendment protections against unreasonable searches.

4. ETHICAL FRAMEWORKS FOR REMOTE SENSING AND COGNITIVE MODELING

Ethical considerations are critical in determining the development and deployment of remote sensing and cognitive modeling technologies. As these technologies evolve, it becomes increasingly vital to establish ethical frameworks that encourage responsible decision-making while mitigating potential risks. In this literature review, we look at existing ethical frameworks for remote sensing and cognitive modeling, specifically how they address issues like privacy, prejudice, justice, and social impact.

4.1 Utilitarianism

Utilitarianism is a significant ethical theory used in remote sensing and cognitive modeling. Utilitarianism holds that activities should be judged on their potential to maximize overall happiness or utility. In the context of modern technologies, utilitarian principles may favor outcomes that maximize society benefit while reducing harm. For example, utilitarian considerations may influence decisions about the deployment of remote sensing technologies for disaster management or environmental monitoring, weighing the potential benefits of early detection and intervention against privacy and surveillance concerns Glennan and Illari, 2018).

4.2 Deontology

In contrast, deontological ethics promotes commitment to moral principles or responsibilities regardless of the consequences. Researchers in remote sensing and cognitive modeling may use deontological ideas to protect fundamental rights and autonomy. For example, deontological considerations may drive data collection and usage decisions in cognitive modeling studies, ensuring that participants' informed consent is gained while also protecting their privacy rights (Mullapudi et al., 2023).

4.3 Virtue Ethics

Virtue ethics is concerned with the character qualities and moral virtues of individuals or societies. In the context of remote sensing and cognitive modeling, virtue ethics may motivate researchers and practitioners to embrace virtues like honesty, integrity, and empathy. Stakeholders can encourage trust and collaboration by encouraging ethical qualities, resulting in more responsible development and deployment of these technologies (Khayal, 2019).

4.4 Feminist Ethics

Feminist ethics provides a critical perspective on power dynamics, inequality, and marginalization, all of which are pertinent to remote sensing and cognitive modeling. Feminist researchers advocate for an intersectional approach that recognizes the different experiences and views of people and communities impacted by these technologies. Feminist ethics, by emphasizing underrepresented voices and addressing concerns of gender, racism, and class, can help to achieve more equitable outcomes in the design and implementation of remote sensing and cognitive modeling systems.

Ethical frameworks are useful guides for traversing the complicated ethical terrain of remote sensing and cognitive modeling. By applying ideas from utilitarianism, deontology, virtue ethics, feminist ethics, and other ethical perspectives, stakeholders can make informed decisions that promote justice, transparency, and social responsibility. However, it is critical to understand that ethical concerns are broad and ever-changing, necessitating continual discourse, reflection, and involvement with many stakeholders to ensure ethical best practices in the development and deployment of these technologies.

5. PRIVACY CONCERNS AND DATA PROTECTION USER

Privacy and data protection are critical considerations in the development and deployment of remote sensing and cognitive modeling technology. These technologies have the potential to capture massive volumes of personal and sensitive data, which raises important ethical and legal concerns about privacy rights, informed permission, data anonymization, and regulatory compliance. In this literature review, we look at the key privacy risks and data protection methods related to remote sensing and cognitive modeling, using academic research and legislative frameworks to provide a thorough analysis.

5.1 Privacy Concerns

One of the key privacy risks with remote sensing and cognitive modeling is the indiscriminate collecting of personal information. Remote sensing technology, such as satellites and drones, have the ability to acquire high-resolution imagery of people, properties, and natural settings, frequently without their knowledge or agreement. This raises worries about monitoring, stalking, and the erosion of personal rights, especially in countries with little regulatory supervision or public understanding of the ramifications of remote sensing activities (Mehmood et al., 2020).

Furthermore, cognitive modeling techniques rely on the analysis of big datasets containing personal information gleaned from sources such as social media, electronic health records, and online transactions. While these databases provide useful insights into human behavior and decision-making, they also jeopardize personal privacy and confidentiality. Unauthorized access, data breaches, and algorithmic biases can all result in the misuse or exploitation of personal data, causing harm and prejudice against vulnerable groups (Kindt, 2013).

5.2 Data Protection Measures

To address these privacy concerns, a variety of data protection mechanisms have been developed and applied in remote sensing and cognitive modeling. One key concept is the necessity for informed consent, which ensures that persons are given clear and transparent information about the purpose, extent, and potential hazards of data collection and analysis. Informed consent enables individuals to make autonomous decisions regarding their participation in research or data-sharing activities while protecting their privacy rights (Gostin et al., 2018).

Another essential data protection strategy is data anonymization, which includes deleting or obscuring personally identifiable information from datasets in order to prevent individuals from being identified. Aggregation, masking, and encryption are examples of anonymization approaches that serve to reduce privacy risks while maintaining the utility and integrity of data for research and analysis (El Emam and Arbuckle, 2013). However, it is crucial to highlight that total anonymization is not always possible, particularly in the context of cognitive modeling, where individual actions and traits can be deduced from aggregated data patterns (Ohm, 2009).

5.3 Regulatory Frameworks

In addition to voluntary measures like informed permission and data anonymization, legal frameworks play an important role in protecting privacy rights and encouraging responsible data practices in remote sensing and cognitive modeling. In the United States, the Health Insurance Portability and Accountability Act (HIPAA) and the Children's Online Privacy Protection Act (COPPA) create standards for protecting health information and children's online privacy, respectively (Hodge et al., 1999). Similarly, the European Union's General Data Protection Regulation (GDPR) establishes stringent criteria for personal data processing and transmission, including regulations for data minimization, purpose limitation, and individual rights (Rossi, 2019).

Privacy and data protection are essential considerations while developing and deploying remote sensing and cognitive modeling technology. By addressing these concerns through informed consent, data anonymization, and adherence to legal frameworks, stakeholders can protect privacy rights, increase transparency, and reduce the risks of damage and discrimination associated with data-driven technology. However, it is critical to remember that privacy is a dynamic concept that necessitates continual attention, dialogue, and collaboration among researchers, politicians, and the general public to guarantee that privacy safeguards keep up with technology improvements and shifting cultural norms.

6. BIAS AND FAIRNESS IN ALGORITHMIC DECISION-MAKING

Bias and fairness in algorithmic decision-making have become more important problems as machine learning and artificial intelligence technologies are used in a variety of fields such as recruiting, lending, criminal justice, and healthcare. Biases in training data, algorithmic design, and decision-making processes can result in unequal outcomes, perpetuate socioeconomic disparities, and discriminate against vulnerable groups. In this literature review, we look at the major concepts, difficulties, and mitigation measures linked to bias and fairness in algorithmic decision-making, using scholarly research and case studies to provide a thorough overview.

6.1 Bias

Bias refers to systemic flaws or departures from the truth in data, algorithms, or decision-making processes that cause unfair treatment or discrimination against specific individuals or groups (McIntosh et al., 2023). Fairness, on the other hand, is a multidimensional term that includes the values of equality, impartiality, and justice. Fairness in algorithmic decision-making is avoiding discrimination based on protected characteristics such as race, gender, or ethnicity (Hodge et al., 1999).

6.1.1 Types of Bias

1 **Sample Bias:** Occurs when training data does not adequately reflect the underlying population, resulting in skewed or unrepresentative results.
2 **Algorithmic bias**: Algorithmic bias results from the design or implementation of algorithms that systematically favor or disadvantage specific groups based on irrelevant or discriminating criteria.
3 **Measurement Bias**: The result of mistakes or biases in data measurement or collecting, which lead to incorrect conclusions or predictions.

4 **Historical bias:** Historical bias is a reflection of historical injustices and inequalities encoded in training data or society standards, which perpetuates unfair outcomes in algorithmic decision-making.

6.2 Fairness Measures

To detect and minimize bias in algorithmic decision-making, researchers and practitioners have created a variety of fairness measures and metrics. This includes:

1 **Statistical Parity:** Ensures that decision outcomes are evenly distributed among distinct demographic groups, regardless of protected characteristics.

2 **Equal Opportunity:** Ensures that people from different backgrounds have equal chances of achieving positive outcomes based on relevant characteristics such as qualifications or performance.

3 **Disparate Impact:** Prohibits choices that have disproportionately negative consequences for protected groups, even if the decision criteria are used similarly across all categories.

4 **Individual Fairness**: Requires that similar people receive the same treatment or decisions, independent of their group membership or attributes (Kim et al., 2018).

6.3 Mitigation Strategies:

To mitigate bias and promote fairness in algorithmic decision-making, technical, legal, and ethical solutions are necessary. This may include:

1. **Data Preprocessing:** Using approaches such as data cleaning, sampling, and augmentation to reduce biases in training data and ensure its representativeness.

2 **Algorithmic Fairness Constraints:** To avoid discriminatory outcomes, algorithms should be designed and optimized with fairness constraints or aims.

3 **Bias Audits:** Conducting audits or reviews of algorithmic systems to detect and mitigate biases in decision-making processes.

4 **Transparency and Accountability:** Increasing transparency and accountability in algorithmic decision-making by providing data sources, algorithms, and decision criteria to stakeholders (Barocas and Selbst, 2016).

7. ETHICAL DECISION-MAKING AND RESPONSIBLE RESEARCH PRACTICES

Ethical decision-making and ethical research procedures are critical components of scientific investigation, assuring integrity, transparency, and societal effect. Ethical issues include values like autonomy, beneficence, nonmaleficence, and fairness. In this literature review, we will look at the principles, problems, and best practices surrounding ethical decision-making and responsible research conduct, drawing on scholarly research and regulatory requirements to present a thorough understanding.

7.1 Ethical Decision-Making Frameworks

Ethical decision-making frameworks offer guidelines and principles to help researchers negotiate challenging ethical quandaries. Let's look into some of these frameworks and compare their approaches:

7.2 Challenges and Considerations

Researchers frequently face several obstacles and ethical quandaries in their work. Let's evaluate some of these difficulties and consider potential solutions to them:

Table 1.

Framework	Description
Belmont Report	Core principles: respect for persons, beneficence, justice
Responsible Conduct of Research	Emphasizes integrity, honesty, and accountability in all aspects of research
Institutional Review Boards (IRBs)	Review research protocols, assess ethical risks, and ensure compliance with regulatory guidelines
Research Ethics Training	Provides education and training in research ethics, including workshops, seminars, and online resources

Table 2.

Challenge	Comparison
Conflicts of Interest	Disclosure vs. Management: Researchers can either disclose potential conflicts of interest or implement management strategies to mitigate their influence.
Pressure to Publish	Quality vs. Quantity: Researchers may prioritize the quality of their work over the quantity of publications, focusing on rigorous methods and reproducibility.
Competing Demands	Prioritization vs. Collaboration: Researchers may prioritize projects based on their ethical significance or collaborate with colleagues to address multiple demands.

7.3 Best Practices and Recommendations

Adopting best practices and suggestions is necessary to promote ethical decision-making and responsible research procedures. Let's compare some of these practices and think about the implications.

8. FUTURE DIRECTIONS AND EMERGING ETHICAL CHALLENGES

As technology advances at a rapid pace, the landscape of ethical concerns in numerous disciplines shifts. In this literature review, we look at the future trends and rising ethical concerns in several disciplines such as artificial intelligence, biotechnology, digital privacy, and environmental sustainability. Drawing on scholarly research and professional viewpoints, we investigate the potential consequences of these difficulties and discuss ways for dealing with them in the next years.

8.1 Artificial Intelligence and Machine Learning

AI and machine learning (ML) technologies are set to transform industries ranging from healthcare and finance to transportation and education. However, with the potential of innovation comes a slew of ethical issues. One significant difficulty is ensuring the fairness and accountability of AI systems, especially in high-stakes areas like criminal justice and healthcare (Barocas and Selbst, 2016), Bias in training data and algorithmic decision-making can produce discriminatory results, worsening socioeconomic inequities and weakening public trust in AI systems.

Furthermore, the rising autonomy and complexity of AI systems raises issues of duty and culpability. Who should be held liable if AI systems make mistakes or hurt people? How do we ensure that AI systems are consistent

Table 3.

Practice	Comparison
Institutional Review Boards (IRBs)	Centralized vs. Decentralized: Institutions may establish centralized IRBs or delegate review responsibilities to departmental or specialized committees.
Research Ethics Training	Standardized vs. Tailored: Research ethics training programs may adopt standardized curricula or tailor content to the specific needs of different disciplines.
Data Management and Sharing	Restricted vs. Open Access: Researchers may restrict access to data to protect confidentiality or adopt open access policies to promote transparency and collaboration.

with human values and priorities? These questions underline the importance of interdisciplinary collaboration and ethical frameworks for the responsible and transparent development, deployment, and regulation of AI technology (Jobin et al., 2019).

8.2 Biotechnology and Genetic Engineering

Advances in biotechnology and genetic engineering have the potential to transform healthcare, agriculture, and environmental conservation. However, these technologies present fundamental ethical concerns regarding safety, justice, and the manipulation of life itself. One rising ethical concern is the ethical application of gene editing technologies such as CRISPR-Cas9, which allows for precise alterations to the human genome (Jasanoff, 2016). While gene editing has the potential to treat genetic illnesses and improve crop resilience, it also raises worries about unintended consequences such as off-target mutations and unanticipated ecological effects.

Furthermore, the growing monetization of genetic information and biotechnological products creates issues of equity and access. Who should have control over genetic information and technology? How can we ensure that biotechnology's benefits are dispersed evenly among different people and communities? These questions highlight the significance of ethical deliberation, public participation, and regulatory control in influencing the future of biotechnology and genetic engineering (Gyngell et al., 2019).

8.3 Digital Privacy and Data Governance

In an increasingly networked and data-driven society, digital privacy and data governance have emerged as critical ethical issues. The expansion of digital technology and online platforms has resulted in unprecedented volumes of personal data being gathered, analyzed, and monetized by governments and businesses. This raises concerns about surveillance, data breaches, and the loss of privacy rights (Zuboff, 2023).

One emerging ethical dilemma is balancing the benefits of data-driven innovation against the protection of individual privacy and autonomy. As data becomes more commodified and valuable, challenges arise regarding who owns and controls personal information, as well as how it should be utilized and protected (Floridi, 2016). Furthermore, the advent of AI and machine learning raises issues about algorithmic discrimination and predictive analytics, in which individuals may be unfairly targeted or disadvantaged based on their data profiles (O'Niel, 2016).

8.4 Environmental Sustainability and Climate Change

The growing concerns of climate change and environmental degradation pose significant ethical issues to global society. As temperatures rise, sea levels rise, and ecosystems deteriorate, vulnerable groups bear disproportionate risks and burdens. One emerging ethical dilemma is how to balance short-term economic goals with long-term environmental sustainability (Gardiner, 2011). The goal of economic expansion and development frequently sacrifices environmental conservation and social equality, aggravating climate change and widening inequities.

Furthermore, the global scope of environmental concerns needs collaborative action and international cooperation. How can we ensure that climate mitigation and adaptation activities are fair and inclusive, especially for marginalized groups? How can we hold governments and corporations accountable for their contributions to environmental degradation and climate injustice? These questions highlight the significance of ethical leadership, intergenerational justice, and planetary stewardship in dealing with the existential risks posed by climate change and environmental degradation (Gupta and Mason, 2014).

9. CONCLUSION

The ethical considerations in remote sensing and cognitive modeling show the fine line between technical progress and ethical duty. Remote sensing technologies provide unparalleled data collecting and analytic capabilities; therefore, ethical frameworks must govern the responsible use of this information, ensuring privacy, consent, and equitable benefit distribution. Similarly, in cognitive modeling, where algorithms attempt to emulate human thought, transparency and accountability are critical in addressing concerns about bias and justice. Looking ahead, stakeholders must maintain continuing communication and collaboration to handle growing ethical problems, stressing ethical principles to protect individual dignity and promote social well-being. Finally, by respecting ethical standards and cultivating an ethical culture, remote sensing and cognitive modeling can help to create a more equitable, transparent, and sustainable future for everyone.

REFERENCES

Barocas, S., & Selbst, A. D. (2016). Big data's disparate impact. *California Law Review*, *104*, 671.

Brennan, T. J., & Macauley, M. K. (1995). Remote sensing satellites and privacy: A framework for policy assessment. *Information & Communications Technology Law*, *4*(3), 233–248. doi:10.1080/13600834.1995.9965723

Cole, M. (1987). *Remote Sensing: Principles and Interpretation*. Academic Press.

Davis, D. S., & Sanger, M. C. (2021). Ethical challenges in the practice of remote sensing and geophysical archaeology. *Archaeological Prospection*, *28*(3), 271–278. doi:10.1002/arp.1837

El Emam, K., & Arbuckle, L. (2013). *Anonymizing health data: case studies and methods to get you started*. O'Reilly Media, Inc.

Faisal, M., Ali, I., Khan, M. S., Kim, J., & Kim, S. M. (2020a). Cyber security and key management issues for internet of things: Techniques, requirements, and challenges. *Complexity*, *2020*, 1–9. doi:10.1155/2020/6619498

Faisal, M., Ali, I., Khan, M. S., Kim, S. M., & Kim, J. (2020b). Establishment of trust in internet of things by integrating trusted platform module: To counter cybersecurity challenges. *Complexity*, *2020*, 1–9. doi:10.1155/2020/6612919

Faisal, M., Attiq-Ur-Rehman, S. N., & Perveen, Z. (2020). Security architecture of cloud network against cyber threats. *Science International (Lahore)*, *32*(1), 63–67.

Fernandez-Diaz, J. C., Cohen, A. S., González, A. M., & Fisher, C. T. (2018). *Shifting perspectives and ethical concerns in the era of remote sensing*. Academic Press.

Foody, G. M. (1996). Relating the land-cover composition of mixed pixels to artificial neural network classification output. *Photogrammetric Engineering and Remote Sensing*, *62*(5), 491–498.

Gabrynowicz, J. I. (1993). (in press). Remote Sensing Law: Obstacle or Opportunity for Geographic Information Systems. *National Center for Geographic Information and Analysis and the Center for the Arizona State University College of Law*, 275–278.

Gardiner, S. M. (2011). *A perfect moral storm: The ethical tragedy of climate change*. Oxford University Press. doi:10.1093/acprof:oso/9780195379440.001.0001

Glennan, S., & Illari, P. M. (Eds.). (2018). *The Routledge handbook of mechanisms and mechanical philosophy*. Routledge.

Gostin, L. O., Halabi, S. F., & Wilson, K. (2018). Health data and privacy in the digital era. *Journal of the American Medical Association, 320*(3), 233–234. doi:10.1001/jama.2018.8374 PMID:29926092

Gupta, A., & Mason, M. (2014). Transparency and international environmental politics. In *Advances in international environmental politics* (pp. 356–380). Palgrave Macmillan UK.

Gyngell, C., Bowman-Smart, H., & Savulescu, J. (2019). Moral reasons to edit the human genome: Picking up from the Nuffield report. *Journal of Medical Ethics, 45*(8), 514–523. doi:10.1136/medethics-2018-105084 PMID:30679191

Hodge, J. G. Jr, Gostin, L. O., & Jacobson, P. D. (1999). Legal issues concerning electronic health information: Privacy, quality, and liability. *Journal of the American Medical Association, 282*(15), 1466–1471. doi:10.1001/jama.282.15.1466 PMID:10535438

Jasanoff, S. (2016). *The ethics of invention: Technology and the human future.* WW Norton & Company.

Jobin, A., Ienca, M., & Vayena, E. (2019). The global landscape of AI ethics guidelines. *Nature Machine Intelligence, 1*(9), 389–399. doi:10.1038/s42256-019-0088-2

Kennicott, P. C. (1974). *Bibliographic Annual in Speech Communication 1973.* Academic Press.

Khayal, O. (2019). *Human Factors and Ergonomics.* Academic Press.

Kim, M., Reingold, O., & Rothblum, G. (2018). Fairness through computationally-bounded awareness. *Advances in Neural Information Processing Systems,* 31.

Kindt, E. J. (2013). Privacy and data protection issues of biometric applications. In *A Comparative Legal Analysis* (Vol. 12). Springer. doi:10.1007/978-94-007-7522-0

Koplow, D. A. (1992). Overflying a Country without Overlooking the Constitution: Legal Implications of Aerial Overflights. *Open Skies, Arms Control and Cooperative Security,* 93-112.

Latin, H. A., Tennehill, G. W., & White, R. E. (1976). Remote sensing evidence and environmental law. *California Law Review, 64*(6), 1300. doi:10.2307/3480040

Lillesand, T., Kiefer, R. W., & Chipman, J. (2015). *Remote sensing and image interpretation.* John Wiley & Sons.

McIntosh, T. R., Susnjak, T., Liu, T., Watters, P., & Halgamuge, M. N. (2023). From google gemini to openai q*(q-star): A survey of reshaping the generative artificial intelligence (ai) research landscape. *arXiv preprint arXiv:2312.10868.*

Mehmood, G., Khan, M. Z., Abbas, S., Faisal, M., & Rahman, H. U. (2020). An energy-efficient and cooperative fault-tolerant communication approach for wireless body area network. *IEEE Access : Practical Innovations, Open Solutions*, *8*, 69134–69147. doi:10.1109/ACCESS.2020.2986268

Miller, R. B., & Small, C. (2003). Cities from space: Potential applications of remote sensing in urban environmental research and policy. *Environmental Science & Policy*, *6*(2), 129–137. doi:10.1016/S1462-9011(03)00002-9

Mullapudi, A., Vibhute, A. D., Mali, S., & Patil, C. H. (2023). A review of agricultural drought assessment with remote sensing data: Methods, issues, challenges and opportunities. *Applied Geomatics*, *15*(1), 1–13. doi:10.1007/s12518-022-00484-6

O'Niel, C. (2016). *Weapons of math destruction*. Crown/Archetype.

Ohm, P. (2009). Broken promises of privacy: Responding to the surprising failure of anonymization. *UCLA l. Rev.*, *57*, 1701.

Rossi, A. (2019). *Legal design for the general data protection regulation. A methodology for the visualization and communication of legal concepts*. Academic Press.

Slonecker, E. T., Shaw, D. M., & Lillesand, T. M. (1998). Emerging legal and ethical issues in advanced remote sensing technology. *Photogrammetric Engineering and Remote Sensing*, *64*(6), 589–595.

Ullah, F., Khan, M. Z., Faisal, M., Rehman, H. U., Abbas, S., & Mubarek, F. S. (2021). An energy efficient and reliable routing scheme to enhance the stability period in wireless body area networks. *Computer Communications*, *165*, 20–32. doi:10.1016/j.comcom.2020.10.017

Zuboff, S. (2023). The age of surveillance capitalism. In *Social theory re-wired* (pp. 203–213). Routledge. doi:10.4324/9781003320609-27

Chapter 5

A Review of Capsule Network Limitations, Modifications, and Applications in Object Recognition

Mahmood Ul Haq
 https://orcid.org/0000-0002-1514-0300
University of Engineering and Technology, Peshawar, Pakistan

Muhammad Athar Javed Sethi
 https://orcid.org/0000-0001-7847-831X
University of Engineering and Technology, Peshawar, Pakistan

Atiq Ur Rehman
 https://orcid.org/0000-0003-0248-7919
Hamad Bin Khalifa University, Qatar

ABSTRACT

Modern computer vision and machine learning technologies have enabled numerous advances in a variety of domains, including pattern recognition and image classification. One of the most powerful machine learning methods is the capsule network, which encodes features based on their hierarchical relationships. A capsule network is a sort of neural network that uses inverted graphics to represent an item in distinct sections and see the existing link between these pieces, as opposed to CNNs, which lose most of the evidence relating to spatial placement and require a large amount of training data. As a result, the authors give a comparison of various capsule network designs utilized in diverse applications. The fundamental contribution of this study is that it summarizes and discusses the major current published capsule network topologies, including their advantages, limits, modifications, and applications.

DOI: 10.4018/979-8-3693-2913-9.ch005

INTRODUCTION

The capsule network neural architectures are a sort of artificial neural network found in machine learning systems (Khan et al., 2023). It is especially noticeable when describing a hierarchical connection and closely resembling biological neural networks (Haq et al., 2019). The capsules network's development is based on the concept of expanding the convolution network (Haq et al., 2023) in order to reuse the end results in order to uncover more consistent and advanced exemplification of the developing capsules. The capsule network has been designed as an alternative for the convolutional neural networks, as the CNN shows few limitations in accomplishing the applications of computer vision despite its efforts in managing the accuracy in the areas where it is applied, as it is a novel architecture in neural networks and an enhanced approach of the prevailing neural network model, particularly for the tasks in computer vision (Ahmad & Adnan, 2015; Ahmad et al., 2018; Anwar, Wang, Khan et al, 2020; Campus, n.d.). Convolutional neural networks, which are defined as the foundation of image processing in a deep learning context (Hosni et al., 2018; Munawar et al., n.d.; Rahim, Zhong, Ahmad, Ahmad, & ElAffendi, 2023; Sohail et al., 2023), were initially developed with the goal of classifying images by utilizing consecutive convolution layers and pooling layers (Anwar, Wang, Ahmad et al, 2020). Despite its ability to achieve accuracy, the convolution neural network caused some performance degradation due to the reduction in the data dimension for acquiring spatial invariance, resulting in a loss of information (location, rotation, various features related to scale and position) that may be required in the process of segmentation, and proper object localization (Fatima et al., 2022). This makes segmentation and detection more difficult (Patrick et al., 2022). The alternative techniques, employing the end to end connected layer (Haq et al., 2024) and utilizing reinforcement learning (Krizhevsky et al., 2012) developing advanced training and designing techniques for the convolutional neural network (Ullah et al., n.d.) to reduce the difficulties in the process of segmentation and detection, to gain accuracy in the classification of the images, were tedious but did not show any improvements (Tahsin et al., 2023), leading to the development of the new convolutional neural network architecture. Geoffrey Hinton developed this approach as a solution to the shortcomings of the convolutional neural network. Figure 1 depicts a conventional convolutional neural network. The input image is scanned in the convolutional layer in order to extract low-level features like edges. To reduce computing complexity and make the model more nonlinear, utilize the RELU function. Down-sampling, or pooling layer, is a technique used to save memory and identify the same object in several images. Different types of pooling, such as max pooling, min pooling, average pooling, and sum pooling, are utilized depending on the requirements. These pooling methods are shown in Figure 2, and the ReLU activation methodology is shown in Figure 3.

Figure 1. CNN structure (Anwar, Wang, Ahmad et al, 2020)

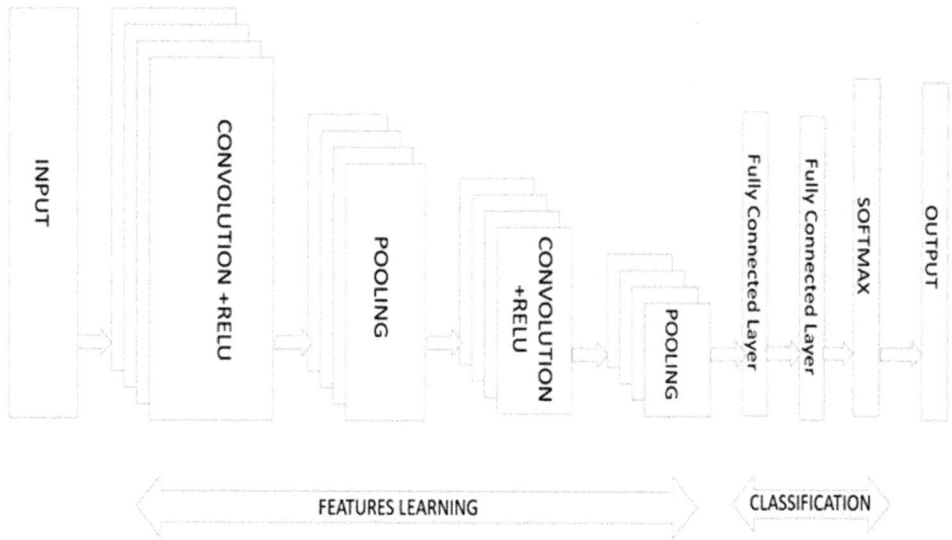

Figure 2. Pooling techniques

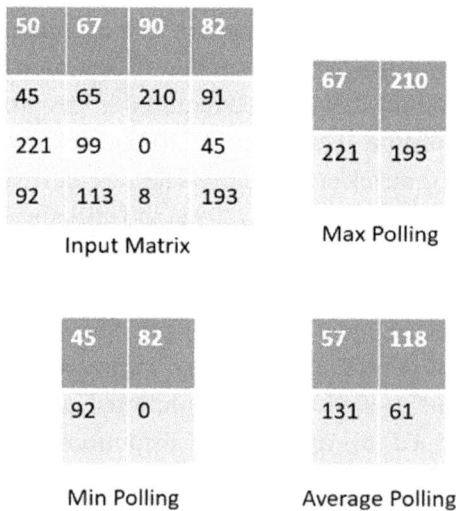

50	67	90	82
45	65	210	91
221	99	0	45
92	113	8	193

Input Matrix

67	210
221	193

Max Polling

45	82
92	0

Min Polling

57	118
131	61

Average Polling

The capsules in the technique represent a collection of neurons that contain all of the minute information about the spatial location of the object in order to reduce difficulties during the segmentation and detection processes (Wahab et al., 2023). In order to depict the image, the capsule network employs the inverse steps of computer

Figure 3. ReLU function of CNN

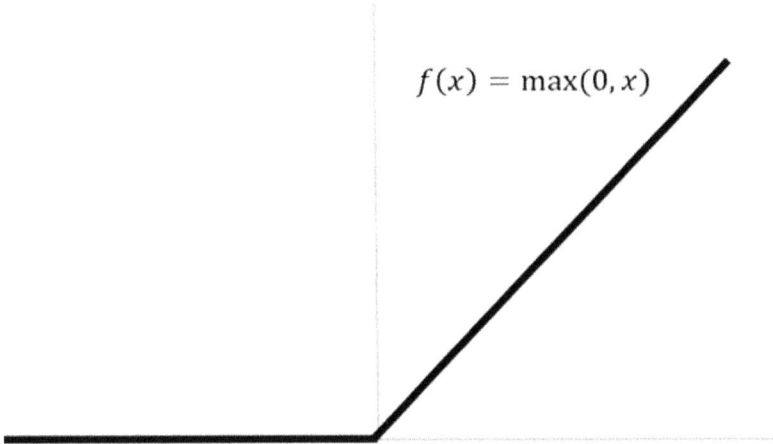

$$f(x) = \max(0, x)$$

graphics. For example, while detecting an object, the object is subdivided inwardly into many parts, and a relationship (hierarchical relationship) is created between all the sub parts of the object to represent the thing. Figure 4 depicts the architecture of the capsule neural network. The capsule neural network architecture is made up of three main parts: the input layer, the hidden layer, and the output layer. The hidden layer is made up of three more layers: the convolutional layer, primary capsules (lower and upper layers, such as digi-caps).

CapsNet's improved resilience is due to its ability to collect data on the relationships between several features in an input image. CapsNet employs "capsules," which are groups of neurons that work together to represent specific characteristics of a picture. These capsules can then be used to generate an output image that is more accurate to the original while also being less vulnerable to adversarial attacks.

When leveraging capsule networks, tasks like vote assaults (Gu et al., 2021), object identification (Hu & Li, 2022), intrusion detection (Devi & Muthusenthil, 2022), and adversarial attacks (Marchisio et al., 2023) have the potential to be very useful in terms of security. When capsule networks include the spatial relationships between objects in an image, they are more effective at detecting unusual or suspicious objects or behaviors (Wang et al., 2023).

This paper's primary contributions are:

- To inspire researchers, it presents cutting-edge capsule models.
- To investigate potential areas for further study
- To provide a comparison analysis of the most advanced CapsNet topologies available today.

Figure 4. Structure of capsule network

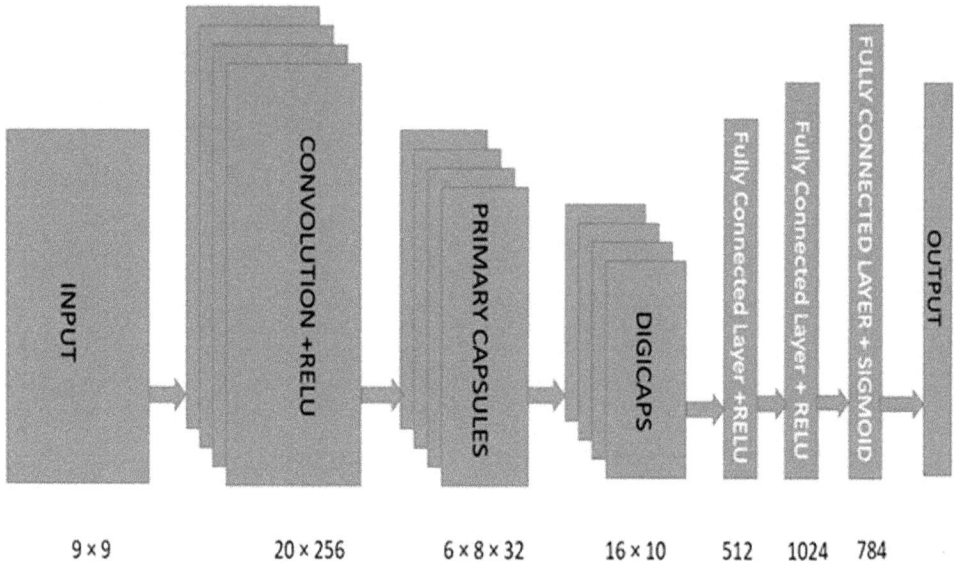

| 9 × 9 | 20 × 256 | 6 × 8 × 32 | 16 × 10 | 512 | 1024 | 784 |

- To provide a comparison analysis of the most advanced routing algorithms for CapsNet topologies as of right now.
- To investigate the elements influencing the functionality and alterations of capsule neural network topologies.

MATHEMATICAL MODEL OF CAPSNET

Input to the CapsNet

An image or a group of photos is used as the input to a CapsNet.

Convolutional Layer

Convolutional layer output is given by:

$$z_{ij,k} = \Sigma\Sigma\Sigma W_{i,k,k,1} X_{i,j,1} + b_{i,j,k} \tag{1}$$

Where $z_{ij,k}$ reflects the activation of the kth feature map in the output at position (i,j). This is the output of a convolution operation on the input. However $W_{i,j,k,1}$ is the weight associated with the ith input channel in the filter of the kth feature map at point (i,j). These weights are acquired through training. Furthermore, $X_{i,j,1}$ represents the activation of the ith input channel at location (i, j) in the input. It represents the convolutional layer's input data. The bias term $(b_{i,j,k})$ for the kth feature map at point (i, j) are extra parameters that are added to the weighted sum of the input to provide the model more flexibility.

Primary Capsule Layer

The primary capsule layer's output is given by:

$$V_{i,j,k} = \text{Squash}(\Sigma W_{i,k,k,1} u_{i,j,1}) \tag{2}$$

Where $u_{i,j,1}$ is the input vector connected to the ith detected feature or part at position (i,j) in the input; $W_{i,j,k,1}$ is the weight matrix connecting the ith input vector to the kth primary capsule; Squash is a non-linear activation function that guarantees that the output vector has a length between 0 and 1.

Routing by Agreement

The higher-level capsule layer's output is provided by:

$$S_j = \sum_i c_{ij} \hat{u}_{j|i} \tag{3}$$

Where c_{ij} is the coupling coefficient, which indicates the likelihood that the ith primary capsule should be routed to the jth higher-level capsule, and S_j is the output of the jth higher-level capsule. Additionally, $\hat{u}_{j|i}$ is the predicted output vector of the jth higher-level capsule based on the input from the jth primary capsule.

Capsule Output

The following is the capsule network's ultimate output (y_k):

$$y_k = squash(s_k) \tag{4}$$

Why CapsNet Generally Outperforms CNN

For image identification and classification applications, two popular deep learning architectures are utilized: convolutional neural networks (CNNs) and capsule networks (CapsNets). When compared to CNNs, CapsNets were said to be superior because of their enhanced capacity to identify and categorize images with changes in orientation, scale, and distortion. The capacity to capture hierarchical relationships between features is one of CapsNets' advantages over CNNs, and it can be very helpful in applications like object recognition. This does not imply, however, that CapsNets are always more reliable than CNNs.

Indeed, some studies have demonstrated that CNNs can perform better than CapsNets under specific conditions, such as working with tiny datasets, adversarial attacks (Rahim, Zhong, Ahmad, Ahmad, Pławiak et al, 2023), or having fewer parameters. Small, skillfully constructed perturbations are added to an input data point in an adversarial assault to trick the model into generating an incorrect prediction. Additionally, CNNs have shown to be very reliable and effective in a variety of real-world applications, including self-driving automobiles (Badue et al., 2021) and facial recognition systems (Weng et al., 2022).

Research and discussion on whether CapsNets outperform CNNs in most situations are still ongoing, though. In certain scenarios, CapsNets may outperform CNNs for several reasons as presented in Table 1:

FACTORS IMPACTING THE PERFORMANCE OF CAPSNET

A dataset is necessary for an algorithm to function properly (Haq et al., 2022). On the MNIST dataset, CapsNet initially produced encouraging results. But this dataset is rather easy in comparison to more complex datasets that have different backgrounds, sizes, colors, noise, and many objects in a single sample. On increasingly complicated datasets, CapsNet outperforms CNN and produces encouraging results (Gordienko et al., 2018). CapsNet performed less well than CNN (Netzer et al., 2011) but still produced a superior result on the high intra-class variation and background noise datasets from SVHN (Krizhevsky & Hinton, 2009) and CIFAR10 (Zhang, 2021), as compared to more sophisticated algorithms like VGG NET (Doon et al., 2018) and CNN (Mukhometzianov & Carrillo, 2018). Furthermore, there are situations in which changing the number of iterations has no effect on accuracy. For a thorough analysis, readers should consult (Nair et al., 2021). Table 2 presents the positive and negative aspects of these factors

MODIFICATION IN CAPSNET

Authors in (Amer & Maul, 2020) conducted an evaluation of CapsNet's routing-by-agreement mechanism and discovered that it is not always guaranteed to link multiple

Table 1. Comparison of CNN and capsule network

Property	CNN	Capsule Network
Occlusion	Tends to exhibit low performance	Exhibits robustness against occlusion
Pose	Occasionally loses pose information	Possesses pose invariance
Data Efficiency	Typically requires large labeled datasets	Can achieve complexity with less labeled data
Small Dataset	Demonstrates lower performance	Performs well even with smaller datasets
Parameters	Generally requires more parameters	Requires fewer parameters compared to CNN
Overlapping Objects	Struggles to identify overlapping objects	Effectively identifies overlapping objects
Computational Efficiency	Often computationally efficient	Requires more computational resources due to dynamic routing

lower-level capsules to a higher-level capsule in order to create a parse tree. Rather than using the original routing-by-agreement method, which requires lower-level capsules to transmit their outputs to all higher-level capsules, an alternate way enables a lower-level capsule to select a single parent. The network's depth and resilience to hostile white-box assaults are increased by this enhancement (Peer et al., 2018).

To determine if a picture is real or artificially made (fake), several researchers have investigated combining a generative adversarial network (GAN) with a high-performing capsule-based discriminator (Jaiswal, 2018). As GAN discriminators, CapsNets may perform better than CNNs since they preserve crucial information without requiring pooling, claims (Saqur & Vivona, 2020).

The original CapsNet's equal-weight initialization routing tended to impede convergence and reduce accuracy. A better approach is to describe the original routing weights as trainable parameters and train them using backpropagation (Ramasinghe et al., 2018). It can also improve the performance of multi-label classification tasks to consider the non-independence of primary capsule predictions.

It has been shown that concentrating on a capsule's length rather than its individual outputs is a more effective approach for entity detection (Zhang et al., 2018). The length of the capsule indicates the presence of an entity; the orientation of the capsule represents the pose properties, which include position, size, orientation, deformation, velocity, albedo, hue, and texture.

Capsules are a potential approach to neural networks, however there are several implementation and performance problems. SoftMax is often used to calculate the assignment probabilities between capsules in neighboring layers; nonetheless, it suffers from the limitation of converging to uniform probability in routing iterations. A solution to this issue was proposed by (Zhao et al., 2019) using the MaxMin function, which permits scale-invariant normalization and for lower-level capsules to assume independent values. Performance is improved with this feature.

Box 1. Literature review

Ref No	Problem Description	Architecture and Parameters	Dataset	Accuracy (%)	Compared with	Comments
(Saqur & Vivona, 2018)	3D image Generation for GAN Discriminators	Dynamic routing: Convolution 1 layer: 32×32 input, 9×9 kernel of stride 1. Primary Capsule Layer: 32 channels each 8×8×8, 8D. Output Capsule layer: 16×1.	MNIST and Small NORB	-	DCGAN (Radford et al., 2015)	The MNIST used in this paper is simplistic; additional experiments needed using complex datasets. CapsGAN has the ability to capture geometric transformations.
(Teto & Xie, 2019)	Identify animals in the wilderness	C-CapsNet.	Serengeti Dataset	96.48	CNN	-
(Singh et al., 2019)	Low Resolution Image Recognition	DirectCapsNet with targeted reconstruction loss and HR-anchor loss.	CMU Multi-PIE dataset (Sim et al., 2002), SVHN dataset, and UCCS dataset.	95.81	Robust Partially Coupled Nets, LMSoftmax, L2Softmax, and Centerloss for VLR.	In future, VLR FR can be done in the presence of aging, adversarial attacks, and spectral variations.
(Afshary et al., 2018)	Brain Tumor Classification	Same as traditional CapsNet. Instead of 256 feature maps in Conv layer, they used 64	Dataset used in (Cheng et al., 2016)	-	CNN	-
(Tiwari & Jain, 2021)	Detect COVID-19 disease through radiography images	VGG-CapsNet	2905 images having 219, 1345, and 1341 images of COVID-19 patients available at (Kaggle, n.d.)	97% for classifying COVID-19, non-COVID samples, and pneumonia with 92% accuracy.	CNN-CapsNet VGG-CapsNet	Proposed approach can be used for clinical practices.
(Mazzia et al., 2021)	Capsule network with self-attention routing	Non-iterative, parallelizable routing algorithm instead of dynamic routing.	MNIST, smallNORB, and MultiMNIST.	-	CNN	Achieved higher accuracy with a considerably lower number of parameters.

continued on following page

Box 1. Continued

Ref No	Problem Description	Architecture and Parameters	Dataset	Accuracy (%)	Compared with	Comments
(Yang et al., 2022)	Identify Pneumonia-related compound	Five Layer Capsule Network.	88 positive samples and 264 negative samples.	An improvement of 1.7–12.9% in terms of AU.	SVM, gcForest, RF, and forgeNet.	-
(Afshar et al., 2018)	Brain tumor type classification	Primary layer: $64 \times 9 \times 9$ convolution filters and stride of 1. Primary Capsule layer: $256 \times 9 \times 9$ convolutions with strides of 2. Decoder: Fully connected layers with $512 \times 1024 \times 4096$ neurons.	Dataset used in (Cheng et al., 2015).	86.56	CNN presented by (Justin et al., 2017).	Accuracy can be increased by varying the number of feature maps.
(Chao et al., 2019)	Emotion Recognition	Features: Feature Matrix Convolution1 layer: $16 \times 16 \times 256$ channels. Primary Capsule layer: $7 \times 7 \times 256$D vector. Class Capsule layer: $2 \times 49 \times 32$D. Dynamic routing.	DEAP dataset (Koelstra et al., 2011).	Arousal=0.6828 Valence=0.6673 Dominance=0.6725.	Compared with five versions of capsule networks, SVM, Bayes classifier and Gaussian naive Bayes.	To verify comprehensively, the proposed method will be tested on more datasets of emotion recognition.
(Kumar & Sachdeva, 2021)	Cyberbullying Detection	CapsNet with dynamic routing and CNN.	10000 comments taken from YouTube, Twitter, and Instagram.	97.05	KNN, SVM, and NB.	High-dimensional, skewed, cross-lingual and heterogeneous data are the limitations of the proposed approach. Future work can be directed to detect and recognize wordplay, creative spellings, slangs words.

continued on following page

Box 1. Continued

Ref No	Problem Description	Architecture and Parameters	Dataset	Accuracy (%)	Compared with	Comments
(Heidarian et al., 2021)	COD-19 patient detection through Chest CT scan images	U-net based segmentation model, Capsule Network.	COVID-CT-MD.	90.82.	-	Proposed algorithm has been tested using a simple dataset having 171 images of COVID-19 positive patients, 60 patients with Pneumonia and 76 normal patients.
(Mukhometzianov & Carrillo, 2018)	Comparative analysis of CapsNet with Fisherfaces, LeNet, and ResNet	256 feature maps, using a 9x9 kernel and valid padding.	Yale face database B, MIT CBCL face dataset, Belgium TS traffic sign dataset and CIFAR-100.	95.3% on Yale dataset, 99.87% on on MIT CBCl, 92% on Belgium TS traffic sign dataset and 18% accuracy on CIFAR-100.	Fisherfaces, LeNet, and ResNet.	With more training iterations CapsNet may have better results.
(Manoharan, 2021)	Hierarchical multi-label text Classification	The features encoded in capsules and routing algorithm are combined.	BGC and WOS datasets.	-	SVM, LSTM, ANN, and CNN architectures.	Proposed algorithm performs efficiently. Future work should include cascading capsule layers.
(Ma et al., 2017)	Classify ultrasonic data for self-driving car	Dynamic routing by agreement. Convolution: 256 kernels of 6×6 size, Activation Function: ReLU. Primary Capsule: 32 channels of 6×6×8. Digit Caps: 16×4.	A dataset of 21,600 measurements.	99.6% using complex CapsNet as compared to CNN (98.9).	Complex CNN	A lot of research is needed to make ultrasonic technology appropriate for self-driving vehicles.
(Janakiramaiah et al., 2021)	Military Vehicle Recognition	Multi-level CapsNet With class capsule layer.	3500 images collected from the Internet.	96.54%	CNN	-

continued on following page

Box 1. Continued

Ref No	Problem Description	Architecture and Parameters	Dataset	Accuracy (%)	Compared with	Comments
(Afshar et al., 2021)	Lung nodule malignancy prediction	MIXCAPS.	LIDC dataset (Armato et al., 2011) and IDRI dataset (Clark et al., 2013).	92.88% with sensitivity of 93.2% and specificity of 92.3%.	-	The proposed approach is independent from pre-defined hand-shaped features and does not require fine annotation.
(Jain, 2019)	Audio Classification	Agreement-based dynamic routing, Flattening Capsule Network, Bi-GRU layer, SoftMax function, and Embedding Layer.	WASSA Implicit Emotion Shared Task (Klinger et al., 2018).	50%	GRU + CNN. GRU + Hierarchical Attention.	-
(Rathnayaka et al., 2018)	Emotion of tweets Prediction	Embedding Layer, Bi-GRU layer with Capsule Network. Flattening with SoftMax layer. Dynamic routing by agreement.	WASSA 2018.	F1 score= 0.692.	GRU with Hierarchical Attention, GRU with CNN. GRU with CapsNet.	-
(Hinton et al., 2018)	Object Detection	NASGC-CapANet	MS COCO val 2017 dataset	43.8% (box mAP)	Faster R-CNN	When compared to the present attention mechanism, the performance is improved by incorporating the capsule attention module into the highest level of FPN.
(Khan et al., 2022)	Hate Speech Detection	HCovBi-Caps (Convolutional, BiGRU and capsule Network)	DS1 dataset and DS2	Training Accuracy= 0.93 and validation accuracy= 0.90	DNN, BiGRU, GRU, CNN, LSTM, etc.	The proposed model detects the date propagation in speech only.

Table 2. Factors impacting the performance of CapsNet

Factors	Positive Impact	Negative Impact
Dataset Quality and Diversity	Enhanced generalization with diverse and high-quality datasets.	Overfitting risks with limited or biased datasets.
Model Architecture	Well-designed architecture improves the network's ability to learn features.	Poorly designed architectures may limit the network's effectiveness.
Routing Algorithm	Efficient routing algorithms (e.g., dynamic routing) facilitate better information flow.	Inefficient routing mechanisms can hinder model convergence.
Number of Parameters	Good performance with fewer parameters, leading to better generalization.	Too few parameters may limit the model's capacity; too many can cause overfitting.
Training Data Size and Complexity	Larger, diverse datasets improve generalization.	Insufficient or overly simple datasets may result in under fitting.
Reconstruction Loss	Inclusion of reconstruction loss aids in learning informative features.	Exclusion may lead to less effective feature learning and higher reconstruction errors.
Transfer Learning	- Leveraging pre-trained models for transfer learning can improve performance in new tasks.	- Transfer learning may not be effective if the pre-trained model is not suitable for the target task.
Learning Rate and Optimization Algorithm	- Proper tuning of learning rates and choice of optimization algorithms accelerate convergence.	- Poorly chosen rates or algorithms may result in slow convergence or suboptimal solutions.
Regularization Techniques	- Integration of regularization techniques (e.g., dropout, weight decay) prevents overfitting.	- Inadequate regularization may lead to overfitting, especially with limited data.

For CapsNets, densely linked convolutional layers can enhance the learning of discriminative feature maps; but, when network depth is increased, this may lead to the vanishing gradient problem. To mitigate this problem, dense connections between layers can be added via feature concatenations or ResNet-style skip connections (Larsson et al., 2016; Phaye et al., 2018).

One weakness in the present CapsNet routing technique is that the training process is not adequately integrated into the system. The optimal number of routing iterations must be manually determined, which might not guarantee convergence (Chen & Crandall, 2018). On datasets other than MNIST or smallNORB, CapsNets have demonstrated potential with additional settings, such as higher Conv and FC layers (Nguyen & Ribeiro, 2019). These modifications are further explained in table 3.

Table 3. Capsnet modifications

Modification	Description	Proposed Problem	Advantages	Disadvantages
Dynamic Routing with RBA	Variant of dynamic routing known as Routing-by-Agreement (RBA).	Enhancing capsule agreement and adaptability to variations.	- Strengthens capsule agreement. - Enhances adaptability to variations in position, rotation, and scale.	- May be computationally intensive. - Sensitive to noise in input data.
Aff-CapsNets	Introduces Affine CapsNets for improved resilience to affine transformations.	Improved resilience to affine transformations.	- Markedly improves resilience to affine transformations. - Achieves this improvement with fewer parameters.	- May not be as effective for non-affine transformations.
Transformation-Aware Capsules	Capsules explicitly designed to handle affine transformations.	Handling affine transformations explicitly.	- Explicitly designed to handle affine transformations. - Learns to identify and apply appropriate transformations.	- May struggle with other types of transformations. - Requires careful training.
Capsule-Capsule Transformation (CCT)	Introduces adaptive transformations between capsules.	Effective handling of varying degrees of affine transformations.	- Enables effective handling of varying degrees of affine transformations.	- Increased computational complexity. - Requires careful parameter tuning.
Margin Loss Regularization	Adds margin loss terms during training to encourage larger margins between capsules.	Increasing resistance to affine transformations.	- Encourages larger margins between capsules, increasing resistance to affine transformations.	- Adds complexity to the loss function. - May lead to slower convergence during training.
Capsule Routing with EM Routing	Utilizes an EM-like algorithm for capsule routing.	Improving capsule agreement process and feature learning.	- Improves capsule agreement process and feature learning.	- Increased computational complexity due to iterative updates.
Self-Routing	A supervised, non-iterative routing method.	Eliminating the need for capsule agreement.	- Alters capsule routing process, eliminating the need for capsule agreement.	- May result in suboptimal routing decisions. - Limited ability to capture global context.
Adversarial Capsule Networks	Combines CapsNet with adversarial training techniques.	Learning robust features by training against adversarial transformations.	- Helps the network learn robust features by training against adversarial affine transformations.	- Increased computational cost due to adversarial training.
Capsule Dropouts	Applies dropouts to capsules during training.	Enhancing generalization and robustness by reducing co-adaptations.	- Enhances generalization and robustness by reducing capsule co-adaptations.	- May lead to information loss. - Requires careful tuning of dropout rates.

continued on following page

Box 1. Continued

Modification	Description	Proposed Problem	Advantages	Disadvantages
Capsule Reconstruction	Augments CapsNet with a reconstruction loss term.	Encouraging the preservation of spatial information.	- Encourages the preservation of spatial information, improving robustness to affine transformations.	- Adds complexity to the network architecture. - May require additional computational resources.
Capsule Attention Mechanism	Incorporates attention mechanisms into capsules.	Improving focus on informative features and aiding robustness.	- Improves focus on informative features, aiding robustness against affine transformations.	- Increased computational complexity. - Requires careful attention mechanism design.

LIMITATIONS AND DIFFICULTIES WITH CAPSULE NETWORK

A relatively new kind of neural network design called capsule networks was presented as an enhancement over conventional convolutional neural networks (CNNs). While CapsNets exhibit encouraging outcomes in specific applications, they are not without limitations and constraints. These are a handful:

Cost of Computation

Because of the increased complexity of the routing-by-agreement algorithm used to calculate the activations of the capsules, CapsNets require more computing power than CNNs. CapsNet training may become slower and more resource-intensive as a result.

Limited Comprehension

Since CapsNets are a relatively new idea, researchers are still trying to figure out how to best utilize them and understand how they operate. Regarding how CapsNets represent and handle information, there are still a lot of unanswered concerns.

Complexity of the Model

CapsNet design is more intricate than those of conventional convolutional neural networks (CNNs). They are more challenging to train and optimize due to their increased complexity.

Few Applications in the Real World

Although CapsNets have demonstrated encouraging performance on some image identification tasks, more research needs to be done to determine how well they work on other real-world issues. The potential applications of CapsNets in a wider range of complicated areas remain largely unanswered.

Limited Pre-Trained Model Availability

The lack of pre-trained CapsNet models compared to CNNs can be a challenge for researchers and practitioners utilizing CapsNets in their work.

Receptive to Minor Changes

Because CapsNets are sensitive to even minute changes in the input, they may predict diverse outcomes for things that are visually similar. This could be problematic for assignments that call for a high degree of accuracy and consistency.

ROUTING ALGORITHMS

A class of neural network architecture known as Capsule Networks seeks to overcome some of the shortcomings of traditional convolutional neural networks (CNNs) in tasks such as natural language processing and object recognition. Capsule networks rely heavily on routing techniques. In Capsule Networks, routing techniques are used to calculate coupling coefficients, which regulate the amount of data transferred between capsules at various layers. Several popular routing algorithms for capsule networks are shown in Table 4. This comparison sheds light on the traits, advantages, and disadvantages of several capsule network routing methods. The needs of the application, the availability of computing power, and the type of input data all influence the routing algorithm that is selected.

CONCLUSION

The capsule network was designed to overcome the difficulties encountered by the classic CNN algorithm, and it has so far shown commendable performance. However, more comprehension of the aforementioned technique is necessary to fully realize it's potential. As a result, this research offers a comparison of the effectiveness of several algorithms found in influential literature. We have examined the use of the

Table 4. Popular routing algorithms

Ref	Routing Algorithm	Description	Advantages	Limitations	Special Applications
(Li et al., 2024)	Dynamic Routing with RBA	Variant of dynamic routing known as Routing-by-Agreement (RBA).	- Strengthens capsule agreement. - Enhances adaptability to variations in position, rotation, and scale.	- May be computationally intensive. - Sensitive to noise in input data.	General-purpose image recognition.
(Sabour et al., 2018)	EM Routing	Utilizes an Expectation-Maximization (EM) algorithm for capsule routing.	- Improves capsule agreement and feature learning. - Helps capsules reach consensus.	- Requires careful initialization. - Computationally demanding.	Medical image analysis, pose estimation.
(Sabour et al., 2017)	Hinton's Dynamic Routing	Original dynamic routing algorithm proposed by Geoffrey Hinton.	- Simple and conceptually clear. - Allows capsules to compete for routing.	- May suffer from gradient vanishing/exploding. - Limited routing capacity.	Image recognition tasks.
(Shi et al., 2022)	Sparse Routing	A variation of dynamic routing that introduces sparsity in the coupling coefficients.	- Reduces computational cost. - Improves scalability.	- May lead to information loss. - Requires careful tuning of sparsity parameters.	Large-scale datasets, efficiency-focused applications.
(Boruah & Das, 2024)	Capsule Forests	Ensembles of capsule networks with routing trees.	- Enhanced generalization through ensemble learning. - Improved robustness.	- Increased computational complexity. - Requires more training data.	Robust classification, anomaly detection.
(Choi et al., 2019)	Routing-by-Agreement with Attention (RBA-Attention)	Integrates attention mechanisms into capsule routing.	- Improved focus on informative features. - Enhanced robustness.	- Increased computational complexity. - Requires careful attention mechanism design.	Tasks with complex and variable structures, attention-driven applications.

current capsule network design and provided insight into its limitations, adjustments, and implementation outcomes. We have also examined the serval routing methods that have been written up in books. The results of this survey will be useful in helping the computer vision community to better understand the shortcomings and achievements of capsule networks and to develop strong machine vision algorithms through additional study.

REFERENCES

Afshar, P., Mohammadi, A., & Plataniotis, K. N. (2018, October). Brain tumor type classification via capsule networks. In *2018 25th IEEE international conference on image processing (ICIP)* (pp. 3129-3133). IEEE. 10.1109/ICIP.2018.8451379

Afshar, P., Naderkhani, F., Oikonomou, A., Rafiee, M. J., Mohammadi, A., & Plataniotis, K. N. (2021). MIXCAPS: A capsule network-based mixture of experts for lung nodule malignancy prediction. *Pattern Recognition*, *116*, 107942. doi:10.1016/j.patcog.2021.107942

Afshary, P., Mohammadiy, A., & Plataniotis, K. (2018). *Brain tumor type classification via capsule networks*. arXiv preprint arXiv:1802.10200. doi:10.1109/ICIP.2018.8451379

Ahmad, S., & Adnan, A. (2015, July). Machine learning based cognitive skills calculations for different emotional conditions. In *2015 IEEE 14th International Conference on Cognitive Informatics & Cognitive Computing (ICCI* CC)* (pp. 162-168). IEEE. 10.1109/ICCI-CC.2015.7259381

Ahmad, S., Li, K., Eddine, H. A. I., & Khan, M. I. (2018). A biologically inspired cognitive skills measurement approach. *Biologically Inspired Cognitive Architectures*, *24*, 35-46.

Amer, M., & Maul, T. (2020). Path capsule networks. *Neural Processing Letters*, *52*(1), 545–559. doi:10.1007/s11063-020-10273-0

Anwar, M. S., Wang, J., Ahmad, S., Khan, W., Ullah, A., Shah, M., & Fei, Z. (2020). Impact of the impairment in 360-degree videos on users VR involvement and machine learning-based QoE predictions. *IEEE Access : Practical Innovations, Open Solutions*, *8*, 204585–204596. doi:10.1109/ACCESS.2020.3037253

Anwar, M. S., Wang, J., Khan, W., Ullah, A., Ahmad, S., & Fei, Z. (2020). Subjective QoE of 360-degree virtual reality videos and machine learning predictions. *IEEE Access : Practical Innovations, Open Solutions*, *8*, 148084–148099. doi:10.1109/ACCESS.2020.3015556

Armato, S. G. III, McLennan, G., Bidaut, L., McNitt-Gray, M. F., Meyer, C. R., Reeves, A. P., Zhao, B., Aberle, D. R., Henschke, C. I., Hoffman, E. A., Kazerooni, E. A., MacMahon, H., van Beek, E. J. R., Yankelevitz, D., Biancardi, A. M., Bland, P. H., Brown, M. S., Engelmann, R. M., Laderach, G. E., ... Clarke, L. P. (2011). The lung image database consortium (LIDC) and image database resource initiative (IDRI): A completed reference database of lung nodules on CT scans. *Medical Physics*, *38*(2), 915–931. doi:10.1118/1.3528204 PMID:21452728

Badue, C., Guidolini, R., Carneiro, R. V., Azevedo, P., Cardoso, V. B., Forechi, A., Jesus, L., Berriel, R., Paixão, T. M., Mutz, F., de Paula Veronese, L., Oliveira-Santos, T., & De Souza, A. F. (2021). Self-driving cars: A survey. *Expert Systems with Applications*, *165*, 113816. doi:10.1016/j.eswa.2020.113816

Boruah, M., & Das, R. (2024). MLCapsNet+: A multi-capsule network for the identification of the HIV ISs along important sequence positions. *Image and Vision Computing*, *145*, 104990. doi:10.1016/j.imavis.2024.104990

Campus, K. (n.d.). *Deep Frustration Severity Network for the Prediction of Declined Students' Cognitive Skills*. Academic Press.

Chao, H., Dong, L., Liu, Y., & Lu, B. (2019). Emotion recognition from multiband EEG signals using CapsNet. *Sensors (Basel)*, *19*(9), 2212. doi:10.3390/s19092212 PMID:31086110

Chen, Z., & Crandall, D. (2018). *Generalized capsule networks with trainable routing procedure*. arXiv preprint arXiv:1808.08692

Cheng, J., Huang, W., Cao, S., Yang, R., Yang, W., Yun, Z., Wang, Z., & Feng, Q. (2015). Enhanced performance of brain tumor classification via tumor region augmentation and partition. *PLoS One*, *10*(10), e0140381. doi:10.1371/journal. pone.0140381 PMID:26447861

Cheng, J., Yang, W., Huang, M., Huang, W., Jiang, J., Zhou, Y., Yang, R., Zhao, J., Feng, Y., Feng, Q., & Chen, W. (2016). Retrieval of brain tumors by adaptive spatial pooling and fisher vector representation. *PLoS One*, *11*(6), e0157112. doi:10.1371/ journal.pone.0157112 PMID:27273091

Choi, J., Seo, H., Im, S., & Kang, M. (2019). Attention routing between capsules. *Proceedings of the IEEE/CVF international conference on computer vision workshops*.

Clark, K., Vendt, B., Smith, K., Freymann, J., Kirby, J., Koppel, P., Moore, S., Phillips, S., Maffitt, D., Pringle, M., Tarbox, L., & Prior, F. (2013). The Cancer Imaging Archive (TCIA): Maintaining and operating a public information repository. *Journal of Digital Imaging*, *26*(6), 1045–1057. doi:10.1007/s10278-013-9622-7 PMID:23884657

Devi, K., & Muthusenthil, B. (2022). Intrusion detection framework for securing privacy attack in cloud computing environment using DCCGAN-RFOA. *Transactions on Emerging Telecommunications Technologies*, *33*(9), e4561. doi:10.1002/ett.4561

Doon, R., Rawat, T. K., & Gautam, S. (2018). *Cifar-10 classification using deep convolutional neural network. In 2018 IEEE Punecon*. IEEE.

Fatima, R., Samad Shaikh, N., Riaz, A., Ahmad, S., El-Affendi, M. A., Alyamani, K. A., Nabeel, M., Ali Khan, J., Yasin, A., & Latif, R. M. A. (2022). A natural language processing (NLP) evaluation on COVID-19 rumour dataset using deep learning techniques. *Computational Intelligence and Neuroscience, 2022,* 2022. doi:10.1155/2022/6561622 PMID:36156967

Gordienko, N., Kochura, Y., Taran, V., Peng, G., Gordienko, Y., & Stirenko, S. (2018). *Capsule deep neural network for recognition of historical Graffiti handwriting.* arXiv preprint arXiv:1809.06693.

Gu, J., Wu, B., & Tresp, V. (2021). Effective and efficient vote attack on capsule networks. *arXiv preprint arXiv:2102.10055.*

Haq, M. U., Sethi, M. A. J., Ahmad, S., ELAffendi, M. A., & Asim, M. (2024). Automatic Player Face Detection and Recognition for Players in Cricket Games. *IEEE Access : Practical Innovations, Open Solutions, 12,* 41219–41233. doi:10.1109/ACCESS.2024.3377564

Haq, M. U., Sethi, M. A. J., & Rehman, A. U. (2023). Capsule Network with Its Limitation, Modification, and Applications—A Survey. *Machine Learning and Knowledge Extraction, 5*(3), 891–921. doi:10.3390/make5030047

Haq, M. U., Sethi, M. A. J., Ullah, R., Shazhad, A., Hasan, L., & Karami, G. M. (2022). COMSATS Face: A Dataset of Face Images with Pose Variations, Its Design, and Aspects. *Mathematical Problems in Engineering, 2022,* 2022. doi:10.1155/2022/4589057

Haq, M. U., Shahzad, A., Mahmood, Z., Shah, A. A., Muhammad, N., & Akram, T. (2019). Boosting the face recognition performance of ensemble based LDA for pose, non-uniform illuminations, and low-resolution images. *KSII Transactions on Internet and Information Systems, 13*(6), 3144–3164.

Heidarian, S., Afshar, P., Enshaei, N., Naderkhani, F., Rafiee, M. J., Fard, F. B., Samimi, K., Atashzar, S. F., Oikonomou, A., Plataniotis, K. N., & Mohammadi, A. (2021). Covid-fact: A fully-automated capsule network-based framework for identification of covid-19 cases from chest ct scans. *Frontiers in Artificial Intelligence, 4,* 4. doi:10.3389/frai.2021.598932 PMID:34113843

Hinton, G. E., Sabour, S., & Frosst, N. (2018, May). Matrix capsules with EM routing. *International conference on learning representations.*

Hosni, A. I. E., Li, K., & Ahmed, S. (2018). HISBmodel: A rumor diffusion model based on human individual and social behaviors in online social networks. *Neural Information Processing: 25th International Conference, ICONIP 2018, Siem Reap, Cambodia, December 13–16, 2018 Proceedings, 25*(Part II), 14–27.

Hu, X.D. & Li, Z.H. (2022). Intrusion Detection Method Based on Capsule Network for Industrial Internet. *Acta Electonica Sinica, 50*(6), 1457.

Huang, L., Wang, J., & Cai, D. (2021). Graph capsule network for object recognition. *IEEE Transactions on Image Processing, 30*, 1948–1961.

Jain, R. (2019). Improving performance and inference on audio classification tasks using capsule networks. arXiv preprint arXiv:1902.05069.

Jaiswal, A. (2018). Capsulegan: Generative adversarial capsule network. *Proceedings of the European Conference on Computer Vision (ECCV) Workshops.*

Janakiramaiah, B., Kalyani, G., Karuna, A., Prasad, L. V., & Krishna, M. (2021). Military object detection in defense using multi-level capsule networks. *Soft Computing*, 1–15.

Justin, S. P., Andre, J. P., Bennett, A. L., & Fabbri, D. (2017). Deep learning for brain tumor classification. *Proc. SPIE Medical Imaging 2017: Biomedical Applications in Molecular, Structural, and Functional Imaging*, 10137.

Kaggle. (n.d.). https://www.kaggle.com/datasets/andrewmvd/convid19-x-rays

Khan, F., Yu, X., Yuan, Z., & Rehman, A. U. (2023). ECG classification using 1-D convolutional deep residual neural network. *PLoS One, 18*(4), e0284791. doi:10.1371/journal.pone.0284791 PMID:37098024

Khan, S., Kamal, A., Fazil, M., Alshara, M. A., Sejwal, V. K., Alotaibi, R. M., Baig, A. R., & Alqahtani, S. (2022). HCovBi-caps: Hate speech detection using convolutional and Bi-directional gated recurrent unit with Capsule network. *IEEE Access : Practical Innovations, Open Solutions, 10*, 7881–7894. doi:10.1109/ACCESS.2022.3143799

Klinger, R., De Clercq, O., Mohammad, S.M. & Balahur, A. (2018). *Iest: Wassa-2018 implicit emotions shared task.* arXiv preprint arXiv.

Koelstra, S., Muhl, C., Soleymani, M., Lee, J. S., Yazdani, A., Ebrahimi, T., Pun, T., Nijholt, A., & Patras, I. (2011). Deap: A database for emotion analysis; using physiological signals. *IEEE Transactions on Affective Computing, 3*(1), 18–31. doi:10.1109/T-AFFC.2011.15

Krizhevsky, A. & Hinton, G. (2009). *Learning multiple layers of features from tiny images*. Academic Press.

Krizhevsky, A., Sutskever, I., & Hinton, G. E. (2012). Imagenet classification with deep convolutional neural networks. *Advances in Neural Information Processing Systems*, 25.

Kumar, A., & Sachdeva, N. (2021). Multimodal cyberbullying detection using capsule network with dynamic routing and deep convolutional neural network. *Multimedia Systems*, 1–10.

Larsson, G., Maire, M., & Shakhnarovich, G. (2016). Fractalnet: Ultra-deep neural networks without residuals. arXiv preprint arXiv:1605.07648

Li, X., Liu, J., Xie, Y., Gong, P., Zhang, X., & He, H. (2024). Magdra: A multi-modal attention graph network with dynamic routing-by-agreement for multi-label emotion recognition. *Knowledge-Based Systems*, *283*, 111126. doi:10.1016/j.knosys.2023.111126

Ma, X., Dai, Z., He, Z., Ma, J., Wang, Y., & Wang, Y. (2017). Learning traffic as images: A deep convolutional neural network for large-scale transportation network speed prediction. *Sensors (Basel)*, *17*(4), 818. doi:10.3390/s17040818 PMID:28394270

Manoharan, J. S. (2021). Capsule Network Algorithm for Performance Optimization of Text Classification. *Journal of Soft Computing Paradigm*, *3*(01), 1–9. doi:10.36548/jscp.2021.1.001

Marchisio, A., Nanfa, G., Khalid, F., Hanif, M. A., Martina, M., & Shafique, M. (2023). SeVuc: A study on the Security Vulnerabilities of Capsule Networks against adversarial attacks. *Microprocessors and Microsystems*, *96*, 104738. doi:10.1016/j.micpro.2022.104738

Mazzia, V., Salvetti, F., & Chiaberge, M. (2021). Efficient-capsnet: Capsule network with self-attention routing. *Scientific Reports*, *11*(1), 1–13. doi:10.1038/s41598-021-93977-0 PMID:34282164

Mukhometzianov, R., & Carrillo, J. (2018). CapsNet comparative performance evaluation for image classification. arXiv preprint arXiv:1805.11195.

Munawar, F., Khan, U., Shahzad, A., Haq, M. U., Mahmood, Z., Khattak, S., & Khan, G. Z. (n.d.). *An Empirical Study of Image Resolution and Pose on Automatic Face Recognition*. Academic Press.

Nair, P., Doshi, R., & Keselj, S. (2021). Pushing the limits of capsule networks. arXiv preprint arXiv:2103.08074.

Netzer, Y., Wang, T., Coates, A., Bissacco, A., Wu, B. & Ng, A.Y. (2011). *Reading digits in natural images with unsupervised feature learning*. Academic Press.

Nguyen, H. P., & Ribeiro, B. (2019). Advanced capsule networks via context awareness. In *Artificial Neural Networks and Machine Learning–ICANN 2019: Theoretical Neural Computation: 28th International Conference on Artificial Neural Networks, Munich, Germany, September 17–19, 2019, Proceedings, Part I 28* (pp. 166-177). Springer International Publishing.

Patrick, M. K., Adekoya, A. F., Mighty, A. A., & Edward, B. Y. (2022). Capsule networks–a survey. *Journal of King Saud University. Computer and Information Sciences*, *34*(1), 1295–1310. doi:10.1016/j.jksuci.2019.09.014

Peer, D., Stabinger, S., & Rodriguez-Sanchez, A. (2018). Training deep capsule networks. arXiv preprint arXiv:1812.09707.

Phaye, R., Sikka, Dhall, & Bathula. (2018). Dense and diverse capsule networks: Making the capsules learn better. arXiv preprint arXiv:1805.04001

Radford, A., Metz, L., & Chintala, S. (2015). Unsupervised representation learning with deep convolutional generative adversarial networks. arXiv preprint arXiv:1511.06434.

Rahim, A., Zhong, Y., Ahmad, T., Ahmad, S., & ElAffendi, M. A. (2023). Hyper-Tuned Convolutional Neural Networks for Authorship Verification in Digital Forensic Investigations. *Computers, Materials & Continua*, *76*(2). Advance online publication. doi:10.32604/cmc.2023.039340

Rahim, A., Zhong, Y., Ahmad, T., Ahmad, S., Pławiak, P., & Hammad, M. (2023). Enhancing smart home security: Anomaly detection and face recognition in smart home iot devices using logit-boosted cnn models. *Sensors (Basel)*, *23*(15), 6979. doi:10.3390/s23156979 PMID:37571762

Ramasinghe, S., Athuraliya, C. D., & Khan, S. H. (2018). A context-aware capsule network for multi-label classification. *Proceedings of the European Conference on Computer Vision (ECCV) Workshops*.

Rathnayaka, P., Abeysinghe, S., Samarajeewa, C., Manchanayake, I., & Walpola, M. (2018). Sentylic at IEST 2018: Gated recurrent neural network and capsule network-based approach for implicit emotion detection. arXiv preprint arXiv:1809.01452. doi:10.18653/v1/W18-6237

Sabour, S., Frosst, N., & Hinton, G. (2018, February). Matrix capsules with EM routing. In *6th international conference on learning representations, ICLR* (pp. 1-15). Academic Press.

Sabour, S., Frosst, N., & Hinton, G. E. (2017). Dynamic routing between capsules. *Advances in Neural Information Processing Systems*, 30.

Saqur, R., & Vivona, S. (2018). Capsgan: Using dynamic routing for generative adversarial networks. arXiv preprint arXiv:1806.03968.

Saqur, R., & Vivona, S. (2020). Capsgan: Using dynamic routing for generative adversarial networks. In A*dvances in Computer Vision: Proceedings of the 2019 Computer Vision Conference (CVC),* Volume 2 (pp. 511-525). Springer International Publishing. 10.1007/978-3-030-17798-0_41

Shi, R., Niu, L., & Zhou, R. (2022). Sparse CapsNet with explicit regularizer. *Pattern Recognition, 124*, 108486. doi:10.1016/j.patcog.2021.108486

Sim, T., Baker, S., & Bsat, M. (2002, May). The CMU pose, illumination, and expression (PIE) database. In *Proceedings of fifth IEEE international conference on automatic face gesture recognition* (pp. 53-58). IEEE. 10.1109/AFGR.2002.1004130

Singh, M., Nagpal, S., Singh, R., & Vatsa, M. (2019). Dual directed capsule network for very low-resolution image recognition. In *Proceedings of the IEEE/CVF International Conference on Computer Vision* (pp. 340-349). 10.1109/ICCV.2019.00043

Sohail, M. Z., Zafar, T., Khan, T. A., Asim, M., Ahmad, S., Mairaj, T., & El Affendi, M. A. (2023). Prediction of Time to Failure (TTF) of Power Systems Using a Deep Learning Technique. *Journal of Hunan University Natural Sciences, 50*(12).

Tahsin, M. S., Al Karim, M., Ahmed, M. U., Tafannum, F., & Firoz, N. (2023). An integrated approach for diabetes detection using fisher score feature selection and capsule network. *Journal of Computing Science and Engineering : JCSE, 4*(2), 61–77.

Teto, J. K., & Xie, Y. (2019, March). Automatically Identifying of animals in the wilderness: Comparative studies between CNN and C-Capsule Network. In *Proceedings of the 2019 3rd International Conference on Compute and Data Analysis* (pp. 128-133). 10.1145/3314545.3314559

Tiwari, S., & Jain, A. (2021). Convolutional capsule network for COVID-19 detection using radiography images. *International Journal of Imaging Systems and Technology, 31*(2), 525–539. doi:10.1002/ima.22566 PMID:33821095

Ullah, H., Haq, M. U., Khattak, S., Khan, G. Z., & Mahmood, Z. (n.d.). *A Robust Face Recognition Method for Occluded and Low-Resolution Images.* Academic Press.

Wahab, H., Mehmood, I., Ugail, H., Sangaiah, A. K., & Muhammad, K. (2023). Machine learning based small bowel video capsule endoscopy analysis: Challenges and opportunities. *Future Generation Computer Systems*, *143*, 191–214. doi:10.1016/j.future.2023.01.011

Wang, X., Wang, Y., Guo, S., Kong, L., & Cui, G. (2023). Capsule Network With Multiscale Feature Fusion for Hidden Human Activity Classification. *IEEE Transactions on Instrumentation and Measurement*, *72*, 1–12. doi:10.1109/TIM.2023.3238749

Weng, Z., Meng, F., Liu, S., Zhang, Y., Zheng, Z., & Gong, C. (2022). Cattle face recognition based on a Two-Branch convolutional neural network. *Computers and Electronics in Agriculture*, *196*, 106871. doi:10.1016/j.compag.2022.106871

Wu, H., Mao, J., Sun, W., Zheng, B., Zhang, H., Chen, Z., & Wang, W. (2016, August). Probabilistic robust route recovery with spatio-temporal dynamics. In *Proceedings of the 22nd ACM SIGKDD International Conference on Knowledge Discovery and Data Mining* (pp. 1915-1924) 10.1145/2939672.2939843

Yang, B., Bao, W., & Wang, J. (2022). Active disease-related compound identification based on capsule network. *Briefings in Bioinformatics*, *23*(1), bbab462. doi:10.1093/bib/bbab462 PMID:35057581

Zhang, L., Edraki, M., & Qi, G.-J. (2018). Cappronet: Deep feature learning via orthogonal projections onto capsule subspaces. *Advances in Neural Information Processing Systems*, 31.

Zhang, X. (2021). The AlexNet, LeNet-5 and VGG NET applied to CIFAR-10. In *2021 2nd International Conference on Big Data & Artificial Intelligence & Software Engineering (ICBASE)* (pp. 414-419). IEEE. 10.1109/ICBASE53849.2021.00083

Zhao, Z., Kleinhans, A., Sandhu, G., Patel, I., & Unnikrishnan, K. P. (2019). *Capsule networks with max-min normalization.* arXiv preprint arXiv:1903.09662

Chapter 6
Advancements in Machine Learning and Deep Learning

Dina Darwish
Ahram Canadian University, Egypt

ABSTRACT

Among the most important methodologies in the field of modern intelligent technology is data-driven advanced machine learning methodology. In order to find rules, it makes use of data samples that have been observed, and it makes use of regular patterns in order to forecast unknown data in the future. In tandem with the development of artificial intelligence, the field of machine learning is making further strides forward. Due to this, there is a need for increased requirements for the training and applications of models, as well as the enhancement of the algorithm and the improvement of technological capabilities. This chapter discusses the recent technologies and trends in the artificial intelligence field, while giving examples and conclusions at the end of the chapter.

INTRODUCTION

Machine learning, a very promising technique within the realm of artificial intelligence, is rapidly gaining widespread adoption across various industries. The applications of AI span across various disciplines, including computer sciences (e.g., cybersecurity, hardware design, and human-machine interaction), social sciences (e.g., economics and education), and natural sciences (e.g., physics and medical). The machine learning research field is dedicated to the advancement of novel methods, algorithms, and models, in addition to their practical applications. Model explanation techniques are utilized to acquire fresh perspectives when

DOI: 10.4018/979-8-3693-2913-9.ch006

addressing complex or significant societal issues. Advanced machine learning, in contrast to regular machine learning, relies more heavily on data rather than experience to enhance the performance of the system. Data-driven advanced machine learning is a crucial methodology in contemporary intelligent technology. It employs observed data samples to identify laws and utilizes regular patterns to forecast future unobservable data. As artificial intelligence expands, the domain of machine learning continues to advance. This necessitates greater requirements for model training and applications, the enhancement of the algorithm, and heightened computational capability. Recent advancements in this technology have facilitated significant advances that enhance the speed and efficiency of corporate intelligence. These discoveries encompass a wide range of capabilities, including facial recognition and natural language processing. Machine learning programs can be considered as discrete elements, or subroutines, of artificial intelligence that have the ability to function autonomously. The objective of achieving genuine artificial intelligence, which refers to a computer or program that possesses the ability to think and communicate in a manner similar to that of a human being, has not yet been accomplished. Nevertheless, specific machine learning algorithms have undergone training to excel in executing particular tasks that are highly beneficial. Machine learning is commonly referred to as AI due to multiple factors. The integration of diverse machine learning algorithms, functioning as subroutines, holds the potential to facilitate the achievement of genuine artificial intelligence.

Deep learning has experienced substantial progress and observed notable patterns in diverse domains. An important advancement is the fusion of deep learning and cloud technology, which enables organizations to create and apply deep learning solutions by leveraging available resources. Deep learning has demonstrated exceptional efficacy in the field of medical image analysis, particularly in the segmentation of anatomical or diseased structures. This technology has proven to be highly beneficial for physicians, aiding them in the process of diagnosis and surgery planning. Deep learning has proven to be advantageous for dialogue systems, a widely-used task in natural language processing. State-of-the-art models are now being employed in both task-oriented and open-domain systems. In addition, deep learning techniques and structures have been utilized in other fields like healthcare, wearable technology, social networks, and others. Presently, the prevailing directions in deep learning research encompass conditional generative adversarial networks, knowledge distillation, active learning, cross-modality learning, and federated learning. These improvements demonstrate the promise of deep learning in diverse applications and emphasize the necessity for additional study in areas such as model performance evaluation and system design. Following the introduction of ChatGPT in November 2022, the year 2023 represented a significant milestone in the field of artificial intelligence. The advancements in AI over the past year, including a thriving open source community and the development of complex multimodal models, have set the foundation for substantial progress. However, while generative AI still holds the interest of the IT industry, organizations are now adopting a more sophisticated and balanced approach as they move from experimental projects to practical applications. The trends observed this year demonstrate a growing level of complexity and prudence in the development and implementation strategies of AI, with a focus on ethical considerations, safety, and the changing legal environment. This chapter focuses on recent trends in the field of artificial intelligence and deep learning, and gives examples on different technologies in this field. Also, this chapter discusses the major recent technologies and trends in the artificial intelligence field, including Multi-modal AI, Agentic AI, Open source AI, Retrieval-augmented generation, and generative AI, and finally comes conclusion.

Major recent Technologies and Trends in the Artificial Intelligence

Technologies Enabling the Operation of AI Systems That Can Process Multiple Modes of Input

Multimodal AI emerges from the collective expertise in several subdomains of AI. AI practitioners and scholars have achieved significant advancements in the storage and processing of data across many formats and modalities in recent years. Here is a list of the domains that are driving the growth of multimodal AI:

1) *Artificial neural* networks that are capable of learning and making complex decisions based on large amounts of data.

2) *Deep learning* is a specialized area within the study of artificial intelligence that utilizes a specific form of algorithm known as an artificial neural network to tackle intricate tasks. The ongoing revolution in generative AI is driven by deep learning models, namely transformers, which are a specific type of neural architecture. The advancement of multimodal AI will also rely on further developments in this field. Specifically, there is a significant need for research to discover novel methods to enhance the capabilities of transformers, as well as innovative solutions for data fusion.

3) *Natural Language Processing (NLP)* is the field of study that focuses on the interaction between computers and human language. It involves developing algorithms and models that enable computers to understand, interpret, and generate human language in a way that is similar to how humans do. Natural Language Processing (NLP) plays a crucial role in the field of artificial intelligence by connecting human communication with machine comprehension.

Natural Language Processing is a field that combines various disciplines to provide computers the ability to understand, analyze, and produce human language. This allows for smooth communication between humans and machines. Given that text is the dominant medium of communication with machines, it is unsurprising that Natural Language Processing (NLP) plays a crucial role in optimizing the performance of generative AI models, including those that incorporate several modes of input.

1) Computer Vision

Image analysis, or computer vision, refers to a collection of techniques that enable computers to visually see and comprehend an image. The advancements in this field have facilitated the creation of multimodal AI models capable of processing images and videos as both inputs and outputs.

2) Sound manipulation

Advanced generative AI models have the ability to process audio files as both inputs and outputs. Audio processing encompasses a wide range of applications, including the interpretation of spoken messages, simultaneous translation, and music composition.

Utilizations of Multimodal Artificial Intelligence

Multimodal learning enables robots to gain more sensory modalities, hence enhancing their accuracy and interpretation capabilities. These capabilities are enabling a wide range of innovative applications across various sectors and industries, including:

1)   Enhanced generative artificial intelligence

The majority of initial generative AI models were designed to do text-to-text tasks, where they could analyze text prompts from users and generate corresponding text responses. Advanced multimodal models such as GPT-4 Turbo, Google Gemini, and DALL-E offer enhanced capabilities that can enhance user experience in terms of both input and output. The potential of multimodal AI agents appears boundless, as they may absorb inputs in numerous modalities and generate material in various formats.

2)   Self-driving vehicles

Self-driving automobiles heavily depend on multimodal artificial intelligence. The cars are fitted with many sensors to analyze data from the environment in different formats. For autonomous vehicles to make intelligent decisions in real time, it is crucial to have multimodal learning that effectively and efficiently combines several sources of information.

3)   Biomedicine

The growing accessibility of biomedical data from biobanks, electronic health records, clinical imaging, and medical sensors, along with genomic data, is driving the development of multimodal AI models in the medical industry. These models have the ability to handle different data sources that come in different forms in order to help us understand human health and illness better, and also to make informed therapeutic decisions. The study of Earth science encompasses the examination of several aspects of our planet, including its geology, atmosphere, and climate. One significant area of focus within Earth science is the analysis of climate change, which refers to the long-term alteration of Earth's climate patterns due to human activities and natural processes. The proliferation of ground sensors, drones, satellite data, and other measurement tools is enhancing our capacity to comprehend the globe. Utilizing multimodal AI is essential for effectively integrating this information and developing novel applications and tools that can assist us in many activities, including monitoring greenhouse gas emissions, forecasting extreme climate events, and implementing precision agriculture.

Presented here are the foremost trends in the field of artificial intelligence and machine learning that one should be ready for in the year 2024.

1.   *Multi-modal AI*

Multimodal AI surpasses the conventional approach of processing data from a single mode and instead incorporates other sorts of input, including text, images, and sound. This advancement aims to imitate the human capacity to handle a wide range of sensory information. "The interfaces across the globe are multimodal," stated Mark Chen, the leader of frontiers research at OpenAI, during a presentation at the EmTech MIT conference in November 2023. OpenAI's GPT-4 model possesses multimodal capabilities, allowing it to process and react to both visual and aural input. During his presentation, Chen provided an illustration of capturing images of the interior of a refrigerator and requesting ChatGPT to propose a recipe based on the components depicted in the photos. If ChatGPT's speech mode is used, the interaction could also include an audio component while making the request verbally.

Key Principles of Multimodal Artificial Intelligence

Multimodal artificial intelligence models introduce an additional level of intricacy to the most advanced language models. These models utilize a neural architecture known as a Transformer. Transformers, created by Google researchers, utilize the encoder-decoder architecture and attention mechanism to facilitate efficient data processing. This is an intricate procedure that can be challenging to comprehend.

Comparison Between Unimodal and Multimodal AI

Several data fusion approaches can be utilized to tackle difficulties involving several modes of data. Data fusion approaches can be classified into three groups based on the processing level at which fusion occurs.

- Initial integration. The process entails encoding many modalities into the model in order to establish a shared representation space. The outcome of this approach is a unified output that captures the semantic information from all the modalities, regardless of their differences.
- Fusion in the middle. The approach entails integrating many modalities during different preprocessing phases. Data fusion is accomplished by constructing dedicated layers within the neural network that are specifically tailored for this purpose.
- Concatenation of multiple sources of information after they have been processed individually. The approach entails generating numerous models to handle various modalities and merging the results of each model in a novel algorithmic layer.

There is no universally optimal data fusion technique that can be used to all types of settings. However, the technique selected will be determined by the specific multimodal task being performed. Therefore, it is probable that a trial and error approach will be necessary in order to discover the best appropriate multimodal AI pipeline. Large language models (LLMs) have been highly successful in the field of natural language processing (NLP). Models like BERT (Devlin et al., 2018), T5 (Raffel et al., 2020), and GPT (Brown et al., 2020) have the ability to acquire general data representations and common knowledge from extensive collections of text by employing self-supervised learning tasks, such as masked language modelling or next token prediction. The acquired models can be additionally optimized for certain tasks and achieve exceptional performance on those tasks. LLMs have also had success in other domains beyond language, such as computer vision or speech processing. Figure 1 shows unimodal vs. multi-modal AI.

Figure 1. Unimodal vs. multi-modal AI

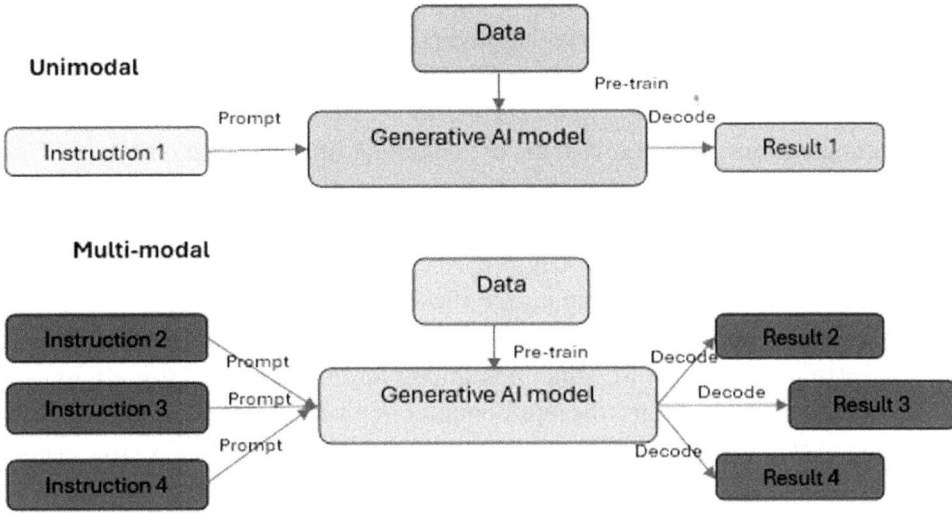

The use of these techniques on various forms of data has led to the emergence of "multimodal AI," which is currently the most popular area of focus in the AI field. In the era of LLMs, the architectural designs of models for many modalities are increasingly converging. Transformers are extensively employed in several contexts, such as text, code, visual, and audio scenarios, to facilitate comprehension and generation tasks. For example, recent LLMs such as ChatGPT or GPT-4 have combined text and code into one model, enabling it to handle tasks involving generating text-only, code-only, text-to-code, and code-to-text. Multi-modal generation models such as DALL·E and NUWA-Infinity are also trained using auto-regressive models like GPT models to generate images and videos. VALL-E, developed by Wang et al. in 2023 (Wang et al., 2023), possesses powerful in-context learning skills. It can be utilized for zero-shot cross-lingual text-to-speech synthesis and zero-shot speech-to-speech translations. VALL-E is built on Transformer and GPT-like models. In addition, we have seen that diffusion models are extensively employed in content creation tasks, including DALL·E 2 and Stable Diffusion (Rombach et al., 2022) for visual generation, as well as NaturalSpeech 2 (Shen et al., 2023) for voice production. However, there is another line of research that seeks to integrate several types of generation models by employing diffusion models. This can also be interpreted as a sign of the convergence of model architectures. There is currently no globally accepted model architecture for multimodal artificial intelligences due to the variations in basic units, data formats, and structures of material across different modalities. Nevertheless, this convergence is undeniably a prominent pattern among the AI field. Transitioning from the comprehension of visual language to the creation of visual content is based on language input. VL pre-trained models are highly representative multimodal artificial intelligences. The objective of these models is to acquire the knowledge of both textual and visual elements simultaneously and facilitate tasks related to visual and textual information, such as finding images, answering questions based on visuals, or generating images based on text.

Over the past few years, there has been a shift in study focus from tasks that involve interpreting visual language to challenges that involve generating visual content. In this section, we will begin by examining the advancements made in models for interpreting visual language. Following that, we will delve into the most recent developments in models that generate visual content based on textual input. There exist three fundamental distinctions among various theories of visual language processing. Firstly, let's discuss the representation of visual inputs. Various models for visual understanding employ different levels of granularity to depict visual elements, including pixels, objects, and patches inside images or videos. Patches have become the prevailing level of granularity in recent times. Furthermore, how can one create visual depictions? Certain models employ convolutional neural network (CNN) architectures like ResNet or Faster R-CNN, while other models utilize Transformers such as ViT, Swin (Liu et al., 2023), and so on. Furthermore, the question arises on how to integrate the representations derived from textual and visual inputs. There

Figure 2. Overview of visual-language models

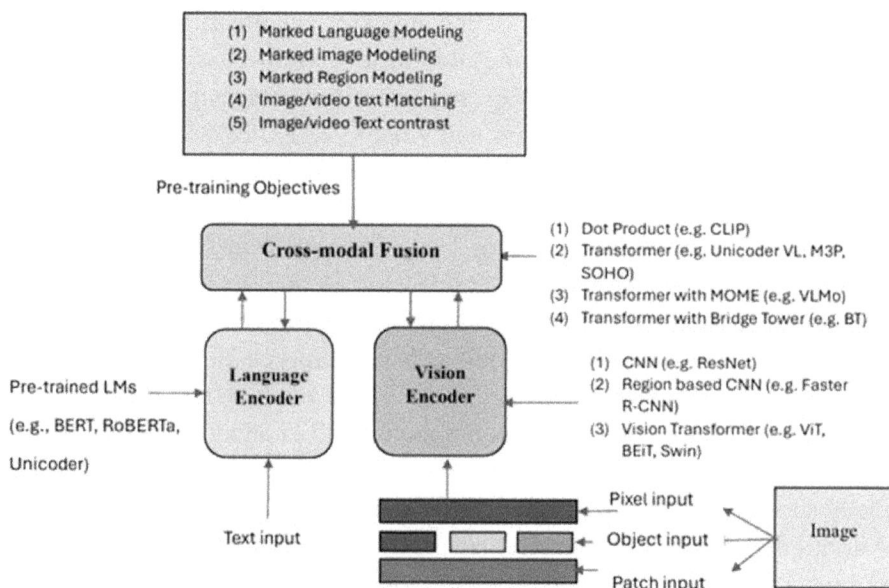

are multiple methods available for doing this assignment. Figure 2 shows overview of visual-language models, and Figure 3 illustrates overview of Visual ChatGPT v2.

The CLIP (Radford et al., 2021) paradigm, for instance, employs a straightforward dot-product component in the fusion process. This design choice significantly reduces computational costs while maintaining a very effective framework for image-text matching. Several first models for visual-linguistic comprehension, such as Unicoder-VL (Li et al., 2022), M3P, and Uniter (Chen et al., 2021), employed Transformers to enhance the integration of textual and visual representations. Mixture-of-Experts are employed to combine representations from several input modalities, including VLMo, hence enhancing the adjustability of model parameters for diverse modalities. Recent studies, such as BridgeTower (Xu et al., 2023) and ManagerTower (Xu et al., 2023), have utilized text or visual representations from several layers to improve the quality of unimodal representations for subsequent visual and linguistic understanding tasks. Currently, the most advanced setup for pre-trained models in the field of visual language (VL) uses patches as the units for visual representation and employs the Transformer model to combine text and visual representations. In addition to photos, the comprehension of videos is equally crucial for the advancement of numerous

Figure 3. Visual ChatGPT v2

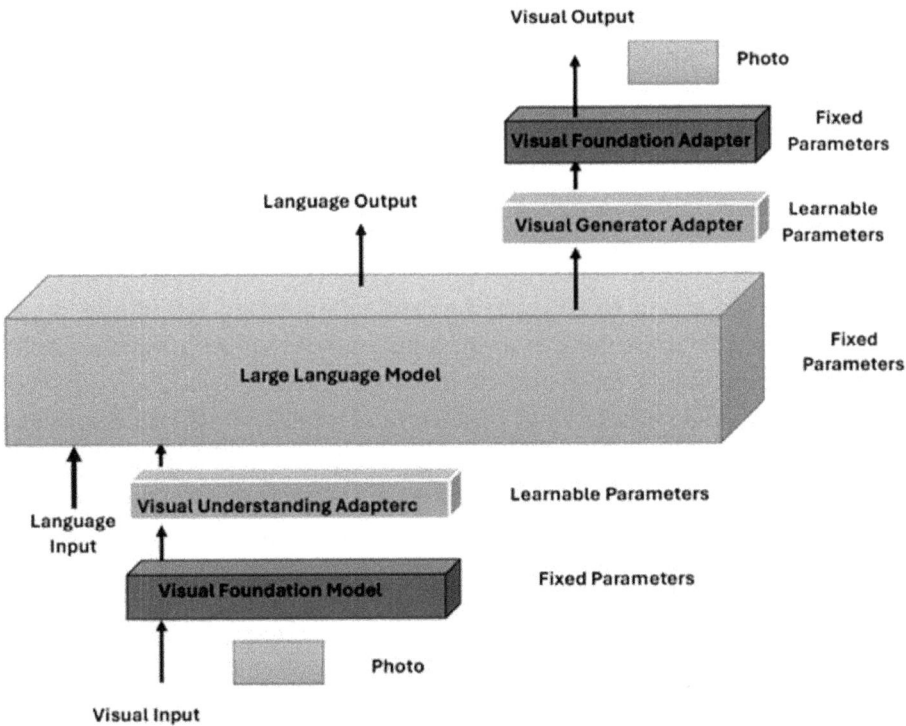

forthcoming AI systems. Currently, image-based visual models are being effectively utilized in video models. Nevertheless, in the future, it will be easy to utilize the extensive video collection directly, enabling the training of more robust multimodal AI models for tasks connected to videos. Multimodal learning is a specialized area of artificial intelligence that aims to enhance the learning capabilities of machines by exposing them to extensive volumes of textual information, together with other forms of data referred to as sensory input, including photos, videos, and audio recordings. This enables models to acquire novel patterns and associations between textual descriptions and their corresponding images, videos, or audio. The utilization of multimodal learning is enabling intelligent systems to explore novel opportunities. Multimodal AI models are well-suited for handling many types of input and producing different types of output due to their ability to incorporate multiple data types during the training process. As an illustration, GPT-4, the fundamental model of ChatGPT, has the capability to process both image and text inputs and produce text outputs. Additionally, OpenAI has just introduced the Sora text-to-video model.

2. *Agentic AI*

Agentic AI represents a notable transition from AI that simply reacts to stimuli to AI that takes initiative and acts proactively. AI agents are self-aware, proactive, and capable of autonomous behaviour. Unlike traditional AI systems, AI agents can perceive their surroundings, set objectives, and act to achieve them autonomously.

Relying on Direct Human Intervention

In environmental monitoring, an AI agent can be trained to collect data, identify patterns, and take proactive efforts to mitigate hazards, such as spotting forest fires early. A financial AI agent can use adaptive approaches to manage an investment portfolio and respond quickly to changing market conditions. According to a blog post by computer scientist Peter Norvig, a fellow at Stanford's Human-Centered AI Institute, AI-enabled chat features will become available in 2023. "By 2024, agents will possess the capability to efficiently accomplish tasks on your behalf." Furthermore, the combination of agentic and multimodal AI has the potential to create new opportunities. Chen's talk included an example of an image recognition and classification application. Previously, establishing a photo recognition application involved training a model and determining the best deployment approach. Using multimodal, agentic models, natural language prompts can fulfil all of these objectives. Agentic AI systems can do long-term actions to achieve goals without relying on predefined behaviors. In popular culture, AI agents are viewed as efficient facilitators capable of doing numerous tasks. The AI, like Samantha in

"Her" and HAL 9000 in "2001: A Space Odyssey," is created to meet the demands of its users. These agents differ from modern AI systems like GPT-4, which are highly intelligent but limited in their ability to execute real-world tasks. There is no clear distinction between "agents" and current AI systems, such as GPT-4. The agentic aspect of AI systems can be understood through multiple dimensions, with additional advancements expected in the field. The degree of agenticness refers to a system's ability to complete complex tasks with minimal supervision. The concept of agenticness encompasses several components, including:

- *Goal complexity:* How difficult would it be for a human to achieve the AI system's goal, and what are the system's capabilities? The aim may include desired levels of reliability, speed, and safety. An AI system that can answer questions on both programming and law has a higher goal-complexity than a text classifier that can only categorize inputs as law or programming.
- *Environmental complexity:* How complicated are the environments where a system can achieve its goals? AI systems that excel at multiple board games have a higher level of environmental complexity compared to those that can only play chess. Other factors to consider include the extent to which they span multiple domains, involve multiple stakeholders, require long-term operation, and utilise external technologies. The first system may survive in a variety of situations, including chess, while the second system is only suitable for chess.
- *Adaptability:* How well can the system modify and respond to novel or unexpected situations? Human customer service people are more adaptable than automated rule-based systems, as they can accommodate unexpected or unique customer requests.
- *Autonomy:* Can the system achieve its objectives with minimum human intervention or oversight?

Level 3 autonomous driving cars, as an example, possess a higher degree of independent execution compared to typical cars that rely on constant human operation. These cars are capable of operating without human involvement in specific situations.

In accordance with recent literature (Chan, et al., 2023), the term "agentic AI systems" can be used to describe systems that demonstrate a significant level of agenticness. This term is used to highlight that agenticness, as defined here, is a characteristic rather than a specific category or classification. However, in certain contexts, we may also use the term "agents" as it is commonly used in the field. This study specifically examines the many impacts and optimal strategies that may be significant as the level of autonomy in systems increases. We want to make it clear that agenticness is a separate idea from awareness, moral patient hood, or self-motivation.

We also differentiate a system's level of agenticness from its anthropomorphism. Agentic AI systems are typically conceptualized as functioning with goals specified by people and in surroundings determined by humans. These systems often work in cooperation with human "teammates" rather than being entirely autonomous and setting their own goals. Agenticness, as we have defined it, is not limited to physicality. In fact, many digital systems exhibit more agenticness than most robots. However, certain forms of "independent execution" that have physical effects, such as in a driverless car, can amplify the risks and opportunities associated with agenticness in specific applications. Finally, it is important to note that agenticness is conceptually separate from an AI system's performance level on a specific job or the breadth of its capabilities. However, enhancements in performance and breadth may enable a system to function as an agent in specific situations (Morris et al., 2023).

The usefulness of agenticness AI systems that possess a high degree of agency can be more advantageous, as long as they are created with safety in mind and adhere to established practices for accountability. Agenticness can enhance the efficacy of a specific system in the following ways:

- ***Enhanced precision and dependability of results***: for instance, a language model with the ability to independently browse the Internet and adapt its queries based on the received outcomes can offer significantly more precise responses to inquiries compared to a system lacking such capabilities. This is especially applicable in cases involving subjects that are constantly changing or events that happened after the model was trained.
- ***Increased time efficiency for users***: for instance, if a user gives an AI system broad instructions on the code they want it to generate, it would be more convenient for the user if the system autonomously performs multiple tasks such as translating the instructions into code, executing the code, presenting the results, evaluating those results, and making necessary code modifications to enhance the outcomes.
- ***Enhanced user preference elicitation***: For instance, a personal assistant AI that can actively communicate with users by sending messages to inquire about specific details using natural language, and does so at opportune moments, may offer a superior user experience compared to an application with multiple intricate settings that are challenging for users to utilize proficiently.
- ***Ability to easily and efficiently handle increasing amounts of work or data***: An AI system with a high level of agency can enable a single user to perform a significantly greater number of acts compared to what they would be able to do otherwise. Additionally, such a system has the potential to benefit a much

broader population of individuals compared to a less agentic version of the same system.

Let's examine the case of radiology. An automated radiology image classification tool could enhance the efficiency of a radiologist to some extent. However, a fully autonomous radiology tool capable of performing specific patient-care tasks without human supervision, such as generating reports on scans and asking patients basic follow-up questions, has the potential to significantly increase a radiologist's efficiency and allow for a greater number of patients to be seen. The Agent AI Paradigm is described in paper (Saenz et al., 2023). It is a framework that involves the use of intelligent agents to perform tasks and make decisions autonomously. This paper explores a novel paradigm and framework for educating Agent AI. The proposed framework aims to achieve multiple objectives:

- Utilize pre-existing pre-trained models and pre-training procedures to efficiently equip the agents with a comprehensive comprehension of crucial modalities, such as textual or visual inputs.
- Provision of adequate long-term task-planning capabilities.
- Implement a memory framework that enables the encoding and retrieval of acquired knowledge at a later time.
- Enable the utilization of environmental feedback to efficiently train the agent in acquiring knowledge about the appropriate activities to undertake.

Figure 4 displays a comprehensive diagram of a new agent, illustrating the significant submodules of the system. Long-lasting memories (LLMs) or very long-lasting memories (VLMs) model can be utilized to initiate the components of the Agent, as illustrated in Figure 4. LLMs have demonstrated strong performance in task-planning (Gong et al., 2023a), possess substantial global knowledge (Yu et al., 2023), and have remarkable logical reasoning ability (Creswell et al., 2022). In addition, Visual Language Models (VLMs) like CLIP (Radford et al., 2021) offer a universal visual encoder that is matched with language and also possess the ability to recognize visual content without prior training. For instance, advanced open-source multi-modal models like LLaVA (Liu et al., 2023) and InstructBLIP (Dai et al., 2023) use fixed CLIP models as visual encoders.

An alternative approach to use frozen LLMs and VLMs for the AI agent is to employ a solitary transformer model that accepts both visual tokens and linguistic tokens as input, such to the Gato model proposed by Reed et al. (2022). Furthermore, alongside visual and linguistic inputs, the authors include a third kind of input called agent tokens. Agent tokens are employed to allocate a distinct subset of the model's input and output space for agentic actions. In the context of robotics or game playing, the input action space of the controller can be considered as a representation. Agent tokens can be utilized for instructing agents to utilize particular tools, such as image-generation or image-editing models, or for other API calls. As depicted in Figure 5, the agent tokens can be merged with visual and verbal tokens to provide a cohesive interface for training multi-modal agent artificial intelligence. Utilizing an agent transformer has various advantages to utilizing huge, proprietary LLMs as agents. Firstly, the model may be readily tailored to very specific agentic tasks that may pose challenges in their representation using natural language, such as controller inputs or other precise actions. Therefore, the agent has the ability to enhance its performance by acquiring knowledge from interactions with the environment and domain-specific data. Furthermore, gaining access to the probabilities of the agent tokens might facilitate a better understanding of the model's decision-making process and the factors that influence its behaviors. Furthermore, some sectors such as healthcare and law are subject to stringent data privacy regulations. In conclusion, a comparatively smaller agent transformer has the potential to be considerably less expensive than a bigger proprietary language model.

3. Open source AI

Building high-capacity generative AI systems, such as comprehensive language models, requires a lot of data and a lot of computing power. Developers can save money and increase access to AI by utilizing an open source approach, which enables them to use previous work. The term "open source AI" describes AI programs that anybody can download and use, typically without paying a dime. Organizations and scholars are able to actively engage and build upon existing code because of this accessibility. According to statistics collected from GitHub, there was a notable uptick in developer interest in artificial intelligence, namely in

Figure 4. the current paradigm for creating multi-modal agents by incorporating a large language model (LLM) with a large vision model (LVM)

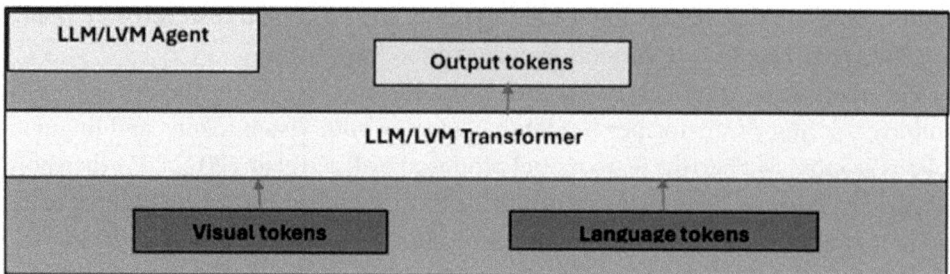

Figure 5. A proposed unified agent multi-modal transformers model for agent systems

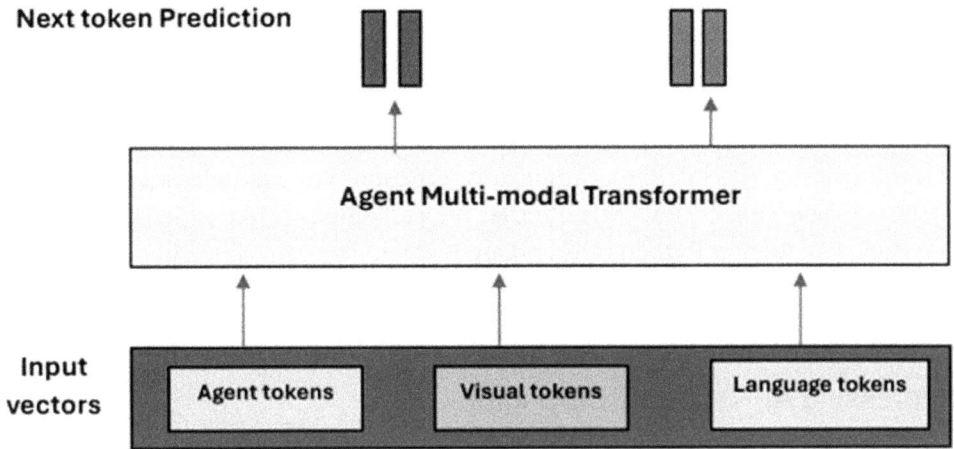

Next token Prediction

Agent Multi-modal Transformer

Input vectors

| Agent tokens | Visual tokens | Language tokens |

generative AI. The most popular code hosting site saw the introduction of generative AI projects in 2023. Many new contributors joined projects such as AutoGPT and Stable Diffusion. Open source generative models were few at the start of the year, and when they were, they often couldn't hold a candle to commercial options like ChatGPT. But the scene grew substantially in 2023 to include major open source rivals like Mistral AI's Mixtral models and Meta's Llama 2. Because of this, smaller, less well-equipped entities may be able to use powerful AI models and technologies that were previously unachievable in 2024, which could change the dynamics of the AI environment. The increased scrutiny of the code increases the likelihood of discovering biases, defects, and security holes, which in turn promotes openness and ethical progress through open source approaches. Nonetheless, scholars have voiced worries about the possibility of open source AI being used to create harmful content, such as disinformation. Building and maintaining open source projects is difficult in general, but it becomes even more so when working with complex and computationally intensive AI models. Rapid progress is being driven in numerous sectors by the convergence of open source software (OSS) with artificial intelligence (AI). In accordance with the EC's proposal for unified AI regulations, software produced using ML, logic-and knowledge-based, or statistical techniques is referred to be an artificial intelligence system (AI system). It has the ability to influence the surroundings it interacts with by producing content, forecasts, suggestions, or decisions based on a set of human-defined objectives. In common parlance, "AI" has always meant machine learning algorithms or software that includes at least one ML algorithm. Machine learning (ML) is a method for analyzing data that entails creating and modifying models. These models enable programs to "learn" from their experiences, usually by training and improving their findings with massive amounts of data. A supervised or unsupervised process of automated incremental improvements is what gives rise to ML, as opposed to AI, which is essentially a one-and-done deal. Emerging as a result of the merging of open source software (OSS) and artificial intelligence (AI), AI-based systems offer developers tools to aid in algorithm design and intelligent application creation. Top examples include Apache SystemDS and TensorFlow. ML libraries are collections of reusable functions that are usually written in Python or R and are ready to be used. Since libraries usually consist of a compilation of functions and procedures that are readily available to us, they prevent developers from writing unnecessary lines of code.

Addressing AI-Related Challenges

When examining the difficulties associated with artificial intelligence (AI), open-source software (OSS) has the capacity to tackle or alleviate certain inherent hazards associated with the use of AI. One instance of this is that open-source AI has the potential to reduce discrimination and bias in machine learning models, in contrast to conventional AI or software, because of the combination of AI and code transparency. Open source toolkits, such the IBM AI Fairness, have been created to enable users to analyze, document, and address prejudice and bias in machine learning models at every stage of the AI application process. Experts in the industry have highlighted that open source solutions greatly enhance the auditability of AI due to their high level of transparency. Furthermore, the field of AI is now controlled by a few large businesses, creating an oligopoly. As a result, the development and use of AI technologies are influenced by a tiny and biassed minority. Applying an open-source software (OSS) approach to artificial intelligence (AI) can effectively address this problem in two clear ways. Firstly, government backing is crucial for the development of a local software sector that can effectively compete with technology giants and enhance the diversity of AI expert sources. For example, the Dutch government demonstrated a greater likelihood of adopting open-source software (OSS) when there was a strong level of political commitment (Hadfield et al., 2023). Furthermore, as demonstrated in the literature, if governments utilise software on platforms other than Windows, the market for these other platforms would experience significant growth. Furthermore, open source software has the ability to enhance transparency when accompanied by a team that renders intricate algorithms comprehensible to the public. This empowers citizens with the means to detect and address any potential biases in artificial intelligence. The open source community provides an additional level of security by continuously scrutinizing software code for defects and vulnerabilities, which is particularly advantageous during the developmental phase of novel and developing technologies such as AI.

The cybersecurity industry utilizes open source tools to conduct risk management and safeguard algorithms against external hackers' manipulation. The decentralized nature of code contributions in OSS fosters innovation by allowing developers from various organizations, industry sectors, and geographical locations to contribute. Due to its nature of welcoming contributions from a wide range of developers, open source software (OSS) communities are more inclined to provide innovative code contributions compared to closed source communities, such as those within a specific organization. It is crucial to include the business sector as well. Businesses that possess the ability to articulate and have confidence in artificial intelligence (AI) can enhance the quantity and precision of AI models being utilized, leading to quantifiable economic benefits. The inclusion of open source is a crucial component of this endeavor. Open-source software (OSS) is a publicly available resource that is based on non-rival use rights. It reduces the barriers to entry in software development and provides cost-effective, reusable tools and resources to public stakeholders. This can stimulate innovation, entrepreneurship, and economic growth. An open-source artificial intelligence (AI) system can also enhance interoperability and prevent difficulties related to vendor lock-in if it is built on open standards. Since the specs are publicly available, it is always feasible to have another entity implement the same solution while complying to the established standards. Furthermore, the utilization of open standards guarantees that data and systems may be evaluated without being dependent on the specific tool that created them. This also enhances control by facilitating migration, so minimizing an organization's dependence on a particular product or supplier (Hendrycks et al., 2023). By fostering competition among different implementations, they can generate value, resulting in reduced pricing and enhanced product quality (Askell, 2019).

Adoption and advancement of artificial intelligence in the private sector The ubiquitous and all-encompassing utilization of technology is altering the day-to-day operations of civilization. The continued development of these technologies has been identified as propelling civilization towards the early stages of the Fourth Industrial Revolution, sometimes referred to as Industry 4.0. AI is a crucial component of Industry 4.0, driving the progress of hyper-automation and hyper-connectivity. Industry 4.0 is being enhanced by Industry 5.0, which leverages new technology to create prosperity beyond employment and economic growth. It also prioritizes environmental limits and focuses on the well-being of workers in the production process. The progress made in AI in recent times has been remarkable, with numerous systems now capable of reaching or surpassing the level of performance achieved by humans in activities such as data analytics, transcription, and picture recognition.

The integration of OSS (Open Source Software) and AI (Artificial Intelligence) is widely regarded as a powerful force that may profoundly reshape and modernize services across several domains in both the corporate and public sectors. Within the European Union, artificial intelligence (AI) presents a significant potential for driving digital transformation in several sectors, including commerce. Presently, only 20% of companies in the EU have achieved a high level of digitalization, while a majority of large industries (60%) and small and medium enterprises (SMEs) (over 90%) are falling behind in terms of digital innovation. Adopting an open-source approach to AI enables firms to utilize the most advanced models and platforms that have already been developed. This allows them to concentrate on enhancing their individual areas of expertise, leading to further acceleration of technical progress. The utilization of open source can contribute to establishing a fair and equitable environment, addressing the challenges that small organizations encounter. This can be achieved by leveraging the open source community as a basis and feedback mechanism for their own technology. Moreover, research on the global economic advancements and consequences resulting from AI indicate that by 2035, AI has the potential to enhance worker productivity by up to 40 percent and double economic growth rates in at least 12 wealthy nations (Munga, 2022). The anticipated advantages include enhanced service quality, higher profitability, business growth, improved operational efficiency, and optimized cost structures (Schneier, 2018). The potential for the use of open-source artificial intelligence (AI) may be observed across several industries owing to its cost-effectiveness and ability to foster innovation. The financial sector utilizes open-source artificial intelligence (AI) technologies to analyze data, fortify security systems, and generate predictions for banking systems. H2O.ai, an open source system, is utilized for the purposes of detecting money laundering, doing credit risk scoring, and performing churn predictive analysis. AI is increasingly being utilized in healthcare for diagnostic purposes and the interpretation of medical images, because to its accuracy, which is comparable to that of qualified clinicians. The fields of computer assisted detection and diagnosis, as well as picture segmentation and registration, have greatly profited from the implementation of AI (Bommasani et al., 2022). OSS AI solutions, like the Tesseract-Medical Imaging (Kleinberg and Raghavan, 2021) are utilized in the medical domain to offer conventional picture viewing and reporting methods. In the cybersecurity field, OSS AI is widely regarded as the future. Open source technologies, such the Adversarial Robustness Toolbox, are now employed in risk management to safeguard algorithms against manipulation by external hackers. Furthermore, aside from the possibility of adoption, there are other prospects for creativity and commerce. The open source environment is very efficient as it has the ability to attract a vast pool of skilled individuals, as compared to a proprietary model, allowing for a greater number of people to contribute as innovators. Programmers with exceptional skills at smaller institutions have the opportunity to attain greater prominence, whereas developers primarily strive to gain notoriety and be highly esteemed in open source groups in order to improve their careers. When contemplating innovation through artificial intelligence (AI), the utilization of open source fosters a beneficial and expansive network of feedback, hence enhancing the pace at which AI algorithms are developed and enhanced. From a business standpoint, Open Source Software (OSS) Artificial Intelligence (AI) promotes economic activity, sustainability, and innovation by enabling enterprises to utilize pre-existing source code and build services upon it. In addition, they provide substantial advantages for organizations and economies by contributing to the rise of productivity, reducing operational expenses, generating income, boosting efficiency, and enhancing the customer experience. When evaluating AI models, it is common for them to be costly to design and necessitate a substantial volume of data for algorithm construction and training. Within the realm of government and innovation capacity, the act of reusing minimizes the creation of incomplete solutions, minimizes waste, and allows for the realization of synergy within the government. If an open source solution is developed in one municipality or country, it can also be advantageous for another community.

Indeed, a significant proportion of the AI ecosystem utilized in numerous applications is presently open source. Python is the primary programming language used in most AI applications, including AI systems that utilize tools like TensorFlow, IBM Watson, Apache Mahout, and others. The determining factor is whether the code is made accessible to the public or not. And this may occur even if the majority of the underlying components are operating systems. An instance of this is the Generative Pre-Trained Transformers (GPT), a pioneering advancement in the field of Natural Language Processing created by OpenAI, an organization specializing in AI research and implementation. The GPT-2 version is a transformer-based language model developed in 2019 using unsupervised deep learning techniques. Its primary objective is to predict the subsequent word(s) in a given sentence. GPT-3 is the third iteration and enhanced version of GPT-2. The GPT model reaches an unprecedented level by being trained on an extensive amount of parameters (specifically, 175 billion), which is more than 10 times larger than its previous version. This training is conducted using an open source dataset called 'Common Crawl,' as well as additional texts from OpenAI, including Wikipedia entries. Crucially, Microsoft declared in September 2020 that it had obtained a 'exclusive' license for utilizing GPT-3. While others may still access the public API to obtain results, only Microsoft possesses authority over the source code. As a result, EleutherAI created its own transformer-based language models based on the GPT architecture. They used their own GPT-Neo to recreate a model of the same size as GPT-3 and made it available to the public as an open-source resource, without any cost.

Utilizing artificial intelligence (AI) in the sphere of digital public administration for landscaping reasons. There has been a lot of talk about how digital technologies, especially self-learning algorithms, are influencing public domain practices including policy analysis, policy formulation, and decision-making supported by evidence through the use of open-source software (OSS) and related data. A number of issues involving AI systems have been the subject of high-level government discussions, including the following: integrating AI into government operations; controlling and encouraging innovation in the corporate environment; enforcing best practices and preventing misuse. A key component of administration is the ability to make decisions. By making previously inaccessible data readily available, AI is reshaping the decision-making process and empowering governments to act intelligently in response to massive data sets pertaining to social conditions. In this setting, it can be used to build and simulate decisions to study possible outcomes of certain laws or to predict specific social dynamics (like crime in public transport), which can help authorities allocate resources more efficiently. People have high hopes for AI systems because of their potential to offer proactive and personalized services. There are three main types of decision-making software: those that just assist decision-makers, those that replace humans entirely, and those that completely change the game by overhauling decision-making processes. There is also a difference in the degree to which the systems can make their own decisions. Some systems are completely autonomous, while others are more like "human-in-the-loop approaches" that involve some human interaction. Decision support systems are those that help humans make decisions. Public procurement is one example of an area where certain decision-support tools are hyper-specialized. Either humans are totally untouched or they are indirectly affected to a very little degree. Information that is directly relevant to the decision-making process is provided by other decision-support technologies. Lastly, software may now make judgements on its own, eliminating the need for humans altogether. A risk-based approach is required in all cases under the new regulatory framework. For AI systems that constitute serious threats to human rights and public safety, the recently released Proposal for the Artificial Intelligence Act lays out a thorough framework that uses a risk-based approach and mandates regulatory duties. The requirements for pre-testing, risk management, and human supervision will also help ensure the protection of other fundamental rights by reducing the likelihood of incorrect or biassed judgements made with the use of AI in important areas such as education and training, employment, law enforcement, and the judiciary.

4.   Retrieval-augmented generation

Retrieval augmented generation (RAG) refers to a process where information is retrieved and used to enhance the development of new content. Despite widespread usage, generative AI technologies still have issues in 2023. Hallucinations are plausible-sounding yet erroneous solutions to consumers' problems. This restriction has hindered enterprise solution adoption as hallucinations in critical business activities or direct interaction with clients might be terrible. RAG is a method developed to reduce hallucinations, which could affect the use of artificial intelligence in business. RAG enhances AI-generated content by combining text generation and information retrieval for greater precision and relevance. LLMs can retrieve external data to generate more accurate and contextual reactions. By avoiding direct interaction putting all knowledge in the LLM reduces model size, improving speed and decreasing costs. RAG can collect plenty of unstructured data, including documents, and input. Enterprise applications that require current and accurate data benefit most from these advantages. Companies can use RAG and foundation models to develop more efficient and informative chatbots and virtual assistants. RAG, introduced by Lewis et al. mid-2020 (Lewis et al., 2020), is a model. LLMs improve generative tasks, demonstrating their capabilities. The Retrieve, Analyze, Generate (RAG) process begins with LLMs requesting an external data source for relevant information. This data is used to answer questions or produce text. This strategy offers information for the following generation and ensures that the accuracy and relevance of responses are considerably improved by retrieving evidence of output. During inference, RAG can retrieve knowledge base information to prevent the creation of false information, or "hallucinations." RAG integration LLMs are commonly used and have greatly enhanced chatbots.

There are four stages of RAG development. When it started in 2017, the initiative focused on Pre-Training Models (PTM) were used to add knowledge to language models. This coincided with Transformer architectural development. In this time, RAG mainly improved pre-training approaches. After this initial phase, inactivity followed. ChatGPT changed everything, propelling LLMs to the forefront. The community focused on using LLMs to increase manage shifting needs. Thus, most RAG efforts concentrated on inference, while a Less time was spent on procedure tweaks. With LLM advancements, The RAG technological landscape changed significantly with GPT-4.

The focus has shifted to a hybrid method combining RAG and fine-tuning, while some still prioritize pre-training optimization. The RAG workflow defines it. The user asks ChatGPT about a recent major event (the abrupt sacking and rehiring of OpenAI's CEO) that generated public debate. ChatGPT is famous and widely utilized, however its pretraining data limits it. In this gap, RAG obtains current document extracts from external sources and knowledge bases. It finds several relevant news stories for the investigation.

The original question and articles are combined to make a better prompt. This example shows RAG's methodology. ability to improve model responses with real-time input. Technology-wise, RAG boosted by varied creative ways that address critical questions about what information to collect, when to retrieve and use the data. When researching "what to retrieve," Basic token and entity retrieval have improved (Khandelwal et al., 2019; Nishikawa et al., 2022) to pieces and knowledge (Ram et al., 2023). These research have focused on retrieval granularity and degree of data structure. More granularity increases information but decreases precision.

Organized text retrieval offers more information but is less efficient. The issue of Many methods have been developed to determine retrieval timing, including single, adaptive, and multiple (Wang et al., 2023b, Shi, 2023) retrieval approaches (Izacard et al., 2022). Integration approaches have been used to use data, and have been developed at several model structure stages, including input (Khattab et al., 2022), intermediate (Borgeaud et al., 2022), and output (Liang, 2023). While "intermediate" and "output layers" are more effective, their training and low efficacy limit them. The RAG paradigm enhances LLMs by using external information. The LLM Synergistic information retrieval improves performance. This framework responds to user queries with relevant information. The biggest benefit of RAG is eliminating certain jobs require LLM retraining. Also, developers can add external knowledge.

A RAG architecture has been popular in LLM systems due to its usability and low entrance hurdle. Many RAG is used for most conversational goods. There are three RAG workflow steps. Vector indices are first applied to corpus segments made with an encoder. Furthermore, RAG uses vector similarity to identify and find chunks like the query and indexed chunks. Ultimately, the model uses contextual information from retrieved chunks to respond. These stages structure the RAG process, enabling information retrieval, and create context-aware content. The RAG research paradigm is evolving, and this part basically describes its growth. The classification includes Naive, Advanced, and although RAG was cost-effective and outperformed native LLM, they also showed several shortcomings. Advanced and Modular RAG were created in response to unique Naive RAG flaws. The research paradigm was a method when ChatGPT got popular. Naive RAG follows tradition in indexing, retrieval, and creation. Another name for it is the "Retrieve-Read" framework (Ma et al., 2023a). Indexing is a crucial data preparation procedure. This begins with data indexing, data cleanup and extraction. File formats like PDF, HTML, Word, etc Markdown is converted to plaintext. To meet language model limitations, Chunking breaks this substance into manageable pieces. The embedding model is chosen to turn chunks into vector representations. This allows comparisons and similarities during retrieval. Finally, a text fragment index is created.

Upon For user inquiries, the system employs the same encoding model as indexing. It then calculates similarity scores between indexed corpus vectorized chunks and query vector. System ranks and returns K most similar chunks to the query. These segments are then used, and expanded contextual framework for user inquiry response. Given query and chosen a complex language receives a cohesive prompt from multiple publications. Task-specific criteria may affect the model's response. letting it use its parametric knowledge or limit its answers to the data provided. The model may participate in multi-turn debates because to its discussion history.

5. Generative AI

Generative AI includes computer methods that generate novel and significant content, such as text, photos, or audio from training data. The widespread adoption of this technology, by Dall-E 2, GPT-4, and Copilot is changing how we operate. Generative AI systems can create text that mimics authors and artist-inspired images. These systems will also be intelligent and assistive question-answering systems. Several fields use generative AI. Incorporating IT helps centers, where it aids transitional knowledge work and tedious tasks giving medical and culinary advice. Industry reports say the Generative AI might boost global GDP by 7% and eliminate 300 million knowledge jobs (Goldman Sachs, 2023). Without doubt, these ramifications affect Business & Information Systems significantly Engineering (BISE) community. We will face groundbreaking chances and hurdles. To responsibly use technology, we must address and govern it in sustainable way.

Model-Level Perspective

Generative AI models use AI algorithms to create fresh data instances of data using training data patterns and relationships. The AI model is important but imperfect because it needs to be refined for specific jobs, applications and systems. Deep neural networks generate data well, especially because they can simulate multiple data types with different designs, such as spatial data like photos or sequential data like human language (Janiesch et al., 2021; Kraus, 2020). For example, diffusion probabilistic models generate images from text, transformer architecture, and LLMs for text generation. A popular language model is the generative pre-trained transformer (GPT). LLMs are usually used to write. GPT is utilized in the creation of talkative agents like ChatGPT, and comprehensive and flexible generative AI models.

Foundation models are model output in specific domains or data types (Bommasani et al., 2021). These big models show two key things; emergence, meaning their behavior is often implicitly induced expressly generated (GPT models can create calendar entries even though they were not trained for this), and homogeneity, which permits one unified model to power many systems and applications (Copilot generates sources in several programming languages).

These models were pre-trained with lots of data. For instance, GPT-4, represents the latest generative model. The AI model behind OpenAI's popular conversational bot ChatGPT (OpenAI, 2023) can process picture and text inputs to output text. Similarly, Midjourney creates graphics using both. In order to accomplish this, generative AI model types include unimodal and multimodal. Unimodal models receive text-based instructions. In contrast, multimodal models can receive input from several sources and output in various formats. Multimodal models exist for text, picture, and audio. Examples include Stable Diffusion (Rombach et al., 2022) for image production. Text, MusicLM (Agostinelli et al., 2023) for text-to-music, Codex, and AlphaCode (Li et al., 2022) generates code from text, and GPT-4 text-from-image and text-to-text creation (OpenAI, 2023).

Methods of Training

Methods of training utilized in generative AI models vary greatly. Generative adversarial networks (GANs) are trained with two competing goals (Goodfellow et al., 2014). One generate fresh synthetic samples, whereas the other is to identify real and fake training samples. Distribution of synthetic samples is the goal.

Conversational models like ChatGPT use human-feedback reinforcement learning. Human-feedback reinforcement learning ChatGPT uses three-step RLHF. It generates demonstration data first. Second, users rate prompt output quality. Finally, policy is learned via reinforcement learning to provide optimal output that ranks high (Ziegler et al., 2019). The system-level view of every system includes linked aspects. Generative AI includes more than the above including generative AI model, infrastructure, user-facing components, and modality as well as rapid data processing. This includes incorporating GitHub Copilot integrates advanced deep learning models like Codex (Chen et al., 2021) into a comprehensive and interactive interface. The integration lets users develop programs with more efficiency. Midjourney's picture generation uses hidden X-to-image generation letting users create graphics with Discord bots. Generative AI includes ability of the mathematical model to produce a user-interactive interface. This stage improves model practicality and applicability, making it more effective in real life. situations. Deep learning model integration into generative AI systems often involves several factors; scale (using distributed computing resources), deploy (for varied contexts, and gadgets), and usability (intent recognition and easier interface). With constant introduction of pretrained open-source alternatives to commercial models, it is becoming more important than ever is making these models available to users, whether organizations or people. The emphasis is on constant model monitoring, and unexpected drop in model performance over time, whether open-source or not closed-source (Chen et al., 2023).

There were powerful text-generating models earlier to ChatGPT's November 2022 debut and universal adoption praised for its easy-to-use UI for non-experts.

Furthermore, at the generative AI system can combine or link its components, platforms or databases with specialist knowledge. Many generative AI models exist. They are only trained on past data up to a certain point. So they don't save data afterward. Information is another drawback. Compression during training causes generative AI models to forget info saw (Chiang, 2023). Adding real-time data to the model reduces both limitations, retrieving abilities, which can boost accuracy and utility. In text generation, Online language modelling trains obsolete models with current data. Thus, these models can know recent occurrences better than static models. Generative AI applications by application relate to AI systems developed to handle specific commercial problems and meet needs giving value to organizational stakeholders. Human-task technology or information systems that use generative AI technology can improve human performance, skills in specific tasks. Many practical applications exist for this level of generative AI in numerous fields. These include SEO content creation (Reisenbichler et al., 2022), synthetic movie generation, AI music generation, and natural language-based Software development (Chen et al., 2021). Generative AI will offer novel approaches, working methods enabled by technology. Consumers learn about these unique Trust, distrust, and utilization of applications will rise. In the future, Applications may move away from writing letters, from booking dinner to seeking medical or legal advice guidance. Significant decisions may demand moral assessment (Krü'gel et al., 2023). Growing popularity of generative AI applications necessitates development of standards for reliable designs. It is also vital to consider user effect and alter trust. Continuous use and examining socio-technical aspects are crucial for identifying the potential for innovation and acceptance of these applications by individuals and organizations.

The definition and scope of AI are continually being enhanced, but human intelligence is the standard. Berente et al. (Berente et al., 2021) report discusses stable intelligence. Previously, AI was known for its analytical abilities, making it ideal for decision-making tasks. AI can now generate chores, making it ideal for content creation. Even though content creation partly analytical due to probabilistic nature, yet creative or artistic results when generative AI innovates part integration. Additionally, IT objects were considered inert used directly by individuals. Reconsider the idea that human agency is paramount, and considering LLM-powered agentic IT artefacts (Baird and Maruping, 2021) (Park et al., 2023). How we create the human-AI interaction will depend on this reevaluation. Future AI capacity models may need to organize, clarify, direct, and limit AI operations and uses. The current research focus lies in studying human-AI connections. In particular, analytic AI scholars Delegation can be used to create a decision-making hierarchy (Baird and Maruping, 2021). People prompt an AI system in generative AI. The AI system then analyses human intent and provides feedback to predict further requests. Initially, this appears to follow a delegation pattern. The following technique doesn't work since the AI's output can affect and help others participate later, intentionally or unconsciously. The creation process usually uses a co-creation approach, which involves multiple roles synchronizing and providing different perspectives to guide design (Ramaswamy and Ozcan, 2018). From the viewpoint of agentic AI artefacts, initiation is not human-only. The above interactions also alter our understanding of hybrid intelligence as humans and AI combining, using their distinct benefits. Hybrid intelligence integrates intuition, creativity, Combining empathy with AI systems' computing capacity, precision, and scalability to overcome Type-specific intelligence constraints. This combo improves decision-making and problem-solving talents (Dellermann et al., 2019). Generative AI and co-creation are generating a stir changing our view of collective intelligence. It may be necessary to create new models and patterns for human-AI interaction to understand and control both Humans and AI. This will make AI applications more efficient and effective assuring AI integration and capability. This transformation in human behavior is theoretical. AI emphasizes that the notion of mind is well-established from psychological notion that explains how people understand and anticipate thoughts feelings, and intentions of others (Carlson et al., 2013; Baron-Cohen, 1997; Gray, 2007). Mastering this aptitude helps people understand and share, making it crucial for social relationships. Additionally, offering AI mind-controlled system can boost its use (Hartmann et al., 2023). The Human theory of mind construction is unconscious and occurs over time. As AI interfaces and output become more realistic, the demand for Human-computer interaction theory of mind grows. Research is underway on the development of AI systems with theory-of-mind awareness to improve (Rabinowitz et al., 2018; Çelikok et al., 2019). However, AI systems indicate interactions minimally. Thus, humans lack a complete theory to convey their understanding of AI systems' intelligent conduct, which is vital. Artificial-mind theory that explains how people perceive and infer AI system states and reasoning to improve their cooperation may solve these challenges.

Debates Over the Relevance of Scientific Approach in Machine Learning

The methodological approaches in ML research can be examined with greater precision. Authors (Forde and Paganini, 2019) emphasizes a disparity between current machine learning approaches and the conventional scientific process, specifically in the domains of hypothesis development and statistical testing. The authors propose that machine learning research frequently lacks the methodical rigor observed in other scientific disciplines.

Their advocacy involves the integration of empirical scientific approaches into machine learning, which encompasses the use of controlled, reproducible, and verifiable experimental designs. According to their argument, this would not only strengthen the fundamental science of deep learning but also improve the quality and dependability of practical machine learning research. Krenn and his colleagues (Krenn et al., 2022) investigate the potential of artificial intelligence (AI) in enhancing scientific comprehension. The research introduces a framework based on the philosophy of science to assess the function of AI in scientific comprehension. It highlights three areas where AI can be beneficial: as a computational microscope, as a source of inspiration, and as a means of enhancing understanding. The examination of each component involves the use of many instances, which emphasize the potential of AI to uncover fresh scientific insights, stimulate innovative ideas, and potentially generate new scientific understanding independently. The research highlights the differentiation between scientific discovery and comprehension, emphasizing the contribution of AI in promoting scientific understanding.

Van Calster and his colleagues (Van Calster et al., 2019) emphasize the importance of calibrating prediction algorithms in clinical decision making. Incorrect calibrations can result in unrealistic anticipations and incorrect choices, underscoring the importance of accurate prediction models that are properly calibrated for the particular population and context. Continuous monitoring and updating of models is necessary due to the variability of patient groups and changing healthcare dynamics.

The primary objective is to improve the effectiveness of predictive analytics in collaborative decision-making and patient counselling.

Varoquax and Cheplygina (Varoquaux and Cheplygina, 2022) present a thorough set of principles for the efficient implementation and assessment of machine learning in the field of medical imaging. This study highlights the crucial importance of maintaining data integrity by emphasizing the early separation of test data to prevent any unauthorized access or leaking of data. The document emphasizes the need of having a well-defined methodology for selecting model hyperparameters, while simultaneously cautioning against using test data for this purpose. The authors emphasize the relevance of having a test dataset that is sufficiently large to assure statistical validity. They recommend using hundreds or preferably thousands of samples, and suggest that performance measures should be accompanied by confidence intervals.

This research supports the use of varied data sets that effectively represent the diversity of patients and diseases across numerous institutions, encompassing a broad spectrum of demographic characteristics and illness conditions. The authors advocate for robust baseline comparisons that encompass not only cutting-edge machine learning approaches but also conventional clinical methodologies. Additionally, they emphasize the necessity of engaging in a critical discourse on the fluctuation of outcomes, taking into account stochastic factors and data origins. Furthermore, it is recommended to utilize a range of quantitative measures to encompass many facets of the clinical issue, connecting them to relevant clinical performance indicators, and making well-informed choices on the balance between false detections and missed detections. Furthermore, it is advisable to incorporate qualitative observations and engage the groups that are most affected by the application in the development of assessment criteria.

Bouthillier and his colleagues (Bouthillier et al., 2019) examine the equilibrium between empirical and exploratory research in the domain of machine learning, specifically in deep learning. The authors emphasize the significance of empirical research in constructing a strong knowledge foundation and guaranteeing consistent scientific advancement.

The authors (Besiroglu et al., 2024) offered an analytical survey that utilizes data to examine the impact of the compute divide on the development of machine learning research. They demonstrated that a computational division has coincided with a diminished portrayal of exclusively academic matters. Research teams focusing on computationally difficult research issues, particularly foundation models. The authors contended that academics is expected to have a diminished impact on the progress of the related approaches, including their critical review, scrutiny, and dissemination. Simultaneously with this shift in research focus, there is a noticeable trend in academic research towards adopting open source, pre-trained models created by the industry. In order to tackle the difficulties that arise from this phenomenon, particularly the decreased examination of prominent models, the authors suggested implementing strategies that focus on carefully broadening scholarly knowledge. By combining nationally-funded computing infrastructure with open science initiatives, academic access to computational resources can be effectively enhanced. This approach would prioritize research in the areas of interpretability, safety, and security. Structured access programs and third-party auditing can facilitate objective external assessment of industry systems.

On the other hand, exploratory research is highly regarded for its ability to broaden the scope of research by uncovering fresh discoveries. Nevertheless, it is crucial to avoid excessive concentration on either strategy, since it might result in weak and unreliable foundations or impede growth by restricting exploration. The research highlights the importance of comprehending the equilibrium between two research methodologies by discussing the recent criticisms in deep learning methodology. The authors utilize the case of batch-norm in deep learning to exemplify the significance of both exploratory and empirical investigation. They propose that a stronger collaboration between these two approaches can result in more effective and logical advancements in the field.

Ethical Challenges in Machine Learning

Machine learning, while revolutionary, presents numerous ethical dilemmas that require meticulous attention. These issues stem from the possible prejudices in algorithms, absence of transparency and comprehensibility, privacy worries, and wider societal implications.

Gaining comprehension and addressing these difficulties are crucial to guarantee the conscientious advancement and implementation of machine learning systems. Some of the primary ethical dilemmas that arise in the field of machine learning are (Mike Stephen et al., 2024):

1. *Algorithmic Bias:* Machine learning algorithms are developed on historical data that may possess inherent biases. Failure to recognize and address these biases might result in algorithms perpetuating prejudice and unfairness. For instance, if a recruiting algorithm is taught with biassed historical data, it may unintentionally exhibit discriminatory behavior towards specific demographic groups. To mitigate algorithmic bias, it is necessary to carefully preprocess the data, utilize diverse and representative training datasets, and continuously evaluate the algorithm's performance to ensure balanced and just outputs.

2. *Transparency and Explainability:* Machine learning algorithms frequently function like complicated "black boxes" that generate outputs without offering explicit justifications for their judgements. The absence of openness can undermine confidence and impede the capacity to hold individuals accountable. Users, particularly those impacted by algorithmic determinations, possess the entitlement to comprehend the rationale behind specific decisions. Improving transparency and explainability can be achieved by creating models that are easy to understand, allowing access to the decision-making processes, and designing user-friendly interfaces that effectively communicate the results of algorithms.

3. *Privacy Concerns:* Machine learning algorithms significantly depend on extensive quantities of personal data for the purposes of training and making predictions. Privacy concerns arise when there is a collection, storage, and processing of sensitive information without sufficient consent or security measures. Implementing strong data security measures, adhering to privacy legislation, and giving priority to data anonymization and encryption techniques are essential for safeguarding individual private rights.

4. *The obligation to take ownership and be responsible for one's actions and decisions:* As machine learning systems gain more autonomy, the issue of accountability becomes crucial. When an algorithm produces a detrimental or prejudiced outcome, who should bear accountability? Establishing clear

lines of accountability and defining the roles and duties of stakeholders, such as developers, organizations, and regulatory agencies, is crucial. Ethical frameworks and guidelines provide a structured approach to responsible decision-making and ensure that machine learning algorithms be held accountable for their outcomes.

5. ***Implications for society:*** Machine learning technology can have extensive societal and economic implications. These developments have the potential to cause employment loss, worsen current inequities, and consolidate power among a small number of organizations or persons. It is essential to comprehend and tackle these wider societal consequences. Proactive efforts, such as implementing retraining and upskilling programs, adhering to inclusive design approaches, and considering the ethical and social implications across the full lifecycle of machine learning systems, are necessary. Interdisciplinary collaboration among researchers, politicians, industry professionals, and ethicists is essential for addressing these ethical concerns. It is imperative to make endeavors to establish and follow ethical principles, advocate for diversity and inclusivity in gathering data and designing algorithms, and encourage continuous discussion and transparency regarding the ethical aspects of machine learning. By adopting this approach, we can promote the development of new and creative ideas while maintaining moral standards and guaranteeing that machine learning systems actively contribute to a just, accountable, and unbiased future.

CONCLUSION

The latest multimodal AI research is summarized in this chapter. First, model architectures are becoming more similar. Second, multimodal understanding models are being replaced by multimodal generation models in research. Finally, it examines the new AI paradigm of mixing Language and Vision Models (LLMs) with external tools and models to complete tasks. Other paths could be explored. Prioritizing video generation could boost ChatGPT and AI. Compositional AI methods for multimodal systems that cover additional modalities while reducing computing costs will also be developed. Autonomous robotics can efficiently perform more practical tasks, boosting human innovation and productivity, hence they must be prioritized.

As highly autonomous AI systems become more common, society may need to take significant steps to ensure their safe and reliable operation and address broader indirect concerns. We expect scholars and practitioners to work together to determine who should use specific approaches and how to make them reliable and affordable for a variety of people and organizations. Consensus on these best practices is unlikely to happen alone. If AI capabilities continue to improve rapidly, society may need to agree on new best practices for each increasingly capable category of AI systems. This is required to stimulate rapid adoption of new techniques that successfully mitigate these systems' increased dangers.

Previous research stresses open source's benefits and potential. Open-source artificial intelligence can utilize community knowledge to counteract the dominance of major technology companies or build on their technologies and data. It's crucial to assess AI's inherent risks and how open-source methods might mitigate them. The recently released AI Act Proposal must be attentively monitored. This act will create a legal framework and trust-based strategy to mitigate risks. Despite its innovation potential, the expert consultation reaffirmed that the EU's non-binding measures have failed to promote open source technology adoption. Based on this study, policy recommendations can be made.

The chapter's summary highlights RAG's LLM capability improvement. Combining parameterized language model information with huge nonparameterized external knowledge sources achieves this. Naive, Advanced, and Modular RAG are the RAG framework's three developmental paradigms. Each paradigm becomes better over time. Advanced architecture components including query rewriting, chunk reranking, and quick summarization make Advanced RAG better than Naive. These advances have created a more complex and adaptive structure that increases LLM efficiency and comprehension. Technical integration of RAG with other AI methods like fine-tuning and reinforcement learning has greatly increased its potential. Content retrieval is improving with a hybrid approach that uses structured and unstructured data. RAG is researching unique ideas including self-retrieval from LLMs and dynamic information retrieval time. RAG technology has advanced, but further study is needed to improve its endurance and capacity to tackle greater scenarios. RAG is being used in multimodal domains to understand and interpret pictures, videos, and code. Expanding RAG shows its real-world effects on AI implementation, which has drawn academic and business interest. RAG-focused AI applications and allied technologies are growing the RAG ecosystem. However, RAG's application field is expanding, thus assessment techniques must be improved to keep up. To effectively assess RAG's contributions to AI research and development, performance assessments must be accurate and representative.

Generative AI can create original text, graphics, and music that increasingly resembles human works. Generative AI can transform industries that rely on creativity, inventiveness, and knowledge processing. It enables the creation of lifelike virtual assistants, customized education and service, and digital art. Generative AI affects BISE practitioners and scholars as an interdisciplinary research community. This framework includes model, system, application, and social-technical viewpoints.

REFERENCES

AgostinelliA.DenkT. I.BorsosZ.EngelJ.VerzettiM.CaillonA.HuangQ.JansenA. RobertsA.TagliasacchiM. (2023) MusicLM: generating music from text. arXiv:2301.11325

Askell, A., Brundage, M., & Hadfield, G. (2019). The role of cooperation in responsible ai development. arXiv preprint arXiv:1907.04534.

Baird, A., & Maruping, L. M. (2021). The next generation of research on IS use: A theoretical framework of delegation to and from agentic IS artifacts. *Management Information Systems Quarterly*, *45*(1), 315–341. doi:10.25300/MISQ/2021/15882

Baron-Cohen, S. (1997). *Mindblindness: an essay on autism and theory of mind.* MIT Press.

Berente, N., Gu, B., Recker, J., & Santhanam, R. (2021). Special issue editor's comments: Managing artificial intelligence. *Management Information Systems Quarterly*, *45*(3), 1433–1450.

Besiroglu, T., Bergerson, S. A., Michael, A., Heim, L., Luo, X., & Thompson, N. (2024). The Compute Divide in Machine Learning: A Threat to Academic Contribution and Scrutiny? *Computers & Society.*

Bommasani, Creel, Kumar, Jurafsky, & Liang. (2022). *Picking on the same person: Does algorithmic monoculture lead to outcome homogenization?* Academic Press.

Bommasani, R., Hudson, D. A., Adeli, E., Altman, R., Arora, S., von Arx, S., Bernstein, M. S., Bohg, J., Bosselut, A., Brunskill, E., Brynjolfsson, E., Buch, S., Card, D., Castellon, R., Chatterji, N. S., Chen, A. S., Creel, K. A., Davis, J., Demszky, D., . . . Liang, P. (2021) On the 123 S. Feuerriegel et al.: Generative AI, Bus Inf Syst Eng opportunities and risks of foundation models. arXiv:2108. 07258. https://10.48550/arXiv.2108.07258

Borgeaud, S., Mensch, A., Hoffmann, J., Cai, T., Rutherford, E., Millican, K., Driessche, G. B. V. D., Lespiau, J.-B., Damoc, B., & Clark, A. (2022). Improving language models by retrieving from trillions of tokens. *International conference on machine learning*, 2206–2240.

Bouthillier, X., Laurent, C., & Vincent, P. (2019). Unreproducible research is reproducible. *Proceedings of the International Conference on Machine Learning (PMLR).*

Carlson, S. M., Koenig, M. A., & Harms, M. B. (2013). Theory of mind. *Wiley Interdisciplinary Reviews: Cognitive Science, 4*(4), 391–402. doi:10.1002/wcs.1232 PMID:26304226

Celikok, M. M., Peltola, T., Daee, P., & Kaski, S. (2019) Interactive AI with a theory of mind. ACM CHI 2019 workshop: computational modeling in human-computer interaction.

Chan, A., Salganik, R., Markelius, A., Pang, C., Rajkumar, N., Krasheninnikov, D., Langosco, L., He, Z., Duan, Y., Carroll, M., Lin, M., Mayhew, A., Collins, K., Molamohammadi, M., Burden, J., Zhao, W., Rismani, S., Voudouris, K., Bhatt, U., ... Maharaj, T. (2023). Harms from increasingly agentic algorithmic systems. *2023 ACM Conference on Fairness, Accountability, and Transparency*, 651-666. arXiv:2302.10329.

ChenL.ZahariaM.ZouJ. (2023) How is chatgpt's behavior changing over time? arXiv:2307.09009

Chen, M., Tworek, J., Jun, H., & Yuan, Q. (2021). *Evaluating large language models trained on code.* arXiv:2107.03374

Chiang, T. (2023). *ChatGPT is a blurry JPEG of the web.* https://www. newyorker. com/tech/annals-of-technology/chatgpt-is-a-blurryjpeg-of-the-web

Creswell, A., Shanahan, M., & Higgins, I. (2022). Selection-inference: Exploiting large language models for interpretable logical reasoning. arXiv preprint arXiv:2205.09712.

Dai, Li, Li, Tiong, Zhao, Wang, Li, Fung, & Hoi. (2023). *Instructblip: Towards generalpurpose vision-language models with instruction tuning.* Academic Press.

Dellermann, D., Ebel, P., So̎llner, M., & Leimeister, J. M. (2019). Hybrid intelligence. *Business & Information Systems Engineering, 61*(5), 637–643. doi:10.1007/s12599-019-00595-2

DevlinJ.ChangM. W.LeeK.ToutanovaK. (2018) BERT: Pretraining of deep bidirectional transformers for language understanding. arXiv:1810.04805.

Forde, J.Z., & Paganini, M. (2019). *The scientific method in the science of machine learning.* Academic Press.

Gong, R., Huang, J., Zhao, Y., Geng, H., Gao, X., Wu, Q., Ai, W., Zhou, Z., Terzopoulos, D., & Zhu, S.-C. (2023). Arnold: A benchmark for language-grounded task learning with continuous states in realistic 3d scenes. *Proceedings of the IEEE/CVF International Conference on Computer Vision (ICCV).* 10.1109/ICCV51070.2023.01873

Goodfellow, I., Pouget-Abadie, J., Mirza, M., Xu, B., Warde-Farley, D., Ozair, S., Courville, A., & Bengio, Y. (2014). Generative adversarial nets. *Advances in Neural Information Processing Systems*, *27*, 2672–2680.

Gray, Gray, & Wegner. (2007). Dimensions of mind perception. *Sci, 315*(5812).

Hadfield, Cuéllar, & Oreilly. (2023). *Its time to create a national registry for large ai models.* Available: https://carnegieendowment.org/2023/07/12/it-s-time-to-create-national-registry-for-large-ai-models-pub-90180]

Hartmann, J., Bergner, A., & Hildebrand, C. (2023). MindMiner: Uncovering linguistic markers of mind perception as a new lens to understand consumer-smart object relationships. *Journal of Consumer Psychology*, *33*(4), 645–667. Advance online publication. doi:10.1002/jcpy.1381

Hendrycks, D., Mazeika, M., & Woodside, T. (2023). *An overview of catastrophic ai risks.* arXiv preprint arXiv:2306.12001.

Izacard, G., Lewis, P., Lomeli, M., Hosseini, L., Petroni, F., Schick, T., Dwivedi-Yu, J., Joulin, A., Riedel, S., & Grave, E. (2022). *Few-shot learning with retrieval augmented language models.* arXiv preprint arXiv:2208.03299.

Janiesch, C., Zschech, P., & Heinrich, K. (2021). Machine learning and deep learning. *Electronic Markets*, *31*(3), 685–695. doi:10.1007/s12525-021-00475-2

Khandelwal, U., Levy, O., Jurafsky, D., Zettlemoyer, L., & Lewis, M. (2019). *Generalization through memorization: Nearest neighbor language models.* arXiv preprint arXiv:1911.00172.

Khattab, O., Santhanam, K., Li, X. L., Hall, D., Liang, P., Potts, C., & Zaharia, M. (2022). *Demonstrate-search-predict: Composing retrieval and language models for knowledge-intensive nlp.* arXiv preprint arXiv:2212.14024.

Kleinberg, J., & Raghavan, M. (2021). Algorithmic monoculture and social welfare. *Proceedings of the National Academy of Sciences of the United States of America*, *118*(22), e2018340118. doi:10.1073/pnas.2018340118 PMID:34035166

Kraus, M., Feuerriegel, S., & Oztekin, A. (2020). Deep learning in business analytics and operations research: Models, applications and managerial implications. *European Journal of Operational Research*, *281*(3), 628–641. doi:10.1016/j.ejor.2019.09.018

Krenn, M., Pollice, R., Guo, S. Y., Aldeghi, M., Cervera-Lierta, A., Friederich, P., dos Passos Gomes, G., Häse, F., Jinich, A., Nigam, A., Yao, Z., & Aspuru-Guzik, A. (2022). On scientific understanding with artificial intelligence. *Nature Reviews. Physics*, *4*(12), 761–769. doi:10.1038/s42254-022-00518-3 PMID:36247217

Kru¨gel, S., Ostermaier, A., & Uhl, M. (2023). ChatGPT's inconsistent moral advice influences users' judgment. *Scientific Reports*, *13*(1), 4569. doi:10.1038/s41598-023-31341-0 PMID:37024502

Li, Y., Choi, D., Chung, J., Kushman, N., Schrittwieser, J., Leblond, R., Eccles, T., Keeling, J., Gimeno, F., Dal Lago, A., Hubert, T., Choy, P., de Masson d'Autume, C., Babuschkin, I., Chen, X., Huang, P.-S., Welbl, J., Gowal, S., Cherepanov, A., ... Vinyals, O. (2022). Competition-level code generation with alphacode. *Science*, *378*(6624), 1092–1097. doi:10.1126/science.abq1158 PMID:36480631

Li, Y., Choi, D., Chung, J., Kushman, N., Schrittwieser, J., Leblond, R., Eccles, T., Keeling, J., Gimeno, F., Dal Lago, A., Hubert, T., Choy, P., de Masson d'Autume, C., Babuschkin, I., Chen, X., Huang, P.-S., Welbl, J., Gowal, S., Cherepanov, A., ... Vinyals, O. (2022). Competition-level code generation with alphacode. *Science*, *378*(6624), 1092–1097. doi:10.1126/science.abq1158 PMID:36480631

Liang, H., Zhang, W., Li, W., Yu, J., & Xu, L. (2023). *Intergen: Diffusion-based multi-human motion generation under complex interactions*. arXiv preprint arXiv:2304.05684.

Liu, P., Yuan, W., Fu, J., Jiang, Z., Hayashi, H., & Neubig, G. (2023). Pre-train, prompt, and predict: A systematic survey of prompting methods in natural language processing. *ACM Computing Surveys*, *55*(9), 1–35. doi:10.1145/3560815

Ma, Y. J., Liang, W., Wang, G., Huang, D.-A., Bastani, O., Jayaraman, D., Zhu, Y., Fan, L., & Anandkumar, A. (2023). *Eureka: Human-level reward design via coding large language models*. arXiv preprint arXiv:2310.12931.

MorrisM. R.Sohl-dicksteinJ.FiedelN.WarkentinT.DafoeA.FaustA.FarabetC.LeggS. (2023). Levels of agi: Operationalizing progress on the path to agi. arXiv:2311.02462.

Munga, J. (2022). *To Close Africa's Digital Divide*. Policy Must Address the Usage Gap. Available: https://carnegieendowment.org/2022/04/26/to-close-africa-s-digital-divide-policy-must-address-usage-gap-pub-86959

Nishikawa, S., Ri, R., Yamada, I., Tsuruoka, Y., & Echizen, I. (2022). *Ease: Entity-aware contrastive learning of sentence embedding*. arXiv preprint arXiv:2205.04260. doi:10.18653/v1/2022.naacl-main.284

OpenA. I. (2023). *GPT-4 technical report*. arXiv:2303.0877

ParkJ. S.O'BrienJ. C.CaiC. J.MorrisM. R.LiangP.BernsteinM. S. (2023) *Generative agents: interactive simulacra of human behavior*. arXiv:2304.03442 doi:10.1145/3586183.3606763

Rabinowitz, N. C., Perbet, F., Song, H. F., Zhang, C., Eslami, S. M. A., & Botvinick, M. M. (2018) Machine theory of mind. *International conference on machine learning, PMLR.* http://proceedings.mlr.press/v80/rabinowitz18a.html

Radford, A., Kim, J. W., Hallacy, C., Ramesh, A., Goh, G., Agarwal, S., Sastry, G., Askell, A., Mishkin, P., & Clark, J. (2021). Learning transferable visual models from natural language supervision. *International conference on machine learning*, 8748–8763.

Ram, O., Levine, Y., Dalmedigos, I., Muhlgay, D., Shashua, A., Leyton-Brown, K., & Shoham, Y. (2023). *In-context retrieval-augmented language models.* arXiv preprint arXiv:2302.00083.

Ramaswamy, V., & Ozcan, K. (2018). What is co-creation? An interactional creation framework and its implications for value creation. *Journal of Business Research, 84*, 196–205. doi:10.1016/j.jbusres.2017.11.027

Reed, S., Zolna, K., Parisotto, E., Colmenarejo, S. G., Novikov, A., Barth-Maron, G., Gimenez, M., Sulsky, Y., Kay, J., & Springenberg, J. T. (2022). *A generalist agent.* arXiv preprint arXiv:2205.06175.

Reisenbichler, M., Reutterer, T., Schweidel, D. A., & Dan, D. (2022). Frontiers: Supporting content marketing with natural language generation. *Marketing Science, 41*(3), 441–452. doi:10.1287/mksc.2022.1354

Rombach, R., Blattmann, A., Lorenz, D., Esser, P., & Ommer, B. (2022). Highresolution image synthesis with latent diffusion models. IEEE/CVF conference on computer vision and pattern recognition, 10684–10695.

Sachs, G. (2023). *Generative AI could raise global GDP by 7%.* https://www.goldmansachs.com/insights/pages/generative-aicould-raise-global-gdp-by-7-percent.html

Saenz, A. D., Harned, Z., Banerjee, O., Abràmoff, M. D., & Rajpurkar, P. (2023). "Autonomous ai systems in the face of liability, regulations and costs. *NPJ Digital Medicine, 6*, 13. PMID:37803209

Schneier, B. (2018). Artificial intelligence and the attack/defense balance. *IEEE Security and Privacy, 16*(02), 96–96. doi:10.1109/MSP.2018.1870857

Shi, W., Min, S., Yasunaga, M., Seo, M., James, R., Lewis, M., Zettlemoyer, L., & Yih, W.-t. (2023). *Replug: Retrievalaugmented black-box language models.* arXiv preprint arXiv:2301.12652.

Stephen, M., Potter, K., & Mohamed, S. (2024). Ethical Considerations in Machine Learning: Balancing Innovation and Responsibility. *Computer Science*.

Van Calster, B., McLernon, D. J., Van Smeden, M., Wynants, L., Steyerberg, E. W., & Collins, P. B. G. S. (2019). Calibration: The Achilles heel of predictive analytics. *BMC Medicine*, *17*(1), 230. doi:10.1186/s12916-019-1466-7 PMID:31842878

Varoquaux, G., & Cheplygina, V. (2022). Machine learning for medical imaging: Methodological failures and recommendations for the future. *NPJ Digital Medicine*, *5*(1), 48. doi:10.1038/s41746-022-00592-y PMID:35413988

Wang, Y., Li, P., Sun, M., & Liu, Y. (2023b). *Self-knowledge guided retrieval augmentation for large language models*. arXiv preprint arXiv:2310.05002. doi:10.18653/v1/2023.findings-emnlp.691

Wang, Z., Cai, S., Liu, A., Ma, X., & Liang, Y. (2023). *Describe, explain, plan and select: Interactive planning with large language models enables open-world multi-task agents*. arXiv preprint arXiv:2302.01560.

Xu, M., Huang, P., Yu, W., Liu, S., Zhang, X., Niu, Y., Zhang, T., Xia, F., Tan, J., & Zhao, D. (2023). *Creative robot tool use with large language models*. arXiv preprint arXiv:2310.13065.

Yu, J., Wang, X., Tu, S., Cao, S., Zhang-Li, D., Lv, X., Peng, H., Yao, Z., Zhang, X., & Li, H. (2023). *Kola: Carefully benchmarking world knowledge of large language models*. arXiv preprint arXiv:2306.09296.

ZieglerD. M.StiennonN.WuJ.BrownT. B.RadfordA.AmodeiD.ChristianoP.IrvingG. (2019). *Fine-tuning language models from human preferences*. arXiv:1909.08593

KEY TERMS AND DEFINITIONS

Convolutional Neural Network (CNN): Is a regularized type of feed-forward neural network that learns feature engineering by itself via filters (or kernel) optimization. Vanishing gradients and exploding gradients, seen during backpropagation in earlier neural networks, are prevented by using regularized weights over fewer connections.

Long-Lasting Memories (LLMs): Can incorporate long-term memory is by utilizing additional memory components, such as third-party databases including vector databases or (knowledge) graph databases. These components store factual data or embeddings that the model can access and use to generate text.

Natural Language Processing (NLP): Combines computational linguistics—rule-based modeling of human language—with statistical and machine learning models to enable computers and digital devices to recognize, understand and generate text and speech.

Open Source Software (OSS): Is computer software that is released under a license in which the copyright holder grants users the rights to use, study, change, and distribute the software and its source code to anyone and for any purpose.

Retrieve, Analyze, Generate (RAG): Is an architectural approach that can improve the efficacy of large language model (LLM) applications by leveraging custom data.

Visual Language (VL): Is a system that communicates through visual elements. It helps users perceive and understand visible signs.

Visual Language Models (VLMs): Combine computer vision and natural language processing to analyze visual and textual medical data.

Chapter 7
COMSATS Face:
An Image Database of Faces With Various Poses, Its Structure, and Features

Mahmood Ul Haq
https://orcid.org/0000-0002-1514-0300
University of Engineering and Technology, Peshawar, Pakistan

Muhammad Athar Javed Sethi
https://orcid.org/0000-0001-7847-831X
University of Engineering and Technology, Peshawar, Pakistan

Aamir Shahzad
COMSATS University Islamabad, Abbottabad campus, Pakistan

Muhammad Shahid Anwar
https://orcid.org/0000-0001-8093-6690
Gachon University, South Korea

ABSTRACT

The human face can appear different depending on the circumstances because of its flexibility and three-dimensional structure. Researchers are facing several obstacles relating to face poses, illumination, facial expressions, head direction, occlusion, hairdo, etc. in the process of developing dependable and efficient algorithms for face detection, face identification, and face expression analysis. To determine the algorithms' effectiveness, they need to be evaluated against a certain set of face image/database benchmarks. This work introduces a dataset of multiple-pose facial photographs. Eight hundred fifty photos from 50 people in 17 distinct stances are included in the collection (0°, 5°, 10°, 15°, 20°, 25°, 30°, 35°, 55°, -5°, -10°, -15°, -20°, -25°, -30°, -35°, -55°). Three distinct lighting conditions are also included in the dataset. Eight resolutions (144 × 256, 200 × 200, 100 × 100, 70 × 70, 50 × 50, 40 × 40, 20 × 20 and 10 × 10 pixels) are available for the dataset. This dataset's facial image content can provide insight on the effectiveness and resilience of upcoming face detection and recognition systems. Additionally, based on the suggested face database, a comparison study of two face recognition methods, such as PAL and PCA, is performed.

DOI: 10.4018/979-8-3693-2913-9.ch007

1. BACKGROUND AND SUMMARY

Face recognition (FR) is an important biometric technique that compares face features of two different images to determine whether they are of same person or not (Anwar, Wang, Ullah et al, 2020). FR is a rapid growing and the most popular research area. Face appearance is dependent to various factors such as illumination variations (Haq et al., 2019), pose variations (Haq et al., 2023), and occlusion (Ahmad et al., 2018). FR systems have come a long way in the last several years because to advances in computer vision and machine learning (Anwar, Wang, Khan et al, 2020). These developments have aided in a variety of applications, from human-computer interaction to security and surveillance (Campus, n.d.). The availability of diverse and high-quality datasets, which provide the basis for model training and testing, is essential to the development and assessment of these systems (Ahmad & Adnan, 2015).

Datasets for facial recognition are essential tools for a wide range of applications in various domains (Hosni et al., 2018; Sohail et al., 2023). These datasets play a key role in security and surveillance by helping to create strong biometric authentication systems (Rahim et al., 2023), improve access control protocols (Munawar et al., n.d.), and support forensic investigations by precisely identifying people from surveillance video or photos taken at crime scenes (Anwar, Wang, Ahmad et al, 2020). Furthermore, in the field of human-computer interaction, facial recognition databases make it possible to design user-friendly interfaces that provide customized experiences for users in applications like emotion detection systems, augmented reality, and virtual reality (Fatima et al., 2022; Haq et al., 2024; Shahid Anwar et al., 2020). These datasets also help in the healthcare industry with the creation of diagnostic tools for facial anomalies, facial expression analysis for mental health evaluation, and patient identification in the management of medical records (Hosni et al., 2020; Ullah et al., n.d.).

Moreover, facial recognition datasets enable retailers and marketers to follow customer activity, assess customer demographics, and customize marketing campaigns (Tahsin et al., 2023). Facial recognition datasets are utilized in a broad range of businesses and domains, contributing significantly to the advancement of technology-driven solutions and the improvement of different facets of human society (Ahmad et al., 2015; Wahab et al., 2023).

One notable addition to this field is the COMSATS Face Dataset, which provides an extensive database of face photos for use in research and development. This dataset offers a wealth of information consisting of 50 unique participants' faces taken in a variety of positions. The dataset captures a wide range of face changes with a total of 17 different postures per participant, making it a reliable testing ground for pose-invariant facial recognition algorithms.

The COMSATS Face Dataset is unique in that it has different lighting conditions. The dataset captures the difficulties caused by different lighting circumstances in real-world applications by including three different illumination conditions, from controlled lighting setups to real-world lighting scenarios. This kind of unpredictability improves the dataset's realism and makes it easier to assess how reliable face recognition systems are in various lighting scenarios.

Moreover, the COMSATS Face Dataset provides images in eight distinct formats, offering diversity in resolution. This flexibility addresses a crucial issue that frequently arises in real-world deployment scenarios by allowing researchers to investigate the effect of image resolution on the efficacy of facial recognition systems.

We provide a thorough examination of the COMSATS Face Dataset in this chapter, covering its production process, salient features, and possible uses. We also do experiments to show how useful the dataset is for assessing how well facial recognition algorithms perform in different positions, lighting scenarios, and resolutions. Our objective is to make a valuable contribution to the field of facial recognition research and expedite the creation of more resilient and dependable systems for practical applications.

The main contribution of this chapter is:

- The COMSATS face dataset is presented in this chapter with updated lighting and resolution variations.
- With a total of 17 distinct poses captured per participant, the dataset provides a broad spectrum of face modifications, making it a trustworthy testing ground for pose-invariant facial recognition algorithms.
- What sets this dataset apart from others is that each image has three distinct lighting conditions.
- The suggested datasets provide photos at eight distinct image resolutions, including 144×256, 200×200, 100×100, 70×70, 50×50, 40×40, 20×20 and 10×10 pixels.

Existing Databases

Facial recognition datasets that are now available cover a wide range of subjects, positions, lighting situations, and resolutions, meeting the various needs of practitioners and researchers in the area. These collections, which range from more specialized collections like CMU Multi-PIE (Gross et al., 2010) and MS-Celeb-1M (Guo et al., 2016) to well-known datasets like LFW (Labeled Faces in the Wild) (Huang et al., 2007) and CelebA (Liu et al., 2015), provide a rich tapestry of facial pictures for facial recognition algorithm testing, training, and benchmarking. Researchers can investigate a broad range of scenarios, including differences in face expressions, occlusions, and environmental conditions, because each dataset brings with it new qualities and problems. Moreover, deep learning-based methods for face recognition tasks can be developed using large-scale datasets, such as VGGFace (Parkhi et al., 2015), VGGFace2 (Zhang et al., 2014), and CASIA-WebFace (Panis et al., 2016), which contain millions of tagged photos. These available datasets are tremendous assets for pushing the boundaries of facial recognition technology, encouraging creativity, and solving practical problems in a variety of fields and applications. Table 1 presents soma available datasets with their key features. Figure 1 presents images of some available datasets.

This chapter briefs the creation of COMSATS face database which is developed in COMSATS University Abbottabad campus. This database contains 850 face images of 50 individuals with seventeen different poses, three illumination conditions and eight resolutions.

Dataset Structure

The fifty participants who took part in the database collection were all male volunteers. Each subject was different in terms of length, color, caste, weight, and age. The permitted age range was 18 to 30 years old. Most of them were students at COMSTAS University Islamabad Abbottabad (Campus), with very few being alumni. All of the subjects for the database collection activity were Pakistani citizens, and it was conducted in the survey lab of the civil engineering department of COMSATS University Islamabad Abbottabad (Campus). Within five months, it was completed. Eight distinct resolutions (144×256, 200×200, 100×100, 70×70, 50×50, 40×40, 20×20 and 10×10 pixels) and three different illumination variations are present in the database, which contains 850 photographs of fifty persons in fifty various positions ($0°$, $5°$, $10°$, $15°$, $20°$, $25°$, $30°$, $35°$, $55°$, $-5°$, $-10°$, $-15°$, $-20°$, $-25°$, $-30°$, $-35°$, $-55°$).

Location (or Observation)

An organized indoor atmosphere has been setup with fluorescent lamps and Natural light. The participant ask to sit the predefined point in front of camera at a distance of (0.5 to 0. 8 meter) and to fallow the predefine structure as shown in fig1 (a). white sheet is placed behind the background to produce uniformity. The camera operator needs to take the images of each individual for the desirable results. The camera operator collects the all dataset images at the end of the experiment. This database consists of 850 images of 50 subjects under 17 different poses ($0°$, $5°$, $10°$, $15°$, $20°$, $25°$, $30°$, $35°$, $55°$, $-5°$, $-10°$, $-15°$, $-20°$, $-25°$, $-30°$, $-35°$, $-55°$) with each subject having different age, weight, height and facial color . All the subjects

Table 1. Available dataset among with their features

Ref	Dataset	Subjects	Images	Resolution	Illumination	Pose	Occlusion	Expression
(Huang et al., 2007)	LFW	5,749	13,000	250 × 250	Variable	Varied	Varied	Neutral
(Gross et al., 2010)	CMU Multi-PIE	337	750,000	High	Varied	Varied	Varied	Various
(Liu et al., 2015)	CELEBA	10,177	202,599	178x218	Varied	Varied	Substantial	Varied
(Panis et al., 2016)	CASIA-WebFace	10,575	500,000	Flexible	Low	Frontal	Diverse	Neutral
(Parkhi et al., 2015)	VGGFace	2,622	2.6 million	Flexible	Low	Varied	Varied	Neutral
(Zhang et al., 2014)	VGGFace2	9,131	3.3 million	Flexible	Low	Varied	Varied	Neutral
(Beveridge et al., 2013)	PaSc	293	9,376	Flexible	Low	Varied	Varied	Neutral
(Bansal et al., 2017)	UMD Faces	8,277	367,888	Flexible	Low	Varied	Varied	Neutral
(Phillips et al., 2015)	IJB-A	2,085	5,712,589	Adjustable	Moderate	Varied	Varied	Varied
(Huang et al., 2012)	CPLFW	500	10,000	Flexible	Low	Varied	Varied	Neutral
(Yang et al., 2016)	WiderFace	392,703	32,203	Flexible	Varied	Varied	Varied	Varied
(Eyiokur et al., 2023)	ISL-UFMD Dataset	-	21,816	Controlled	Significant	Varied	Varied	Varied
(Li et al., 2023)	Chinese Face Dataset	130	7,395	Controlled	Significant	Varied	Varied	Varied
(Haq et al., 2022)	COMSATS Face Dataset	50	850	Significant	Low	Varied	Controlled	Neutral
(Bae et al., 2023)	DigiFace 1M	50,000	1,000,000	Significant	Moderate	Varied	Varied	Neutral
(Medvedev et al., 2023)	YLFW	3,000	10,000	Significant	Low	Varied	Varied	Neutral
(Guo et al., 2016)	MS-Celeb-1M	100,000	10,000,000	Flexible	Low	Varied	Varied	Neutral
(Ferrari et al., 2018)	Extended YTF	3,425	22,000 videos	Significant	Low	Varied	Varied	Neutral
(Muhajir et al., 2023)	USK FEMO Dataset	-	2,250	-	-	-	-	Five different emotions
(Wang et al., 2023)	Masked Face Recognition Dataset	10,000	500,000	-	Significant	Varied	Varied	Varied
(Mekonnen, 2023)	Balanced Face dataset	-	10,000	-	Low	Varied	Varied	Varied

were pakistani nationals and were students of COMSATS University Islamabad, Abbottabad(Campus).The angle measurement process is performed in survay labin

Figure 1. Images of (a) CelebA dataset, (b) LFW dataset, (c) CMU multi-PIE dataset, (d) CASIA-WebFace dataset

(a)   (b)

(c)   (d)

the department of Civil Engineering COMSATS University Islamabad Abbottabad (campus). A consent form has been signed by evey indivisual, which ensures that their face images will be used only for research purpose . Specification of dataset are presented in table 2.

METHODS

Different instruments are used to collect database like Total Station (Trimble M3 DR5), Theodolite (DT-5), Staff Road, stand, permanent Marker, background sheet and a digital camera. The angles were measured using Theodolite and Total station (normally used in Civil Engineering). The staff road of 5-meters is used to find the elevation of angles. The Theodolite is used to measure the angles of vertical

Table 2. Specifications table

Subject Area	Electrical and Electronic Engineering		
More specific subject area	Computer Vision, Image processing, Face recognition, Electronics		
Type of data	Images		
How data was acquired	The data was acquired by using following instruments 1-Camera (cannon 20.6 Mega Pixel (MP)) 2-Theodolite(DT-5) 3-Totla station (Trimble M3 DR5) 4-Flouresent lamp (For lightening)		
Data format	.jpeg		
Experimental factors	Images with17 different possess (-55 to +55) of total 50 individuals were captured with the help of a photographer, lab assistant and lab engineer.		
Experimental features	The database consists of 850 images of fifty subjects with seventeen different poses (0°, 5°, 10°, 15°, 20°, 25°, 30°, 35°, 55°, -5°, -10°, -15°, -20°, -25°, -30°, -35°, -55°) and eight resolutions (144 × 256, 200 × 200, 100 × 100, 70 × 70, 50 × 50, 40 × 40, 20 × 20 and 10 × 10 pixels) with three different illumination variations		
Data source location	COMSATS University Islamabad, Abbottabad (campus), Pakistan		
Data availability	https://www.kaggle.com/datasets/mahmoodulhaq/comsats-face-dataset		

axis and horizontal axis. The accuracy of theodolite and total station are very fine. We use these instruments to find the horizontal and vertical distances. The images were captured by cannon EOS6D in the lightening of fluorescent lamps as shown in Figure 2.

We collect the images of 50 subject each indivisuals under 17 different poses (0°, 5°, 10°, 15°, 20°, 25°, 30°, 35°, 55°, -5°, -10°, -15°, -20°, -25°, -30°, -35°, -55°). The Images of some indivisuals are shown in figure 3. These images were captured closed to real world conditions in duration of five months. Figure 4 shows 17 different poses of each indivisuals. Face images involved in this dataset can reveal the effectiveness and robustness of different face detection and recognition algorithms. Eight distinct resolutions of 144 × 256, 200 × 200, 100 × 100, 70 × 70, 50 × 50, 40 × 40, 20 × 20, and 10 × 10 pixels were created from these photos. Figure 5 shows dataset photos with resolutions. Furthermore to test the effectiveness of state-of-the-art algorithms against illumination variations, the dataset includes images in three different lighting conditions as presented in figure 6.

DATA RECORDS

- This data set will be used for evaluation of performance of different algorithms proposed for security and attendance purpose.

Figure 2. (a) Angle measurement process and predefined structure, (b) measured angles

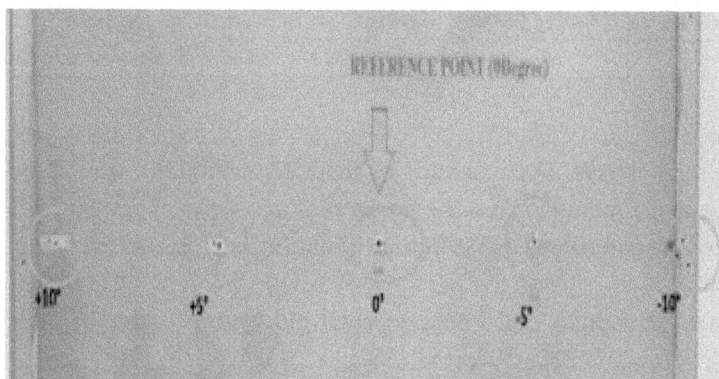

(b)

Table 3. Description of dataset

No of Images	850
No of subjects	50
No of Images(poses) per subject	17
Face poses	0°, 5°, 10°, 15°, 20°, 25°, 30°, 35°, 55°, -5°, -10°, -15°, -20°, -25°, -30°, -35°, and -55°
Image resolution	144 256, 200 200, 100 100, 70 70, 50 50, 40 40, 20 20 and 10 10 pixels
Illumination conditions	3 different illumination conditions
Image Format	.Jpg

Figure 3. Example images of subjects

Figure 4. Seventeen different poses of individual subject

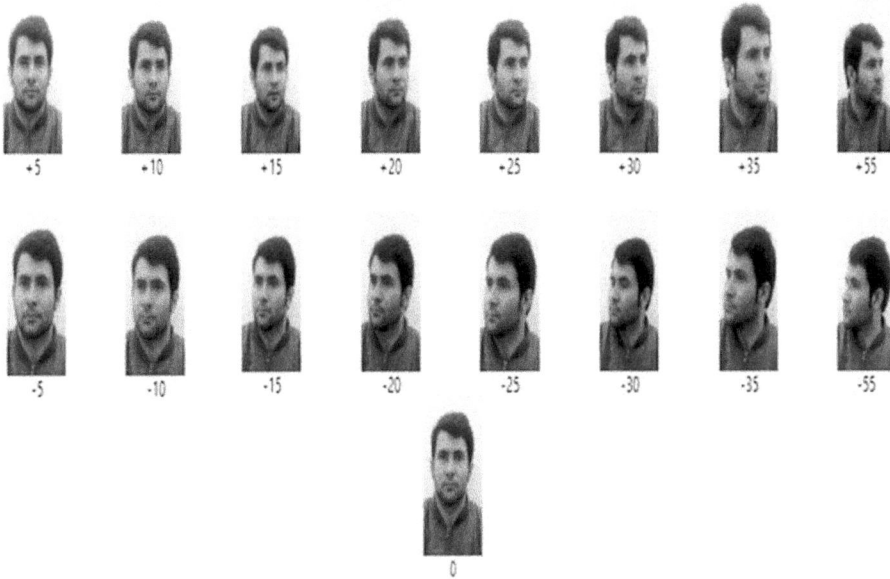

- This data will be source for different algorithms like as LDA (Lu et al., 2003), Local Binary Pattern (Ahonen et al., 2006) Eigen faces (Turk & Pentland,

Figure 5. Images with resolution of (a) 144 × 256, (b) 200 × 200, (c) 100 × 100, (d) 50 × 50, (e) 20 × 20

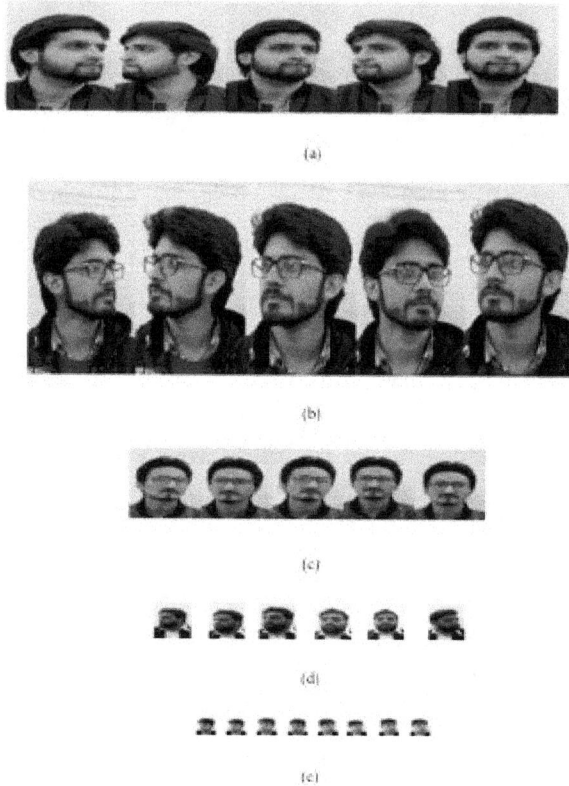

1991) and Deep Learning and will be challenge for recently published face recognition algorithms.

- It includes the poses in the range of (-55 to +55) of all the subjects. These poses are (0°, 5°, 10°, 15°, 20°, 25°, 30°, 35°, 55°, -5°, -10°, -15°, -20°, -25°, -30°, -35°, -55°). This dataset includes fifty subjects having different age of people; the age range is from 18 to 25 years.
- There are three distinct lighting settings in these pictures.
- Eight distinct image resolution formats, including 144 × 256, 200 × 200, 100 × 100, 70 × 70, 50 × 50, 40 × 40, 20 × 20, and 10 × 10 pixels, are available for this dataset.
- These requirements set the dataset apart from other usable datasets

Properties of image i.e dimensios and pixels before and after preprocessing are presented in Figure 7. In database preprocessing we rename all the images of each

Figure 6. Images with different lighting conditions: (a) Illumination 1, (b) Illumination 2, (c) illumination 3

(a)

(b)

(c)

indivisual by there angles.the iamges were resize by using irfan view and Matlab. The images were croped manually to get the specifice portion of image and in resize preocess we change the dimensions of images detaset to get appropriate results.

DATASET TESTING

The FR algorithm comparison in this section is based on the aforementioned database. The investigation was conducted with various image sizes. Four photographs from

Figure 7. Properties of a single image: (a) original image, (b) image after preprocessing

Image	Dimension (Pixels)
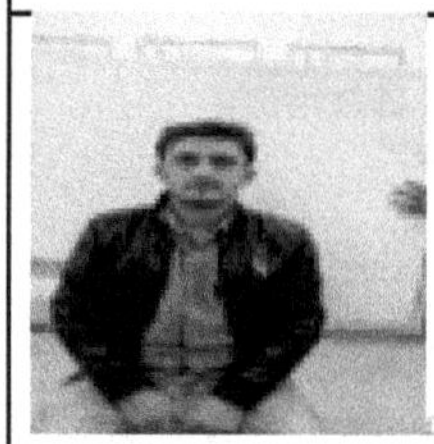	[illegible]

(a)

Image	Dimension (Pixels)
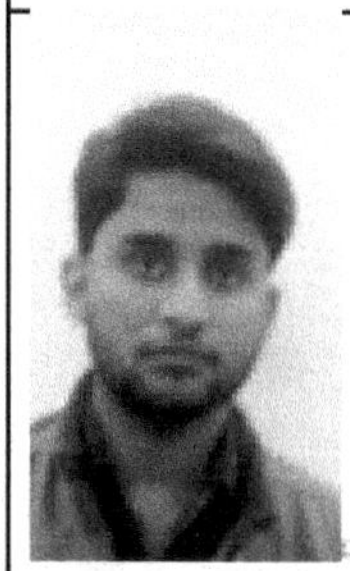	144 × 256
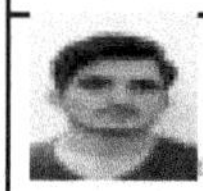	70 × 70
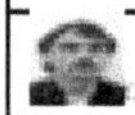	40 × 40
	20 × 20
	10 × 10

lighting 1 were chosen for each person in order to train these algorithms. 51 photos per person (17 from each lighting) were chosen for testing.

PCA-Based FR

A statistical method called principal component analysis (PCA) uses transformation as a set of observed possible. The principle of components is the process that converts correlated data into linearly noncorrelated variables. PCA is an essential component of the face recognition system since it is a highly effective face recognition technique. Similar to PCA, each training set picture is represented as a mixture of weighted eigenfaces, and covariance matrices are computed. Eigenvectors are derived from a training set of pictures' covariance matrix. Weights of eigenvectors are determined by the most significant set of Eigen faces.

PAL-Based FR

PAL based FR model used 68-points face landmark localization for face detection followed by partial PCA and LDA. The pseudo code of PAL based FR model are presented in figure 8.

SIMULATIONS RESULTS

A 32 GB RAM HP Core i7 computer was used for the experiments, and MATLAb 2019 was used as the simulation tool. Numerous experiments were conducted on the suggested database, which contains several face photos with three various conditions, such as face positions, illumination and image resolutions, in order to evaluate the FR algorithms listed above.

Figure 8. Pseudo code of PAL-based FR algorithm

Input: A set of input images $A = \left\{a_{i=1}^{J}\right\}_{i=1}^{I}$ with $I = \{1,2, \ldots\ldots, I\}$ classes and J images of each class.
Do for $i = 1, \ldots, I$
    (2) convert RGB images to grey,
    (3) estimate and crop face $(I_{cropped})$.
    (4) update mean and standard deviation of each image, $I_n = (I_{cropped} - \bar{X}) \times \sigma_{def}/\sigma_i + \bar{X}_{def}$
    (5) calculate mean image of each class, $tr_i = \sum_{j=1}^{J} a_j^i/J$.
Final training images of each class, $T_r = \{tr_1, \ldots\ldots tr_i\}$.
Initialize mislabelled distribution over m, $D_1(i) = 1/m = 1/N(-1)$
Do for $t = 1, \ldots T$:
    (1) if $t = 1$, choose i samples per class for the learner.
    (2) train LDA feature extractor.
    (3) build a g classifier h_t
    (4) calculate pseudo loss, e_t
    (5) calculate $\beta_t = e_t/(1 - e_t)$
    (6) if $\beta_t = 0$, abort the loop
    (7) update the distribution
Final g classifier of training image, $hf(z) = \arg\max(\sum(\log 1/\beta_t)h_t(z,y))$.
Generate a matching score.
Output: Maximum matching score (M_{score}), $I_{recog} = \arg\max(M_{score})$.

Image Resolution Analysis

Face photos of 144 × 256, 200 ×200, 100 × 100, 70 × 70, 50 × 50, 40 × 40, 20 × 20 and 10 × 10 pixels were reexamined in this investigation. The outcomes of the two FR algorithms are listed in Table 4. The following are key explanations taken from Table 4.

- PAL produces the highest recognition rate of 86.66% for face images with resolutions of 70×70 pixels and above, followed by PCA with 78.4% accuracy
- The recognition rate of the PAL method is 71.3%, however the accuracy of PCA has dropped to 32.7% for face resolution of 10×10.
- All of the above specified ranges of image resolutions are where the PAL algorithm performs best.

Illumination Analysis

To test the effectiveness of the FR models toward the illumination variation of proposed dataset, these algorithms were tested on these three illumination conditions (144 ×256 pixels) separately. Figure 9 presents the accuracy of FR on these illumination conditions. The key points of figure 9 are:

- PAL algorithms achieve high accuracy on all these three lighting conditions as compared to PCA based FR algorithm
- The illumination 3 are the toughest condition in this dataset.

Table 4. Recognition accuracy for image resolution

Image Resolution	PCA	PAL
144 × 256	78.48	86.66
200 ×200	78.48	86.66
100 × 100	78.48	86.66
70 × 70	78.48	86.66
50 × 50	77.22	86.66
40 × 40	68.31	86.66
20 × 20	53.46	81.43
10 × 10	45.05	77.75

Figure 9. Recognition accuracy for each lighting condition

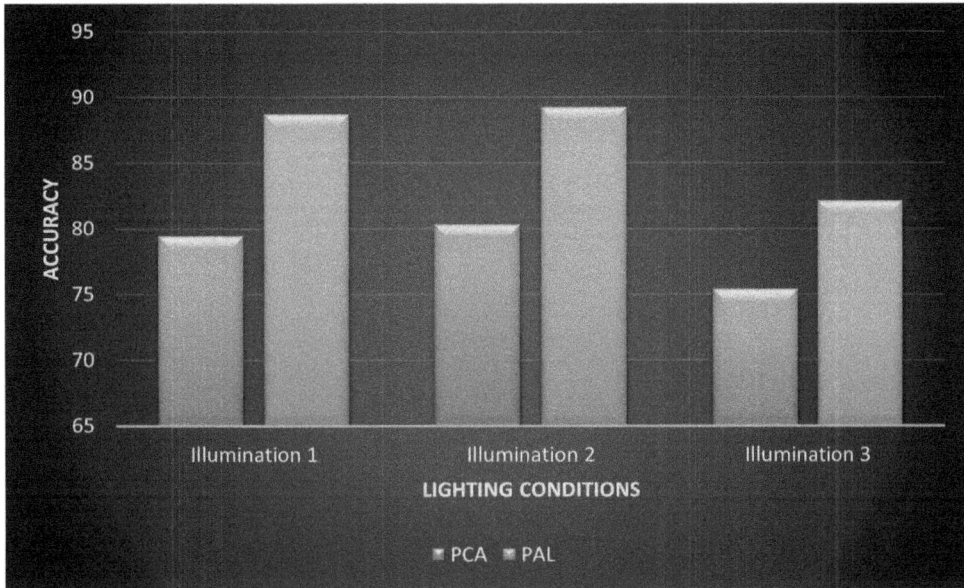

Pose Analysis

Variations in the facial position obscure certain aspects of an individual's face. A strong face recognition algorithm should be able to identify faces from a variety of viewing angles and be resilient to changes in stance. This study examines 17 different facial positions on 50 participants.

Figure 10 illustrates that for frontal face images, 100% recognition accuracy is achieved by the PAL technique, and PCA based face recognition algorithms. Moreover, the PAL approach performs superior to PCA based FR algorithm in terms of pose variation from frontal to ± 55°.

Discussion

Ongoing efforts are being made to create a strong FR algorithm that can replicate the human visual system. Table 5 emphasizes the significance of the FR algorithms even further.

(i)   The PAL and PCA algorithms can be applied to frontal images with incredibly low resolution, as those with 20 by 20 pixels.

(ii)   Only PAL should be utilized for low-resolution non-frontal photos, including crime scenes.

165

Figure 10. Recognition accuracy of presented algorithms for each face pose

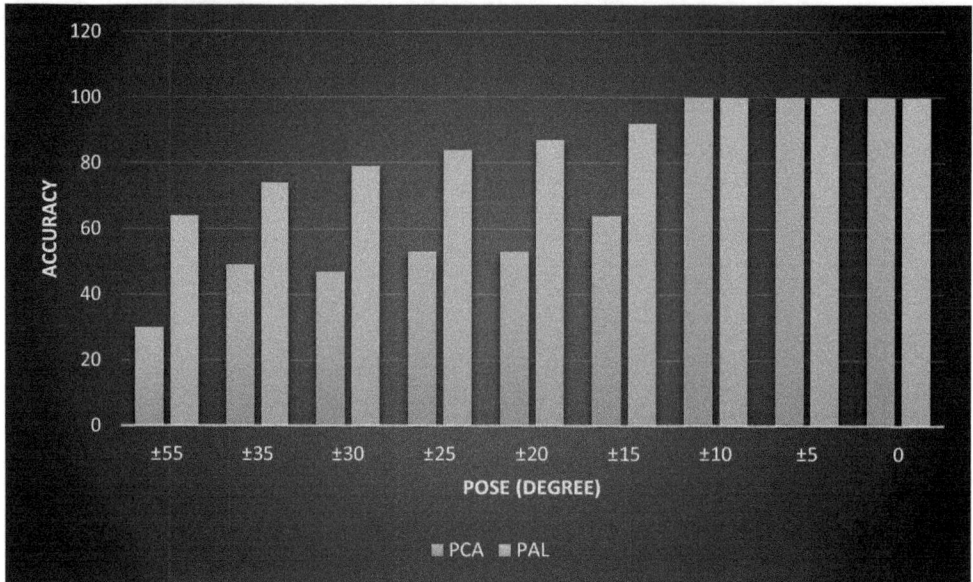

Table 5. FR algorithm selection based on performance

Description	Algorithms
Low Resolution frontal images	PCA (Turk & Pentland, 1991) and PAL (Haq et al., 2019)
Face pose images	PAL (Haq et al., 2019)
Face Illumination variations	PCA (Turk & Pentland, 1991) and PAL (Haq et al., 2019)

CONCLUSION

This study introduces the COMSATS face database, a dataset of face photographs with numerous poses. In the course of five months, these photos were taken on the campus of COMSATS University in Abbottabad, close to actual conditions. This dataset's facial image content can provide insight on the effectiveness and resilience of upcoming face detection and recognition systems. Other study fields that can make use of this database include face modeling, emotion recognition, age estimation, gender classification, and face position recognition.

The following step compares two popular face recognition algorithms using the suggested dataset: (i) PAL face recognition method and (ii) PCA based face recognition (eigenfaces). The PAL face recognition method may be employed with reliability for low resolution photos up to 10×10 pixels, for frontal ($0°$) ranges up to $\pm 55°$ of face posture variation and illumination variations in near real time, according to simulation results using the suggested database.

Our goal for the future study is to create a new face recognition algorithm that can distinguish between poses that vary by $\pm 90°$ and low-resolution face photos up to 5×5 pixels.

REFERENCES

Ahmad, S., & Adnan, A. (2015, July). Machine learning based cognitive skills calculations for different emotional conditions. In *2015 IEEE 14th International Conference on Cognitive Informatics & Cognitive Computing (ICCI* CC)* (pp. 162-168). IEEE. 10.1109/ICCI-CC.2015.7259381

Ahmad, S., Adnan, A., Khan, G., & Mehmood, N. (2015). Emotions, age, and gender based cognitive skills calculations. *International Journal of Computer Theory and Engineering*, *7*(1), 76–80. doi:10.7763/IJCTE.2015.V7.934

Ahmad, S., Li, K., Eddine, H. A. I., & Khan, M. I. (2018). A biologically inspired cognitive skills measurement approach. *Biologically inspired cognitive architectures*, *24*, 35-46.

Ahonen, T., Hadid, A., & Pietikainen, M. (2006). Face description with local binary patterns: Application to face recognition. *IEEE Transactions on Pattern Analysis and Machine Intelligence*, *28*(12), 2037–2041. doi:10.1109/TPAMI.2006.244 PMID:17108377

Anwar, M. S., Wang, J., Ahmad, S., Khan, W., Ullah, A., Shah, M., & Fei, Z. (2020). Impact of the impairment in 360-degree videos on users VR involvement and machine learning-based QoE predictions. *IEEE Access : Practical Innovations, Open Solutions*, *8*, 204585–204596. doi:10.1109/ACCESS.2020.3037253

Anwar, M. S., Wang, J., Khan, W., Ullah, A., Ahmad, S., & Fei, Z. (2020). Subjective QoE of 360-degree virtual reality videos and machine learning predictions. *IEEE Access : Practical Innovations, Open Solutions*, *8*, 148084–148099. doi:10.1109/ACCESS.2020.3015556

Anwar, M. S., Wang, J., Ullah, A., Khan, W., Ahmad, S., & Fei, Z. (2020). Measuring quality of experience for 360-degree videos in virtual reality. *Science China. Information Sciences*, *63*(10), 1–15. doi:10.1007/s11432-019-2734-y

Bae, G., de La Gorce, M., Baltrušaitis, T., Hewitt, C., Chen, D., Valentin, J., Cipolla, R., & Shen, J. (2023). DigiFace-1M: 1 Million Digital Face Images for Face Recognition. In *Proceedings of the IEEE/CVF Winter Conference on Applications of Computer Vision* (pp. 3526-3535). 10.1109/WACV56688.2023.00352

Bansal, A., Nanduri, A., Castillo, C. D., Ranjan, R., & Chellappa, R. (2017, October). Umdfaces: An annotated face dataset for training deep networks. In *2017 IEEE international joint conference on biometrics (IJCB)* (pp. 464-473). IEEE.

Beveridge, J. R., Phillips, P. J., Bolme, D. S., Draper, B. A., Givens, G. H., Lui, Y. M., ... Cheng, S. (2013, September). The challenge of face recognition from digital point-and-shoot cameras. In *2013 IEEE Sixth International Conference on Biometrics: Theory, Applications and Systems (BTAS)* (pp. 1-8). IEEE. 10.1109/BTAS.2013.6712704

Budiarsa, R., Wardoyo, R., & Musdholifah, A. (2023). Face recognition for occluded face with mask region convolutional neural network and fully convolutional network: a literature review. *International Journal of Electrical & Computer Engineering, 13*(5).

Campus, K. (n.d.). *Deep Frustration Severity Network for the Prediction of Declined Students' Cognitive Skills*. Academic Press.

Eyiokur, F. I., Ekenel, H. K., & Waibel, A. (2023). Unconstrained face mask and face-hand interaction datasets: Building a computer vision system to help prevent the transmission of COVID-19. *Signal, Image and Video Processing, 17*(4), 1027–1034. doi:10.1007/s11760-022-02308-x PMID:35910402

Fatima, R., Samad Shaikh, N., Riaz, A., Ahmad, S., El-Affendi, M. A., Alyamani, K. A., Nabeel, M., Ali Khan, J., Yasin, A., & Latif, R. M. A. (2022). A natural language processing (NLP) evaluation on COVID-19 rumour dataset using deep learning techniques. *Computational Intelligence and Neuroscience, 2022*, 2022. doi:10.1155/2022/6561622 PMID:36156967

Ferrari, C., Berretti, S., & Del Bimbo, A. (2018, August). Extended youtube faces: a dataset for heterogeneous open-set face identification. In *2018 24th International Conference on Pattern Recognition (ICPR)* (pp. 3408-3413). IEEE. 10.1109/ICPR.2018.8545642

Gross, R., Matthews, I., Cohn, J., Kanade, T., & Baker, S. (2010). Multi-pie. *Image and Vision Computing, 28*(5), 807–813. doi:10.1016/j.imavis.2009.08.002 PMID:20490373

Guo, Y., Zhang, L., Hu, Y., He, X., & Gao, J. (2016). Ms-celeb-1m: A dataset and benchmark for large-scale face recognition. *Computer Vision–ECCV 2016: 14th European Conference, Amsterdam, The Netherlands, October 11-14, 2016 Proceedings, 14*(Part III), 87–102.

Haq, M. U., Sethi, M. A. J., Ahmad, S., ELAffendi, M. A., & Asim, M. (2024). Automatic Player Face Detection and Recognition for Players in Cricket Games. *IEEE Access: Practical Innovations, Open Solutions, 12*, 41219–41233. doi:10.1109/ACCESS.2024.3377564

Haq, M. U., Sethi, M. A. J., & Rehman, A. U. (2023). Capsule Network with Its Limitation, Modification, and Applications—A Survey. *Machine Learning and Knowledge Extraction, 5*(3), 891–921. doi:10.3390/make5030047

Haq, M. U., Sethi, M. A. J., Ullah, R., Shazhad, A., Hasan, L., & Karami, G. M. (2022). COMSATS Face: A Dataset of Face Images with Pose Variations, Its Design, and Aspects. *Mathematical Problems in Engineering, 2022*, 2022. doi:10.1155/2022/4589057

Haq, M. U., Shahzad, A., Mahmood, Z., Shah, A. A., Muhammad, N., & Akram, T. (2019). Boosting the face recognition performance of ensemble based LDA for pose, non-uniform illuminations, and low-resolution images. *KSII Transactions on Internet and Information Systems, 13*(6), 3144–3164.

Hosni, A. I. E., Li, K., & Ahmad, S. (2020). Minimizing rumor influence in multiplex online social networks based on human individual and social behaviors. *Information Sciences, 512*, 1458–1480. doi:10.1016/j.ins.2019.10.063

Hosni, A. I. E., Li, K., & Ahmed, S. (2018). HISBmodel: A rumor diffusion model based on human individual and social behaviors in online social networks. *Neural Information Processing: 25th International Conference, ICONIP 2018, Siem Reap, Cambodia, December 13–16, 2018 Proceedings, 25*(Part II), 14–27.

Huang, G. B., Ramesh, M., Berg, T., & Learned-Miller, E. (2007). Labeled Faces in the Wild: A Database for Studying Face Recognition in Unconstrained Environments. *Proceedings of the Workshop on Faces in "Real-Life" Images: Detection, Alignment, and Recognition.*

Huang, J., Singh, V., Kumar, S., & Jain, A. K. (2012). *Cross-Pose LFW: A database for studying cross-pose face recognition in unconstrained environments. Technical Report.* Department of Computer Science and Engineering, Michigan State University.

Li, N., Shen, X., Sun, L., Xiao, Z., Ding, T., Li, T., & Li, X. (2023). Chinese face dataset for face recognition in an uncontrolled classroom environment. *IEEE Access : Practical Innovations, Open Solutions*, *11*, 86963–86976. doi:10.1109/ ACCESS.2023.3302919

Liu, Z., Luo, P., Wang, X., & Tang, X. (2015). Deep learning face attributes in the wild. *Proceedings of the IEEE International Conference on Computer Vision*, 3730-3738. 10.1109/ICCV.2015.425

Lu, J., Plataniotis, K. N., & Venetsanopoulos, A. N. (2003). Face recognition using LDA-based algorithms. *IEEE Transactions on Neural Networks*, *14*(1), 195–200. doi:10.1109/TNN.2002.806647 PMID:18238001

Medvedev, I., Shadmand, F., & Gonçalves, N. (2023). Young Labeled Faces in the Wild (YLFW): A Dataset for Children Faces Recognition. *arXiv preprint arXiv:2301.05776.*

Mekonnen, K. A. (2023). Balanced Face Dataset: Guiding StyleGAN to Generate Labeled Synthetic Face Image Dataset for Underrepresented Group. arXiv preprint arXiv:2308.03495.

Muhajir, M., Oktiana, M., Muchtar, K., Fitria, M., Akhyar, A., Pratama, M. D., & Lin, C. Y. (2023, August). USK-FEMO: A Face Emotion Dataset using Deep Learning for Effective Learning. In *2023 2nd International Conference on Computer System, Information Technology, and Electrical Engineering (COSITE)* (pp. 199-203). IEEE. 10.1109/COSITE60233.2023.10249834

Munawar, F., Khan, U., Shahzad, A., Haq, M. U., Mahmood, Z., Khattak, S., & Khan, G. Z. (n.d.). *An Empirical Study of Image Resolution and Pose on Automatic Face Recognition*. Academic Press.

Panis, G., Lanitis, A., Tsapatsoulis, N., & Cootes, T. F. (2016). Overview of research on facial ageing using the FG-NET ageing database. *IET Biometrics*, *5*(2), 37–46. doi:10.1049/iet-bmt.2014.0053

Parkhi, O. M., Vedaldi, A., & Zisserman, A. (2015). Deep Face Recognition. *Proceedings of the British Machine Vision Conference*, 1-12. 10.5244/C.29.41

Phillips, P. J., Beveridge, J. R., Draper, B. A., Givens, G. H., O'Toole, A. J., Bolme, D. S., Dunlop, J., Lui, Y. M., Sahibzada, H., & Weimer, J. A. (2015). IARPA Janus Benchmark A (IJB-A) Face Dataset. *Proceedings of the IEEE Conference on Computer Vision and Pattern Recognition (CVPR).*

Rahim, A., Zhong, Y., Ahmad, T., Ahmad, S., & ElAffendi, M. A. (2023). Hyper-Tuned Convolutional Neural Networks for Authorship Verification in Digital Forensic Investigations. *Computers, Materials & Continua, 76*(2). Advance online publication. doi:10.32604/cmc.2023.039340

Rajeshkumar, G., Braveen, M., Venkatesh, R., Shermila, P. J., Prabu, B. G., Veerasamy, B., ... Jeyam, A. (2023). Smart office automation via faster R-CNN based face recognition and internet of things. *Measurement. Sensors, 27*, 100719. doi:10.1016/j.measen.2023.100719

Rehman, A., Saba, T., Mujahid, M., Alamri, F. S., & ElHakim, N. (2023). Parkinson's disease detection using hybrid lstm-gru deep learning model. *Electronics (Basel), 12*(13), 2856. doi:10.3390/electronics12132856

Shahid Anwar, M., Wang, J., Ahmad, S., Ullah, A., Khan, W., & Fei, Z. (2020). Evaluating the factors affecting QoE of 360-degree videos and cybersickness levels predictions in virtual reality. *Electronics (Basel), 9*(9), 1530. doi:10.3390/electronics9091530

Sohail, M. Z., Zafar, T., Khan, T. A., Asim, M., Ahmad, S., Mairaj, T., & El Affendi, M. A. (2023). Prediction of Time to Failure (TTF) of Power Systems Using a Deep Learning Technique. *Journal of Hunan University Natural Sciences, 50*(12).

Tahsin, M. S., Al Karim, M., Ahmed, M. U., Tafannum, F., & Firoz, N. (2023). An integrated approach for diabetes detection using fisher score feature selection and capsule network. *Journal of Computing Science and Engineering : JCSE, 4*(2), 61–77.

Turk, M., & Pentland, A. (1991). Eigenfaces for recognition. *Journal of Cognitive Neuroscience, 3*(1), 71–86. doi:10.1162/jocn.1991.3.1.71 PMID:23964806

Ullah, H., Haq, M. U., Khattak, S., Khan, G. Z., & Mahmood, Z. (n.d.). *A Robust Face Recognition Method for Occluded and Low-Resolution Images.* Academic Press.

Wahab, H., Mehmood, I., Ugail, H., Sangaiah, A. K., & Muhammad, K. (2023). Machine learning based small bowel video capsule endoscopy analysis: Challenges and opportunities. *Future Generation Computer Systems, 143*, 191–214. doi:10.1016/j.future.2023.01.011

Wang, Z., Huang, B., Wang, G., Yi, P., & Jiang, K. (2023). Masked face recognition dataset and application. *IEEE Transactions on Biometrics, Behavior, and Identity Science, 5*(2), 298–304. doi:10.1109/TBIOM.2023.3242085

Yang, S., Luo, P., Loy, C. C., & Tang, X. (2016). Wider face: A face detection benchmark. In *Proceedings of the IEEE conference on computer vision and pattern recognition* (pp. 5525-5533). IEEE.

Zhang, J., Shan, S., Kan, M., & Chen, X. (2014). WebFace: A Scalable Face Image Dataset with Varying Pose and Age. *Proceedings of the IEEE Conference on Computer Vision and Pattern Recognition (CVPR)*.

Zheng, X., Fan, Y., Wu, B., Zhang, Y., Wang, J., & Pan, S. (2023). Robust physical-world attacks on face recognition. *Pattern Recognition*, *133*, 109009. doi:10.1016/j. patcog.2022.109009

Chapter 8
The Development, Applications, Challenges, and Analysis of a Cricket Player Face Recognition Dataset

Mahmood Ul Haq

(iD) https://orcid.org/0000-0002-1514-0300

University of Engineering and Technology, Peshawar, Pakistan

Muhammad Athar Javed Sethi

(iD) https://orcid.org/0000-0001-7847-831X

University of Engineering and Technology, Peshawar, Pakistan

Subhan Ullah

University of Malakand, Pakistan

Abd Ullah

University of Malakand, Pakistan

ABSTRACT

Over the last decade, face recognition technology has played a critical role in various circumstances, such as airport boarding, security applications, biometric verification, and smart homes. Along with the major role of face recognition in the areas above, we must recognize the important role of face recognition in various sports (i.e., cricket and football). The importance of proper player surveillance and identification in sports, particularly cricket, cannot be overstated. Articles are saturated with many deep-face evaluation systems; however, they are not up to mark due to the lack of significant face posture data. To address the black box in facial expression datasets, this chapter presents a comprehensive cricket player facial recognition dataset. The authors have a wide selection of cricket player images from various teams, playing styles, and backgrounds. It includes images taken during games, practices, and official team photos, providing a diverse range of facial

DOI: 10.4018/979-8-3693-2913-9.ch008

changes and challenges for facial recognition systems. Furthermore, they evaluate the efficacy of cutting-edge facial recognition algorithms on our dataset, providing insights into the effectiveness of current methodologies as well as potential areas for development. Eventually, the extensive experimental analyses demonstrate that the current work is significant in addressing the black box in facial expression datasets.

1. INTRODUCTION

Over the last decade, images focusing on face posture have been used to train face recognition algorithms. An algorithm trained with accurate face posture data can be used for various applications in critical areas with the aim of authentication and verification. So, the applications of such systems vary from security (internal and border security) to internal bank transactions and ATM transactions. Such a system focuses on the human face posture with a deeply featured algorithm to identify one identity (Sohail et al., 2023; Ullah et al., 2019). Facial expression is the fundamental mode of communication and recognition among individuals, and it forms the basis of our social interactions. Humans have had an innate ability to recognize and distinguish between faces for different ages (Ahmad & Adnan, 2015; Munawar et al., 2019). The effort to reproduce this intrinsic talent through technology has given rise to the interesting discipline of face recognition in the digital age (Anwar et al., 2024; Haq et al., 2023; Haq et al., 2022; Hosni, Li, Ding et al, 2018).

As face recognition plays a critical in the military and our daily life, we must also recognize the important role of face recognition in various sports, i.e., cricket and football. The significance of proper player management and tagging in sports, particularly cricket, cannot be overstated. Cricket, a sport cherished by millions throughout the world, offers an enthralling blend of athleticism, strategy, and spectacle (Haq et al., 2024; Rahim, Zhong, Ahmad, Ahmad, & ElAffendi, 2023). Cricket's fascination stretches far beyond the playing field, including a diversified and passionate global fan base. It has not only thrived as a source of pleasure in this digital age, but it has also become a focal point for technical innovation (Anwar et al., n.d.; Anwar et al., 2020). Among the numerous improvements, the incorporation of facial recognition technology into the cricketing arena has emerged as a trailblazer, set to transform the way we experience and analyze the game.

Identification of cricket players is essential for several aspects of the game, such as performance evaluation, fan participation, and security measures (Kowalczewski et al., 2023; Rahim, Zhong, Ahmad, Ahmad, Pławiak et al, 2023; Wickramasinghe, 2022). In order to develop tactics and make wise choices, coaches and analysts rely on player tracking and performance information. Fans want a more engaging viewing experience that allows them to interact with their favorite athletes and teams. Accurate player identification is crucial since stadiums are responsible for managing access and guaranteeing security (Chen et al., 2021). The creation of a comprehensive cricket player facial recognition dataset is a critical resource in this scenario. A dataset like this one not only allows academics and developers to enhance the state of the art in facial recognition (Deng et al., 2020), but it also solves the unique issues given by cricket matches as dynamic, uncontrolled settings (Yu et al., 2019).

Literature shows numerous deep-face evaluation systems; nevertheless, they need to be up to mark due to the lack of significant face posture data. The other reason for the insignificant performance is the small dataset for training deep models. Insignificant datasets may lead deep models towards deterioration in prediction. To answer these multifarious requirements, this work introduces a fresh cricket player face recognition dataset that has been rigorously curated to encompass the diverse tapestry of cricket. This dataset contains a diverse mix of facial photos from various cricket teams, formats, and match circumstances. It addresses the difficulties and complexities of recognizing players in dynamic, uncontrolled contexts such as cricket stadiums, which have fluctuating lighting conditions, various facial expressions, and probable occlusions.

This paper's key contribution is:

- This study describes the development of the cricket player face database. This database comprises 850 face images of 50 different cricket players in two different image resolutions.
- Images have been acquired in a very comparable setup. Separate sessions were performed for each process of dataset collection.
- The images were collected from the internet under several conditions, such as positions, expressions, lighting conditions, occlusion, and game situations.
- By varying the picture resolution, the suggested dataset was tested on three baseline FR algorithms: PCA, PAL, and LDA. The extensive experimental analyses show that the current work is significant in addressing the highlighted issues in facial expression datasets.

2. LITERATURE REVIEW

For a variety of reasons, FR is a popular problem in computer vision (Tang et al., 2018). First, because individuals are labelled with their names, it is simple to create well-posed problems and gather data. Second, it's worth looking at because it's a shining example of fine-grained classification (Najibi et al., 2018). Third, the facial recognition problem is critical and can be used to a wide range of applications (Wu et al., 2021). Face recognition has been a hot topic in the vision community for a variety of reasons (Phillips et al., 1998).

Face recognition tasks are often classified into two types. One is face identification, which means that given a gallery set and a query set, we want to find the most similar face in the gallery set to a given image in the query set and use the identity of the similar face as the query image's identity. Face verification, on the other hand, assesses whether two images are of the same person.

Early face datasets were almost entirely collected in controlled environments such as PIE (Sim et al., 2002) and FERET (Huang et al., 2008), resulting in very good performance on these constrained datasets. However, due to the complexities of real-world faces, most models built from these datasets do not perform well in actual applications. The dataset focus eventually switched from controlled to uncontrolled contexts to boost the generalization of face recognition systems. As a result, in 2007, the landmark dataset Labelled Faces in the Wild (LFW) (Zhang et al., 2014) was established. The most significant distinction between LFW and the previous benchmark dataset is that the photos were gathered via the Internet rather than acquired in a variety of pre-defined situations.

Numerous new face recognition databases have recently been collected to research face recognition and verification. CASIA database (Kemelmacher-Shlizerman et al., 2016), Megaface (Phillips et al., 2015), IJB-A (Cao et al., 2018), FaceScrub and VGGFACE (Karpathy et al., 2014) were among them. The CASIA dataset has 494,414 photos of 10,575 different persons. FaceScrub comprises 106,863 photos of 530 celebrities gathered from the internet. Each person has roughly 200 photos on average. Though the percentage of correct labels is difficult to calculate, these vast and deep databases can help researchers train facial recognition systems with sophisticated frameworks.

Existing facial recognition datasets, such as LFW, focus on athletes from a variety of sports. While these datasets have helped to advance facial recognition technology in sports, they may not fully capture the complexities and problems of cricket, which has its own set of scenarios, lighting conditions, and player expressions.

NBA Faces and soccer player databases specialize to individual sports, allowing researchers to construct basketball and soccer-specific apps and algorithms. These datasets have shown the value of domain-specific datasets in sports face recognition. However, due to the unique characteristics of cricket, a dedicated dataset is required to properly leverage the potential of face recognition technology in this sport. Table 1 presents comparative analysis of some sports face recognition dataset based on no of images/subject and conditions.

Table 1. Comparative analysis of some sports face recognition dataset

Ref	Dataset	No. of Images	No. of Subjects	Face Poses	Illumination	Occlusion	Expressions
(Barman et al., 2018)	Sports-1M	1 million+	6,000+	Multiple	Varied	Minimal	Varied
(Kolyperas et al., 2019)	FIFA 2018 (Sports-100)	9,425	100	Frontal	Varied	Minimal	Neutral
(Palmer & King, 2006)	NBA Faces (Casual and Panini)	450,000+	4,000+	Multiple	Varied	Minimal	Varied
(Pappalardo et al., 2019)	MLB Faces	6,000+	300+	Multiple	Varied	Minimal	Varied
(Feng et al., 2020)	UEFA Champions League (2018)	7,000+	200+	Frontal	Varied	Minimal	Neutral
(Yu & Yang, 2001)	SSET	Varies	Varies	Multiple	Varied	Minimal	Varied

There is currently a noteworthy lack of a cricket player facial recognition dataset. Given the global popularity of cricket and the necessity for precise player identification in numerous elements of the sport, such as fan engagement, performance analysis, and security, the development of a dedicated dataset for cricket players becomes critical.

This work proposes the creation of a Cricket Player Face Recognition Dataset with the goal of capturing a varied range of facial photographs of cricket players in various playing styles, match circumstances, and lighting conditions. A dataset like this would be a great resource for scholars, coaches, and cricket fans, providing the foundation for powerful facial recognition algorithms in cricket.

In conclusion, while there are existing sports facial recognition datasets, the lack of a cricket-specific dataset presents a research and development potential. The proposed Cricket Player Face Recognition Dataset can address this need and help to progress face recognition technology in cricket, ultimately improving the cricketing experience for fans and assisting with numerous cricket-related applications.

3. METHODOLOGY

The method of creating the cricket player face recognition dataset can be divided into the following steps: These steps are presented in figure 1.

3.1. Collecting Raw Internet Images

In 1st step of dataset collection, images of different cricket players were collected. Google is used to search for this purpose. Keeping in mind several conditions such as positions, expression, lighting conditions and occlusion, several images were selected for the proposed dataset.

3.2. Player Detection

After collecting dataset, players were detected and cropped. For this purpose a group of five image processing experts were suggested. These experts have selected the visible players to be added in dataset from collected images.

3.3. Face Detection

The next stage is to detect faces; because conventional detection methods do not perform well in huge poses and occlusion, we manually detect the faces. The image is then cropped and rescaled (as detailed below) before being saved as a separate JPEG file.

3.4. Rescaling and Cropping the Detected Faces

We use the following approach to construct the photos in the proposed dataset. The region obtained by human for the given face is multiplied by two based on the maximum length and width values. If the expanded region falls outside the image's original region, a new image of the necessary size will be created by filling up the area outside the original image with black pixels. The extended image is then downsized to 100×100 and 50×50 pixels using MATLAB's imresize function. Finally, the image is saved as a JPEG file.

3.5. Remove Duplicate Images

Before we can remove duplicate photos, we must first define what constitutes a duplicate. The two images are numerically comparable at each pixel, according to the most basic concept. The criterion, however, ignores many scenarios in which faces in photographs are indistinguishable to human eyes since images taken by volunteers may have been re-cropped, rescaled, renormalized, or differently compressed. If we do not remove these face photographs, we may end up with positive pairs that are visually equivalent but differ quantitatively. So, in accordance with (Wang et al., 2004), duplicates are photographs that are determined to share a common original source photograph. For eliminating duplicate photos, we have two methods.

To begin, all possible pairs of pictures from the same identity are compared using a structural similarity metric (Ahmed et al., 2022; Hosni, Li, & Ahmed, 2018; Rehman et al., 2023; Ullah et al., 2022). Only pairs with a high degree of similarity are examined, and the low-quality version is deleted. We then thoroughly examine each individual's photo to confirm that there are no duplicate images in the dataset.

3.6. Determining Whether or Not Labels Are Correct

For each subject, we take great care to personally determine whether the scraped photographs are actually about the player or not. The abundant information on the original page improves labelling quality, especially in difficult circumstances.

4. DATASET DESCRIPTION

4.1. Image Specification

The database includes 850 images of 50 subjects with a resolution of 100×100, and 50×50 pixels as presented in Figure 2.

Figure 1. Cricket player face recognition dataset methodology

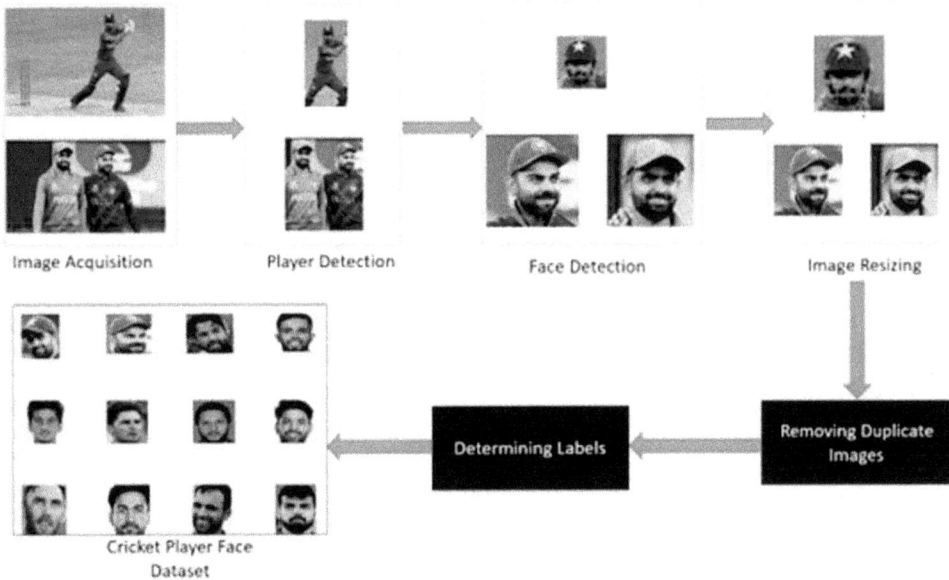

Figure 2. Dataset images of (a) 100 × 100, (b) 50 × 50 pixels

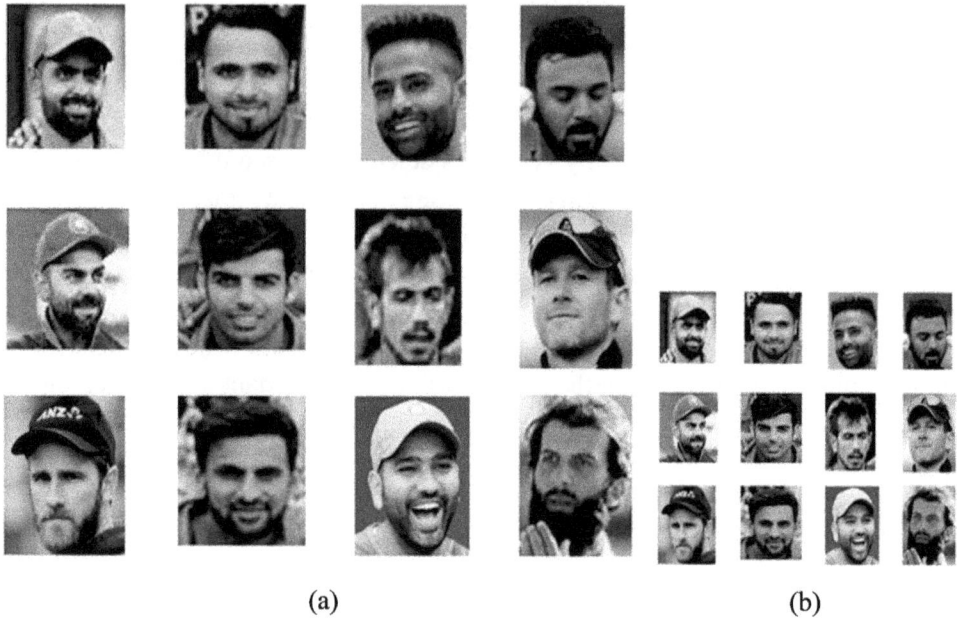

(a)   (b)

4.2. Dataset Structure

The database contains 850 photos of fifty different persons in several distinct face poses, illumination, and occlusion and lighting conditions. The images of individual in dataset are presented in figure. Figure 3 presents the images of three subjects under several conditions while figure 4 presents some sample images of dataset. Table 2-3 presents the data specification and challenging conditions added in dataset.

4.3. Data Records

This dataset will be used to assess the performance of various algorithms provided for security, monitoring and entertainment purposes

• This data will be used as a source for various methods such as LDA (Chen et al., 2013), Local Binary Pattern (Kshirsagar et al., 2011), Eigenfaces (Haq et al., 2019), PAL (Hammouche et al., 2022) and Deep Learning, as well as a challenge for recently released face recognition algorithms (Boutros et al., 2022; Deng et al., 2019; Huang & Chen, 2022; Jeevan et al., 2022; Karpagam et al., 2022).

- The dataset includes images of each subject have two images of frontal face, two face images in helmets, two images in caps, with different illumination and poses
- The dataset have images in 100×100, and 50×50 pixels.

5. APPLICATIONS OF CRICKET PLAYER FACE RECOGNITION DATASET

The applications of cricket player face recognition dataset demonstrate the versatility and importance of a cricket player face recognition dataset in the world of cricket, benefiting not just players and teams but also fans, broadcasters, and other stakeholders.

5.1. Identification and Tracking of Players

Accurate player identification and tracking during matches is one of the key applications of a cricket player facial recognition dataset. Even in dynamic and fast-paced scenarios, this technology can assist viewers and analysts in identifying players on the pitch.

Figure 3. Images of individual in proposed dataset

Figure 4. Images of three subjects under several conditions

Table 2. Data specification table

Subject Area	Computer Vision, Machine Learning
Specific Subject Topic	Face Detection and recognition
Data Acquired	Internet
Data Format	Images (JPG)
Image Resolution	100 × 100 and 50 × 50 pixels
Experimental Factors	Dataset were collected under the supervision of computer vision experts
Experimental Features	Dataset includes images of cricket players under several conditions such as poses, occlusion and illumination
Conditions	Pose variations, Occlusion, Illumination variations, Expressions
Data Source Location	UET Peshawar
Data Availability	Dataset will be available on request

Table 3. Description of dataset based on challenging conditions

Dataset	Conditions					
	Face Pose	Occlusion	Illumination	Resolution	Facial Expressions	Tilted Face
Cricket Player Face Recognition Dataset	Yes	Yes	Yes	100 × 100 and 50 × 50 pixels	Yes	Yes

5.2. Increased Fan Engagement

Cricket fans are extremely loyal to their favorite players and teams. Cricket broadcasters and apps can use face recognition to deliver personalized material and statistics about players, improving the overall fan experience.

5.3. Performance Evaluation

Face recognition can be used by coaches and analysts to track player performance over time. They can analyze statistics and patterns by recognizing players in training sessions and matches, assisting in strategy building and player improvement.

5.4. Statistics and Records for Players

Face recognition can be used to keep accurate player statistics and records, lowering the possibility of errors in documenting players' career accomplishments such as runs scored, wickets taken, and milestones reached.

5.5. Controlling Access and Security

Face recognition technology can be used to improve security at cricket stadiums and facilities. It can assist limit access to restricted areas, detect unauthorized persons, and improve player and spectator safety.

5.6. Apps for Fan Engagement

Face recognition tools can be included in cricket-specific mobile apps and websites. Fans can post selfies or images and receive information on recognized players, their stats, and forthcoming matches featuring their favorite clubs in real time.

5.7. Social Media Inclusion

Face recognition technology can be used by social media platforms to automatically tag cricket players in user-uploaded images and videos, making it easier for fans to share their cricket experiences and interact with their favorite athletes.

5.8. Merchandising and Marketing

Face recognition data can be used by cricket teams and sponsors to create tailored marketing campaigns. They can tailor marketing and item suggestions to fans' favorite athletes.

5.9. Highlight Reel Production

Face recognition can be used by content developers and broadcasters to automatically construct highlight reels starring certain athletes. This can help to speed up the video editing process and provide fans rapid access to noteworthy moments.

5.10. Branding of Players and Teams

Face recognition data can help players and teams establish and reinforce their branding. It can be used in promotional materials, team apparel, and marketing documents to identify players.

5.11. Protocols for Player Safety and Health

Face recognition can help enforce health protocols for players and ensure compliance with safety measures during tournaments and practice sessions in the light of global health concerns.

5.12. Broadcast Player Recognition

Face recognition can be used by broadcasters to display player information, statistics, and career highlights during live broadcasts, improving the viewing experience for viewers.

5.13. Injury Evaluation and Rehabilitation

Face recognition can help to monitor players' physical health, analyze injuries, and track the progress of rehabilitation programs.

5.14. Analysis of Game Strategy

Face recognition can be used by coaches and analysts to monitor player movements and positions during games, assisting in the development of game strategies and tactics.

6. CHALLENGES AND CONSIDERATIONS

To make the proposed dataset robust, several consideration has been made keeping in mind the challenges occurred

6.1. Data Gathering and Annotation

The need for access to players during matches, practice sessions and official events made gathering a comprehensive dataset of cricket player facial photos difficult. It took a long time to annotate the dataset with precise player IDs and contextual information.

6.2. Data Variety

It is critical to ensure that the dataset is diverse. Cricket teams, playing formats (e.g., Test, One-Day, T20), and geographical locations are all represented.

6.3. Variations Within a Class

Cricket players' look can vary significantly, including changes in facial hair, hairstyles, and accessories (e.g., helmets, caps). Intra-class variances have make recognition more difficult.

6.4. Similarities Between Classes

Players on the same team may wear jerseys that are visually identical, which can also confound recognition systems. It is necessary to distinguish between players who have similar appearances.

6.5. Occlusion

During matches, cricket players frequently wear helmets or caps, which can partially cover their faces. Headgear in the dataset might dramatically diminish the visible facial features, making it more difficult for face recognition systems to identify the player.

7. EXPERIMENTAL RESULTS

This section compares facial recognition methods based on the aforementioned database. The experiment was carried out with images of 100×100, and 50×50 pixels size. Images from fifty players were chosen for training (three samples per individual), and these algorithms were tested on ten photographs of each subject.

7.1. Face Recognition Algorithm Based on PCA

Principal Component Analysis (PCA) is a statistical procedure that converts a set of observable possibly correlated variables into a set of linearly uncorrelated variables known as the principal of components. Because it is a particularly efficient method for face recognition, PCA plays an essential role in the face recognition system. Like in PCA, all pictures in the training set are represented as a combination of weighted eigenfaces and covariance matrices. Eigenvectors are calculated using the covariance matrix of a training set of images. The weights of eigenvectors are determined by the collection of most relevant eigenfaces. A test image is projected onto the eigenfaces subspace to achieve face recognition. The distance between the test and training images is calculated with help of Equation 1.

$$d_{x_i x_j} = \sum_{a=1}^{n} (b_{k_i} - b_{k_i})^2 \tag{1}$$

where x_i and x_j represent two matrices for training and test samples, respectively, and $(b_{k_i} - b_{k_i})^2$ is the Euclidean distance (ED) between two image components. Test image must have minimum Euclidean distance with a recognized image that exists in the training images.

Where x_i and x_j are training and test sample matrices, respectively, and $(b_{k_i} - b_{k_i})^2$ is the Euclidean distance (ED) between two picture components. The test image must have a minimal Euclidean distance from a recognized image in the training images.

7.2. Linear Discriminant Analysis (LDA)

LDA is offered as an improvement to PCA. LDA creates a discriminant sub-space by minimizing scatter between photos of the same class and increasing dispersion between images of different classes. Let i [X1, X2,..., Xi] be the face classes in the database and each face class Xi have face images xj, where j=1,2,...n. With-in class variance can be calculated as:

$$V_w = \sum_{i=1}^{I} \sum_{j=1}^{J} \left(b_k - \mu_i \right) \left(b_k - \mu_i \right)^T \tag{2}$$

Where b_k represents the jth image for all classes (i) and μi denotes the mean of it^h class. The μi_c an be calculated as:

$$\mu_i = \frac{\sum b_j}{k} \tag{3}$$

Similarly, the scatter matrix between classes can be computed as follows:

$$V_b = \sum_{i=1}^{I} N_i \left(\mu_i - \mu \right) \left(\mu_i - \mu \right)^T \tag{4}$$

Where μ signifies the average of all classes and can be determined as follows:

$$\mu = \frac{\sum \mu_i}{i} \tag{5}$$

Readers are directed to (Barman et al., 2018) for a more in-depth examination.

7.3. PAL Face Recognition Algorithm

The PAL FR approach detects 68 distinct points on training and testing faces after face detection using a machine learning algorithm. Following that, all of these faces are cropped using these 68 landmarks. To eliminate error due to lighting differences, the mean and standard deviation of each face image are calculated and updated based on the relationship indicated in (6).

$$I_n = \frac{\left(I - \bar{X} \right) \sigma_d}{\sigma_i + \bar{X}_d} \tag{6}$$

Where $\bar{X}$ and σi represent the mean and standard deviation of it^h class while $\bar{X}_d$ and σd represent the predefined suggested mean and standard deviation. This technique uses a mean image of each class to reduce time complexity, memory requirements, and pose variation mistakes. The average image can be calculated as follows:

$$\bar{X}_m^s = \frac{\sum_{j=1}^{J} I_{nj}^s}{j} \tag{7}$$

Furthermore, these photos are fed into AdaBoost in conjunction with LDA for recognition. The final classifier computes a scoring value for the test image with each class, and the image with the highest scoring value is considered the recognized image with the desired class. For a more in-depth examination, readers are recommended to (Hammouche et al., 2022). The pseudocode for the PAL algorithm is shown in Table 4. Three baseline techniques are used to test the proposed face data-base: LDA (Chen et al., 2013), PCA (Palmer & King, 2006), and PAL (Pappalardo et al., 2019). Figure 5 depicts the total accuracy of the discussed methods on the suggested face database.

Table 4. Pseudocode for the PAL based FR algorithm

Input: A set of input images $A = \left\{a_{i=1}^{j}\right\}_{i=1}^{I}$ with $I = \{1, 2, \ldots \ldots, I\}$ classes and J images of each class.
Do for $i = 1, \ldots, I$
    (2) convert RGB images to grey,
    (3) estimate and crop face ($I_{cropped}$).
    (4) update mean and standard deviation of each image, $I_n = (I_{croped} - \bar{X}) \times \sigma_{def}/\sigma_i + \bar{X}_{def}$
    (5) calculate mean image of each class, $tr_i = \sum_{j=1}^{J} a_j^i / J$.
Final training images of each class, $T_r = \{tr_1, \ldots \ldots tr_i\}$.
Initialize mislabelled distribution over m, $D_1(i) = 1/m = 1/N(-1)$
Do for $t = 1, \ldots T$:
    (1) if $t = 1$, choose i samples per class for the learner.
    (2) train LDA feature extractor.
    (3) build a g classifier h_t
    (4) calculate pseudo loss, e_t
    (5) calculate $\beta_t = e_t/(1 - e_t)$
    (6) if $\beta_t = 0$, abort the loop
    (7) update the distribution
Final g classifier of training image, hf $(z) = \arg\max(\sum(\log 1/\beta_t)h_t(z, y))$.
Generate a matching score.
Output: Maximum matching score (M $_{score}$), I $_{recog}$ = argmax (M$_{score}$).

7.4. Simulation Results

The trials were conducted out using a Super-Server 7047 GR system with 32 GB of RAM and the simulation software MATLAB 2019. To evaluate the FR algorithms indicated above, numerous experiments were performed on the proposed database, which contains various face pictures with two distinct image resolutions. Three photographs of each subject were chosen for training, and these algorithms were trained on 850 images.

Face images with 100×100, and 50×50 pixels with several face poses, illumination and occlusion were reinvestigated in this work. Figure 5 shows the outcomes of the three FR algorithms. The following are the key explanations from Figure 5.

- PAL achieves the highest recognition rate of 75.24% for face photos with a resolution of 100×100, followed by PCA at 63.91% and LDA at 63.21%.
- PAL achieves the highest recognition rate of 72.38% for face photos with a resolution of 50×50 pixels, followed by PCA at 57.73% and LDA at 43.31%.
- The PAL algorithm performs best across all of the image resolution ranges listed above.

Figure 5. Recognition accuracy of FR algorithms on proposed dataset

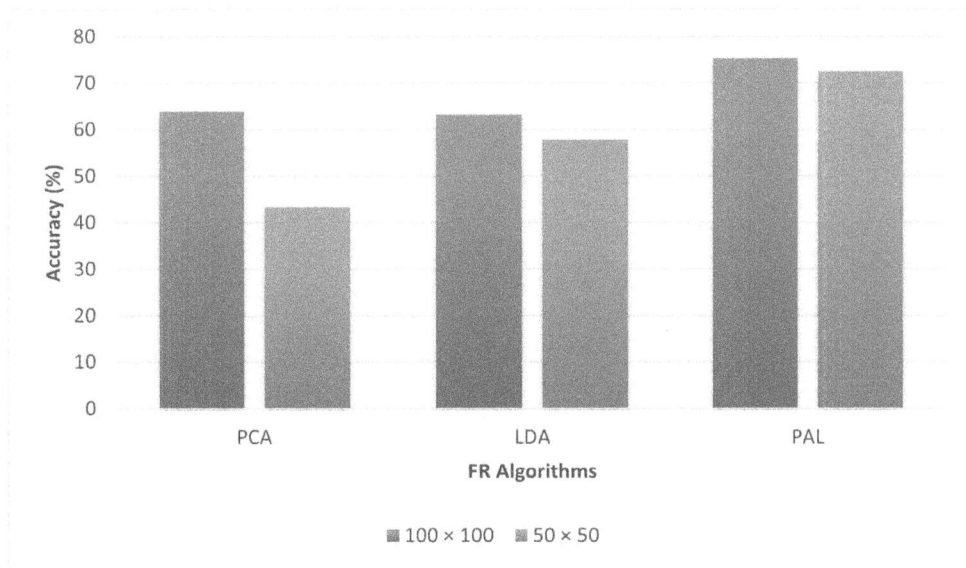

7.5. Complexity of Computation

Figure 6 depicts the method execution timings for various image resolution facial images. The following are some key findings from Figure 6.

- When compared to PCA (Haq et al., 2019) and LDA (Chen et al., 2013), the PAL (Hammouche et al., 2022) technique takes more than 9 seconds for each face image resolution category and is the most computationally complex.
- In case of time, the winner is PCA (Haq et al., 2019) based face recognition algorithm

7.6. Discussion

Continuous efforts are being undertaken to develop a robust FR algorithm capable of simulating the human visual system. Table 5 highlights the importance of the FR algorithms even more.

- Use PAL for low-resolution facial poses, occlusion, and high accuracy.
- Readers are suggested to use the PCA approach for lower computing cost and average accuracy.

Figure 6. Based on the proposed dataset, the computational complexity of FR algorithms

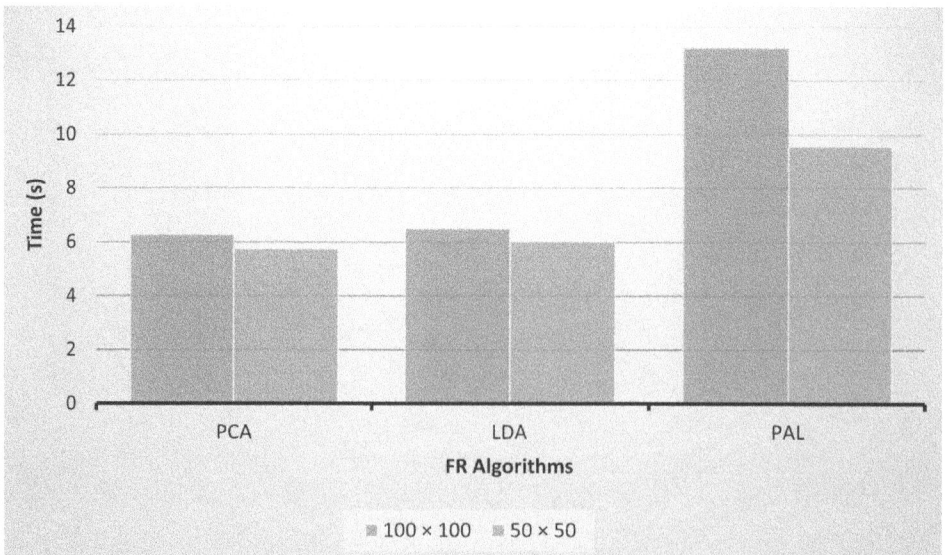

- Based on these findings, it can be stated that the presented dataset is extremely difficult to recognize efficiently using a FR algorithm.

8. CONCLUSION

The development of the Cricket Player Face Recognition Dataset marks an important step forward in the integration of face recognition technology into the world of cricket. This research presents a dataset of cricket player face dataset. This dataset's face photos can demonstrate the efficiency and robustness of future face detection and identification algorithms. This database can be utilized for a variety of research purposes, including cricket player identification and statistics, gender classification, age estimate, emotion identification, face posture recognition, age estimation, and face modelling.

The suggested dataset is then used to compare three well-known face recognition algorithms: (i) PCA based face recognition (eigenfaces), (ii) LDA based face recognition, and (iii) PAL face recognition algorithm. The proposed database simulation results show that PAL based FR recognition have high accuracy in recognizing the proposed dataset.

9. FUTURE WORK

As we look to the future, we invite researchers, cricket fans, and technology pioneers to join us in this fascinating initiative, where the only limit is our collective imagination.

Cricket, with its unique blend of tradition and contemporary, is ready to embrace the digital age, and we are thrilled to be a part of it.

10. AVAILABILITY OF SUPPORTING DATA

The article only evaluates existing research that are available online on various platforms such as Google Scholar.

Table 5. Performance-based FR algorithm selection

Description	Algorithm
Low-resolution facial poses, occlusion and High accuracy	PAL (Hammouche et al., 2022)
Less computational cost and Average Accuracy	PCA (Haq et al., 2019)

11. CONFLICTS OF INTEREST

The authors declare that they have no conflicts of interest to report regarding the present study.

REFERENCES

Ahmad, S., & Adnan, A. (2015, July). Machine learning based cognitive skills calculations for different emotional conditions. In *2015 IEEE 14th International Conference on Cognitive Informatics & Cognitive Computing (ICCI* CC)* (pp. 162-168). IEEE. 10.1109/ICCI-CC.2015.7259381

Ahmad, S., Aoun, N. B., El Affendi, M. A., Anwar, M. S., Abbas, S., & Abd El Latif, A. A. (2022). Optimization of Students' Performance Prediction through an Iterative Model of Frustration Severity. *Computational Intelligence and Neuroscience, 2022,* 2022. doi:10.1155/2022/3183492 PMID:36017453

Ahmed, S., Eisa, Al-Wesabi, Elsafi, Al Duhayyim, Yaseen, Hamza, & Motwakel. (2022). *Parkinson's Detection Using RNN-Graph-LSTM with Optimization Based on Speech Signals.* Tech Science Press. doi:10.32604/cmc.2022.024596

Anwar, M. S., Alhalabi, W., Choi, A., Ullah, I., & Alhudali, A. (2024). Internet of Metaverse Things (IoMT): Applications, Technology Challenges and Security Consideration. In *Future Communication Systems Using Artificial Intelligence, Internet of Things and Data Science* (pp. 133–158). CRC Press.

Anwar, M. S., Ullah, I., Ahmad, S., Choi, A., Ahmad, S., Wang, J., & Aurangzeb, K. (n.d.). Immersive Learning and AR/VR-Based Education: Cybersecurity Measures and Risk Management. In Cybersecurity Management in Education Technologies (pp. 1-22). CRC Press.

Anwar, M. S., Wang, J., Ahmad, S., Khan, W., Ullah, A., Shah, M., & Fei, Z. (2020). Impact of the impairment in 360-degree videos on users VR involvement and machine learning-based QoE predictions. *IEEE Access : Practical Innovations, Open Solutions, 8,* 204585–204596. doi:10.1109/ACCESS.2020.3037253

Barman, N., Zadtootaghaj, S., Schmidt, S., Martini, M. G., & Möller, S. (2018, June). GamingVideoSET: a dataset for gaming video streaming applications. In *2018 16th Annual Workshop on Network and Systems Support for Games (NetGames)* (pp. 1-6). IEEE. 10.1109/NetGames.2018.8463362

Boutros, F., Damer, N., Kirchbuchner, F., & Kuijper, A. (2022). Elasticface: Elastic margin loss for deep face recognition. In *Proceedings of the IEEE/CVF conference on computer vision and pattern recognition* (pp. 1578-1587). 10.1109/CVPRW56347.2022.00164

Cao, Q., Shen, L., Xie, W., Parkhi, O. M., & Zisserman, A. (2018, May). Vggface2: A dataset for recognising faces across pose and age. In *2018 13th IEEE international conference on automatic face & gesture recognition (FG 2018)* (pp. 67-74). IEEE. 10.1109/FG.2018.00020

Chen, J., Kellokumpu, V., Zhao, G., & Pietikäinen, M. (2013). RLBP: Robust Local Binary Pattern. BMVC.

Chen, Z., Luo, P., Wang, Y., Chen, Y., & Tang, X. (2021). BlazeFace: Sub-millisecond Neural Face Detection on Mobile GPUs. *Proceedings of the IEEE Conference on Computer Vision and Pattern Recognition (CVPR).*

Deng, J., Guo, J., Xue, N., & Zafeiriou, S. (2019). Arcface: Additive angular margin loss for deep face recognition. In *Proceedings of the IEEE/CVF conference on computer vision and pattern recognition* (pp. 4690-4699). 10.1109/CVPR.2019.00482

Deng, J., Guo, J., Xue, N., & Zafeiriou, S. (2020). RetinaFace: Single-stage dense face localisation in the wild. arXiv preprint arXiv:1905.00641.

Faheem, M. Y., Wang, X., Ahmad, S., Azeem, M. B., & Essani, I. Y. (2019, December). Ultra-low power small size rf transceiver module for 5g/lte applications. In *2019 4th International Conference on Emerging Trends in Engineering, Sciences and Technology (ICEEST)* (pp. 1-6). IEEE.

Feng, N., Song, Z., Yu, J., Chen, Y. P. P., Zhao, Y., He, Y., & Guan, T. (2020). SSET: A dataset for shot segmentation, event detection, player tracking in soccer videos. *Multimedia Tools and Applications*, *79*(39-40), 28971–28992. doi:10.1007/s11042-020-09414-3

Hammouche, R., Attia, A., Akhrouf, S., & Akhtar, Z. (2022). Gabor filter bank with deep autoencoder based face recognition system. *Expert Systems with Applications*, *197*, 116743. doi:10.1016/j.eswa.2022.116743

Haq, M. U., Sethi, M. A. J., Ahmad, S., ELAffendi, M. A., & Asim, M. (2024). Automatic Player Face Detection and Recognition for Players in Cricket Games. *IEEE Access: Practical Innovations, Open Solutions*, *12*, 41219–41233. doi:10.1109/ACCESS.2024.3377564

Haq, M. U., Sethi, M. A. J., & Rehman, A. U. (2023). Capsule Network with Its Limitation, Modification, and Applications—A Survey. *Mach. Learn. Knowl. Extr.*, *5*(3), 891–921. doi:10.3390/make5030047

Haq, M. U., Sethi, M. A. J., Ullah, R., Shazhad, A., Hasan, L., & Karami, G. M. (2022). COMSATS Face: A Dataset of Face Images with Pose Variations, Its Design, and Aspects. *Mathematical Problems in Engineering*, *2022*, 2022. doi:10.1155/2022/4589057

Haq, M. U., Shahzad, A., Mahmood, Z., Shah, A. A., Muhammad, N., & Akram, T. (2019). Boosting the face recognition performance of ensemble based lda for pose, non-uniform illuminations, and low-resolution images. *KSII Transactions on Internet and Information Systems*, *13*(6), 3144–3164.

Hosni, A. I. E., Li, K., & Ahmed, S. (2018). HISBmodel: A rumor diffusion model based on human individual and social behaviors in online social networks. *Neural Information Processing: 25th International Conference, ICONIP 2018, Siem Reap, Cambodia, December 13–16, 2018 Proceedings*, *25*(Part II), 14–27.

Hosni, A. I. E., Li, K., Ding, C., & Ahmed, S. (2018, November). Least cost rumor influence minimization in multiplex social networks. In *International Conference on Neural Information Processing* (pp. 93-105). Cham: Springer International Publishing. 10.1007/978-3-030-04224-0_9

Huang, G. B., Mattar, M., Berg, T., & Learned-Miller, E. (2008, October). Labeled faces in the wild: A database forstudying face recognition in unconstrained environments. *Workshop on faces in 'Real-Life' Images: Detection, Alignment, and Recognition.*

Huang, Y. H., & Chen, H. H. (2022). Deep face recognition for dim images. *Pattern Recognition*, *126*, 108580. doi:10.1016/j.patcog.2022.108580

Jeevan, G., Zacharias, G. C., Nair, M. S., & Rajan, J. (2022). An empirical study of the impact of masks on face recognition. *Pattern Recognition*, *122*, 108308. doi:10.1016/j.patcog.2021.108308

Karpagam, M., Jeyavathana, R. B., Chinnappan, S. K., Kanimozhi, K. V., & Sambath, M. (2022). A novel face recognition model for fighting against human trafficking in surveillance videos and rescuing victims. *Soft Computing*, 1–16.

Karpathy, A., Toderici, G., Shetty, S., Leung, T., Sukthankar, R., & Fei-Fei, L. (2014). Large-scale video classification with convolutional neural networks. In *Proceedings of the IEEE conference on Computer Vision and Pattern Recognition* (pp. 1725-1732). 10.1109/CVPR.2014.223

Kemelmacher-Shlizerman, I., Seitz, S. M., Miller, D., & Brossard, E. (2016). The megaface benchmark: 1 million faces for recognition at scale. In *Proceedings of the IEEE conference on computer vision and pattern recognition* (pp. 4873-4882). 10.1109/CVPR.2016.527

Khan, W., Hua, W., Ayaz, M., Shahid Anwar, M., & Ahmad, S. (2020). A balanced energy efficient (BEE) routing scheme for underwater WSNs. In *Innovative Mobile and Internet Services in Ubiquitous Computing: Proceedings of the 13th International Conference on Innovative Mobile and Internet Services in Ubiquitous Computing (IMIS-2019)* (pp. 82-93). Springer International Publishing.

Khan, W., Wang, H., Anwar, M. S., Ayaz, M., Ahmad, S., & Ullah, I. (2019). A multi-layer cluster based energy efficient routing scheme for UWSNs. *IEEE Access : Practical Innovations, Open Solutions*, 7, 77398–77410. doi:10.1109/ACCESS.2019.2922060

Kolyperas, D., Maglaras, G., & Sparks, L. (2019). Sport fans' roles in value co-creation. *European Sport Management Quarterly*, 19(2), 201–220. doi:10.1080/16184742.2018.1505925

Kowalczewski, P. Ł., Siejak, P., Jarzębski, M., Jakubowicz, J., Jeżowski, P., Walkowiak, K., Smarzyński, K., Ostrowska-Ligęza, E., & Baranowska, H. M. (2023). Comparison of technological and physicochemical properties of cricket powders of different origin. *Journal of Insects as Food and Feed*, 9(5), 637–646. doi:10.3920/JIFF2022.0030

Kshirsagar, V. P., Baviskar, M. R., & Gaikwad, M. E. (2011, March). Face recognition using Eigenfaces. In *2011 3rd International Conference on Computer Research and Development* (Vol. 2, pp. 302-306). IEEE. 10.1109/ICCRD.2011.5764137

Munawar, F., Khan, U., Shahzad, A., Haq, M. U., Mahmood, Z., Khattak, S., & Khan, G. Z. 2019, January. An empirical study of image resolution and pose on automatic face recognition. In *2019 16th International Bhurban Conference on Applied Sciences and Technology (IBCAST)* (pp. 558-563). IEEE. 10.1109/IBCAST.2019.8667233

Najibi, M., Samangouei, P., Chellappa, R., & Davis, L. S. (2018). SSH: Single Stage Headless Face Detector. *Proceedings of the European Conference on Computer Vision (ECCV).*

Palmer, M. C., & King, R. H. (2006). Has salary discrimination really disappeared from Major League Baseball? *Eastern Economic Journal*, 32(2), 285–297.

Pappalardo, L., Cintia, P., Rossi, A., Massucco, E., Ferragina, P., Pedreschi, D., & Giannotti, F. (2019). A public data set of spatio-temporal match events in soccer competitions. *Scientific Data*, *6*(1), 236. doi:10.1038/s41597-019-0247-7 PMID:31659162

Phillips, P. J., Beveridge, J. R., Draper, B. A., Givens, G. H., O'Toole, A. J., Bolme, D. S., Dunlop, J., Lui, Y. M., Sahibzada, H., Weimer, J. A., & ... . (2015). IARPA Janus Benchmark A (IJB-A) Face Dataset. *Proceedings of the IEEE Conference on Computer Vision and Pattern Recognition (CVPR)*.

Phillips, P. J., Wechsler, H., Huang, J., & Rauss, P. J. (1998). The FERET database and evaluation procedure for face-recognition algorithms. *Image and Vision Computing*, *16*(5), 295–306. doi:10.1016/S0262-8856(97)00070-X

Rahim, A., Zhong, Y., Ahmad, T., Ahmad, S., & ElAffendi, M. A. (2023). Hyper-Tuned Convolutional Neural Networks for Authorship Verification in Digital Forensic Investigations. *Computers, Materials & Continua*, *76*(2). Advance online publication. doi:10.32604/cmc.2023.039340

Rahim, A., Zhong, Y., Ahmad, T., Ahmad, S., Pławiak, P., & Hammad, M. (2023). Enhancing smart home security: Anomaly detection and face recognition in smart home iot devices using logit-boosted cnn models. *Sensors (Basel)*, *23*(15), 6979. doi:10.3390/s23156979 PMID:37571762

Rehman, Saba, Mujahid, Alamri, & Elhakim. (2023). *Parkinson's Disease Detection Using Hybrid LSTM-GRU Deep Learning Modell*. Multidisciplinary Digital Publishing Institute. doi:10.3390/electronics12132856

Sim, T., Baker, S., & Bsat, M. (2002, May). The CMU pose, illumination, and expression (PIE) database. In *Proceedings of fifth IEEE international conference on automatic face gesture recognition* (pp. 53-58). IEEE. 10.1109/AFGR.2002.1004130

Sohail, M. Z., Zafar, T., Khan, T. A., Asim, M., Ahmad, S., Mairaj, T., & El Affendi, M. A. (2023). Prediction of Time to Failure (TTF) of Power Systems Using a Deep Learning Technique. *Journal of Hunan University Natural Sciences*, *50*(12).

Tang, S., Wu, W., Yan, J., & Zhang, X. (2018). PyramidBox: A context-assisted single shot face detector. *Proceedings of the European Conference on Computer Vision (ECCV)*. 10.1007/978-3-030-01240-3_49

Ullah, A., Ullah, A., Ahmad, S., Haq, M. U., Shah, K., & Mlaiki, N. (2022). Series type solution of fuzzy fractional order Swift–Hohenberg equation by fuzzy hybrid Sumudu transform. *Mathematical Problems in Engineering*, *2022*, 2022. doi:10.1155/2022/3864053

Ullah, H., Haq, M. U., Khattak, S., Khan, G. Z., & Mahmood, Z. (2019, August). A robust face recognition method for occluded and low-resolution images. In *2019 International Conference on Applied and Engineering Mathematics (ICAEM)* (pp. 86-91). IEEE. 10.1109/ICAEM.2019.8853753

Wang, Z., Bovik, A. C., Sheikh, H. R., & Simoncelli, E. P. (2004). Image quality assessment: From error visibility to structural similarity. *IEEE Transactions on Image Processing*, *13*(4), 600–612. doi:10.1109/TIP.2003.819861 PMID:15376593

Wickramasinghe, I. (2022). Applications of machine learning in cricket: A systematic review. *Machine Learning with Applications*, *10*, 100435. doi:10.1016/j.mlwa.2022.100435

Wu, Z., Huang, Y., Zhang, L., & Wang, L. (2021). FaceDet: A Lightweight Anchor-Free Face Detector for Mobile Devices. *Proceedings of the IEEE Conference on Computer Vision and Pattern Recognition (CVPR)*.

Yu, H., & Yang, J. (2001). A direct LDA algorithm for high-dimensional data—With application to face recognition. *Pattern Recognition*, *34*(10), 2067–2070. doi:10.1016/S0031-3203(00)00162-X

Yu, S., Jiang, Y., Lu, J., & Zhou, J. (2019). CenterFace: Joint Face Detection and Alignment Using Face as Point. *Proceedings of the IEEE Conference on Computer Vision and Pattern Recognition (CVPR)*.

Zhang, J., Shan, S., Kan, M., & Chen, X. (2014). WebFace: A Scalable Face Image Dataset with Varying Pose and Age. *Proceedings of the IEEE Conference on Computer Vision and Pattern Recognition (CVPR)*.

Chapter 9
Exploring COVID–19 Classification and Object Detection Strategies:
X–Rays Image Processing

Saifullah Jan
City University, Peshawar, Pakistan

Aiman
City University, Peshawar, Pakistan

Bilal Khan
ⓘ https://orcid.org/0000-0002-6816-3776
City University, Peshawar, Pakistan

Muhammad Arshad
City University, Peshawar, Pakistan

ABSTRACT

The overlapping imaging characteristics of COVID-19 viral pneumonia and non-COVID-19 viral pneumonia chest X-rays (CXRs) make differentiation difficult for radiologists. Machine learning (ML) has demonstrated promising outcomes in a range of medical sectors, enhancing diagnostic accuracy through its interaction with radiological tests. The potential contribution of ML models in assisting radiologists in discriminating COVID-19 from non-COVID-19 viral pneumonia from CXRs, on the other hand, deserves further examination and exploration. The goal of this study is to empirically assess ML models' capacity to classify X-ray images into COVID-19, pneumonia, and normal cases. The study evaluates the efficacy of K-nearest Neighbor (KNN), random forest (RF), AdaBoost (AB), and neural networks (NN) with various hidden neuron configurations using a wide range of performance measures. These metrics evaluate the area under the curve (AUC), classification accuracy (CA), F1 score (F1), precision, and recall, resulting in a comprehensive evaluation technique. ROC analysis is used to gain a thorough knowledge of the models' discriminating skills. The results show that NN models, particularly those

DOI: 10.4018/979-8-3693-2913-9.ch009

with 100 and 150 hidden neurons, outperform in all criteria, proving their ability to reliably categorize medical disorders. Notably, the study emphasizes the difficulties in separating COVID-19 from pneumonia, emphasizing the importance of strong classification methods. While the study provides useful insights, its drawbacks include the use of a single dataset, the absence of more sophisticated deep learning architectures, and a lack of interpretability analyses. Nonetheless, the study adds to the developing area of medical picture categorization, directing future attempts to improve diagnosis accuracy and widen the use of machine learning in healthcare. The findings highlight the utility of NN models in medical diagnostics and pave the way for future study in this vital area of technology and healthcare.

1. INTRODUCTION

The infectious disease termed COVID-19 is caused by the severe acute respiratory syndrome coronavirus 2 (SARS-CoV-2). This virus induces pronounced respiratory symptoms and was initially identified in Wuhan, China, in 2019. Following its rapid global dissemination, it was officially designated as a worldwide pandemic (Cucinotta and Vanelli, 2020). As of October 15, 2021, the COVID-19 Dashboard maintained by John Hopkins University has documented an estimated 240 million confirmed cases of infection and approximately 4.8 million recorded fatalities on a global scale (Lauer et al., 2020; Jaiswal et al., 2021). The cardinal manifestations of COVID-19 encompass elevated body temperature, fatigue, respiratory distress, and impairment of gustatory or olfactory senses. In instances of heightened severity, the disease can give rise to respiratory complications, particularly pneumonia, warranting hospitalization and occasionally necessitating admission into intensive care units (Lauer et al., 2020).

Distinguishing between COVID-19 pneumonia and typical pneumonia based on clinical attributes can present challenges. Both conditions exhibit akin signs and symptoms, including fever, fatigue, non-productive cough, and respiratory distress. The elevated morbidity and mortality rates linked to COVID-19 pneumonia have imposed a substantial burden on healthcare infrastructures (Jaiswal et al., 2021). To curb the propagation of the pandemic and optimize the allocation of medical resources, it is imperative to swiftly diagnose and isolate individuals afflicted with either common pneumonia or COVID-19.

Despite the convergence of symptoms and diagnostic complexities, the radiographic representations obtained through Computed Tomography (CT) for general pneumonia and COVID-19 manifest resemblances. This overlapping imagery further compounds the intricacy of effectively discerning between the two maladies (Cheng et al., 2019).

The identification of COVID-19 is commonly accomplished through real-time polymerase chain reaction (RT-PCR) testing for the presence of the SARS-CoV-2 virus. While RT-PCR exhibits commendable specificity for COVID-19, its sensitivity for accurately detecting cases of the disease has been relatively lower (Nishio et al., 2020). Notably, chest computed tomography (CT) scans have demonstrated utility in identifying atypical conditions associated with COVID-19 pneumonia. The distinctive CT findings pertinent to COVID-19 offer a basis for differentiation from viral and bacterial pneumonia presentations (Bai et al., 2020).

Numerous recent investigations have aimed to differentiate between COVID-19 and pneumonia, each harboring inherent limitations and strengths. This study, however, presents an empirical scrutiny of diverse machine learning (ML) models, including K-nearest Neighbor (KNN), Random Forest (RF), AdaBoost (AB), and Neural Networks (NN) with concealed layers containing 100 and 150 neurons. Employing a dataset sourced from the Kaggle repository, these models are evaluated utilizing state-of-the-art performance metrics encompassing the area under the curve (AUC), classification accuracy (CA), F1 Score (F1), precision, and recall.

The study makes a substantial contribution to the domain of medical image classification by addressing the crucial challenge of classifying individuals into COVID-19, Pneumonia, and Normal categories based on X-ray images. The research undertakes a comprehensive empirical analysis to evaluate the effectiveness of several machine learning (ML) models, encompassing K-nearest Neighbor (KNN), Random Forest (RF), AdaBoost (AB), and Neural Network (NN) with various configurations. This meticulous investigation is underpinned by a range of well-established evaluation metrics, including Area under the Curve (AUC), Classification Accuracy (CA), F1 Score, accuracy, and recall. These metrics collectively provide a comprehensive understanding of the strengths and limitations intrinsic to each model's diagnostic capacities for medical conditions.

Moreover, the study employs Receiver Operating Characteristic (ROC) analysis to provide intuitive insights into the discriminatory capabilities of the models across distinct classes. The novel application of these advanced algorithms to healthcare challenges underscores the potency of machine learning in the realm of medical diagnostics. The research also proposes intriguing avenues for future exploration, such as the development of more sophisticated deep-learning architectures and the utilization of larger datasets.

2. LITERATURE REVIEW

The emergence of the SARS-CoV-2 virus, which gives rise to the novel viral pneumonia named COVID-19, was initially documented in China in December 2019. The rapid escalation in the number of individuals affected globally has propelled this ailment into a worldwide public health emergency of substantial magnitude (Guan et al., 2020). In response to the expansive dissemination of COVID-19, nations across the world are implementing comprehensive measures to curtail its propagation and reinstate societal equilibrium. Governments have taken decisive actions, including the imposition of social constraints and extensive public outreach initiatives, as a testament to the gravity posed by the COVID-19 threat (Nishio et al., 2020).

Moreover, the expeditious identification of individuals displaying symptoms is of paramount significance to mitigate future transmission. This necessitates the application of diverse diagnostic strategies, encompassing rapid testing, real-time polymerase chain reaction (RT-PCR) assays, and the utilization of chest X-ray or computed tomography (CT) scans for prompt detection and isolation of COVID-19 cases (Cenggoro and Pardamean, 2023). Across the global spectrum, governments are compelled to enhance their diagnostic capabilities to effectively confront the ongoing epidemic.

While rapid tests hold the potential to detect positive cases, their accuracy in isolation may fall short. Supplementary assessments, such as real-time polymerase chain reaction (RT-PCR) tests, are imperative to substantiate the initial findings. RTPCR currently stands as the definitive diagnostic standard, although its utility is not devoid of constraints, including notable rates of false negatives and considerable time demands. Research findings (Ng et al., 2020) underscore the capability of both chest X-rays and chest CT scans to identify pulmonary anomalies attributed to COVID-19. The execution of such diagnostic procedures necessitates the expertise of skilled radiologists. However, the scarcity of radiological professionals, coupled with the escalating caseload of COVID-19 incidents, has accentuated the exigency for an auxiliary system that can provide support to radiologists and healthcare practitioners (Shankar and Perumal, 2021).

Enhancements in radiologists' diagnostic precision can be facilitated through the integration of computer-aided diagnostic (CAD) technologies. In the realm of detection, scholars employ either manually crafted or synthetically generated features extracted from lung morphology, textural attributes, and anatomical traits. Nonetheless, the pivotal and intricate facet of this strategy persists in selecting an optimal classifier adept at accommodating the unique attributes inherent to lung feature spaces (Hussain et al., 2020).

The cultivation of machine learning methodologies within medical imaging holds promises for fostering a more precise and expedient identification of COVID-19 cases. Such techniques have the potential to complement prevailing COVID-19 detection methodologies, offering the prospect of refined diagnostic outcomes.

In recent epochs, Deep Learning has emerged as a preeminent technique within the purview of machine learning. However, its efficacy is inextricably linked to voluminous datasets, as smaller datasets are susceptible to overfitting. To ameliorate this quandary, scholars have embraced transfer learning, a paradigm that harnesses pre-trained models from the ImageNet dataset (Adhinata et al., 2021). Jaiswal et al. (2021) exemplified this approach through the utilization of pre-trained models such as DenseNet201 (Huang et al., 2017), VGG16 (Simonyan and Zisserman, 2015), ResNet152V2 (He et al., 2016), and InceptionResNetV2 (Szegedy et al., 2017). The acme of performance, attaining a remarkable accuracy of 96.25%, was attained via the employment of DenseNet201.

Panwar et al. (2020) embarked on a distinct trajectory, employing chest X-Ray and chest CT images as inputs. They harnessed the pre-trained VGG19 model for classification, introducing alterations by incorporating 5 MaxPool Layers and 1 SoftMax Layer. This endeavor culminated in the achievement of a precision level reaching 95.61%. Correspondingly, a comparable study that embraced the pre-trained VGG19 model secured an accuracy of 93.01% (Pathak et al., 2022).

Ismael and Sengur (2021) delved into the realm of deep feature extraction via transfer learning, deploying ResNet18, ResNet50, ResNet101, VGG16, and VGG19. The Support Vector Machine (SVM) (Noble, 2006) was enlisted as the classifier, culminating in ResNet50+SVM yielding the pinnacle accuracy of 94.74%.

The prevalence of transfer learning methodologies pervades various investigations. Shah et al. (2021), for instance, adroitly harnessed transfer learning, augmenting VGG19 by excising the final two fully connected (FC) layers and replacing them with two FC layers comprising 4096 neurons each. This adaptation translated to a commendable precision level of 94.50%. In an independent inquiry, scholars undertook the training of pre-existing models such as ResNet18, ResNet50, COVID-Net, and DenseNet121. Notably, among these models, ResNet18 emerged triumphant with the attainment of the highest accuracy rate, standing at 91.00% (Tartaglione et al., 2020).

3. EXPERIMENTAL SETUP AND METHODOLOGY

The principal aim of this research is to introduce an enhanced classification model designed to differentiate between instances of COVID-19 and pneumonia through the analysis of chest X-ray images. The research investigations were systematically conducted on a Windows 10 operating system characterized by a 64-bit architecture. The analytical framework was facilitated by employing Orange3 version 3.27.1-Miniconda-x86_64. The overarching procedural outline followed in this study is delineated in Fig. 1, encompassing a comprehensive trajectory commencing from the procurement of data and extending to the evaluation of the model's performance. Further elucidation of the constituent phases is subsequently provided.

Figure 1. Experimental methodology

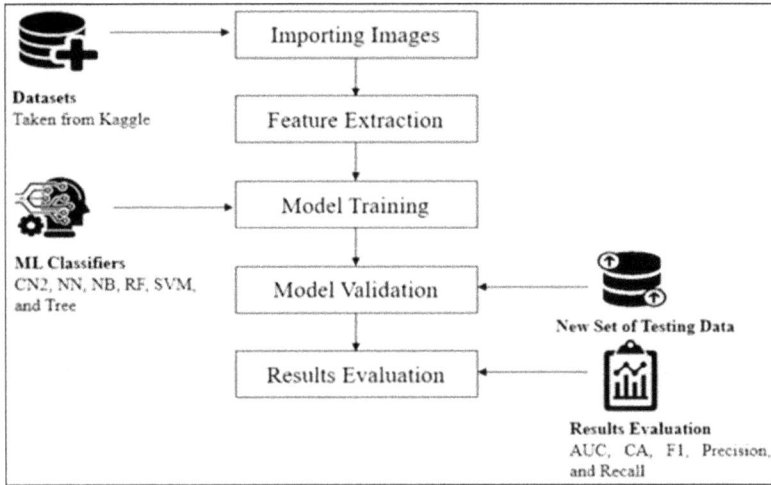

3.1 Data, Training, and Testing

The dataset utilized in this investigation was sourced from Kaggle[1], encompassing a total of 4970 instances characterized by the absence of any missing data. Each instance is characterized by a multitude of 1000 features. The dataset's three designated target features encompass "normal," "pneumonia," and "COVID-19."

To assess the efficacy of each aforementioned technique, a 10-fold crossvalidation methodology was adopted. This entailed the partitioning of the dataset into ten equivalent subsets. During this procedure, nine subsets were allocated for model training, while the remaining subset was dedicated to the evaluation of the trained models' performance.

3.2 Performance Assessments

The evaluative criteria employed for performance assessment within this study encompassed several metrics, namely, AUC, CA, Precision, Recall, and the F1 measure.

A prominent aggregate performance metric employed herein is the Area under the Receiver Operating Characteristic Curve. This metric serves as a comprehensive gauge of a model's capability to effectively differentiate between positive and negative instances across an array of classification thresholds. The Area under the Receiver Operating Characteristic Curve yields a singular value that encapsulates the overarching accuracy of predictions engendered by the classification model (Khan et al., 2021).

Furthermore, the CA metric was enlisted to quantify the proportion of instances within the dataset that were correctly categorized. This straightforward metric serves as an indicator of the model's predictive precision and efficacy. CA can be calculated as follows:

$$CA = TP + TN / TP + FP + TN + FN \tag{1}$$

Here, TP denotes true positives, FP stands for false positives, TN represents true negatives, and FN corresponds to false negatives.

When confronting scenarios characterized by imbalanced classes, the evaluation of a predictor's performance necessitates due consideration of precision and recall as pivotal measures. Recall, in particular, serves as a litmus test for the predictor's proficiency in accurately identifying all pertinent outcomes. In contrast, precision scrutinizes the fidelity of the projected outcomes. These metrics bear pronounced significance contingent upon the nature of the classification task, serving as fundamental yardsticks to discern the model's efficacy in accurately detecting positive instances and concurrently mitigating instances of false positives or false negatives (Naseem et al., 2020).

$$Precision = TP / TP + FP \tag{2}$$

$$Recall = TP / TP + FN \tag{3}$$

The F1 score calculates predictive accuracy by determining a weighted harmonic mean between precision and recall, formulated as follows:

$$F1 = 2 * precision * recall / precision + recall \tag{4}$$

3.3 Employed Machine Learning Models

This study employs a comprehensive selection of four distinct machine learning (ML) models, which encompass the following:

3.3.1 K-Nearest Neighbours

K-Nearest Neighbors is a fundamental supervised machine learning algorithm suited for both classification and regression tasks. It functions by allocating objects within the feature space into categories based on their proximity to the training instances (Al-Areqi and Konyar, 2022). The fundamental variant of the KNN technique arises when K is designated as 1, signifying that the classification of the test sample hinges upon its closest neighbor within the training dataset. In this research, we specifically consider a K value of 5 for the number of neighbors. The measure of distance between two instances, denoted as p and q, is calculated via the utilization of the Euclidean Distance Formula:

$$\text{Euclidian Distance } d\left(p,q\right) = \sqrt{\sum\nolimits_{i=1} k\left(p_i - q_i\right)^2} \qquad (5)$$

3.3.2 Neural Networks

Neural networks (NNs) represent computational constructs designed to emulate the operational principles observed in the natural neural networks within the human brain and other species. These networks comprise interconnected components, facilitating intercommunication among them. By their adaptability, Artificial Neural Networks (ANNs) are capable of dynamically adjusting their configuration during the learning process in response to the data they are processing (Morra et al., 2010).

Specifically, the neural network employed in this context encompasses a concealed layer housing 100 neurons, where the Rectified Linear Unit (ReLu) serves as the designated activation function. The training process is bounded by a maximum of 200 iterations, and regularization is governed by a value of 0.0001. For the hidden layers, the study selects a neuron count of 100 and 150, respectively, and adheres to a uniform maximum iteration count of 200 across both cases.

3.3.3 Random Forest

Renowned for its user-friendliness and propensity to yield favorable outcomes with minimal fine-tuning, this algorithm enjoys popularity. It functions by amalgamating numerous decision trees, each accounting for a limited subset of attributes at individual decision junctures. This algorithm is also characterized by the adoption of a rule that halts the continued subdivision of subgroups containing fewer than five instances (Zamir et al., 2020). In the context of this study, the algorithm's configuration entailed the inclusion of 10 decision trees, and at each bifurcation, a set of 5 attributes were taken into consideration.

3.3.4 AdaBoost

Combining many weak classifiers to build a strong classifier is a powerful and efficient strategy. "Weak" in this context refers to classifiers with comparatively poor accuracy and performance. Classification issues are the main use of this technology. A random subset of the overall training set is chosen as the training set for each new classifier in the process. After each classifier is created, its weights are chosen based on how accurate it is (Morra et al., 2010; Abdullah et al., 2022; Anwar et al., 2019; Ahmad et al., 2018, 2022; Rahim et al., 2023). This can be determined mathematically using Eq. (6), where more precise classifiers are given greater weights.

$$H\left(x\right) = sign\left(\sum_{t=1}^{T} a_1 h_1\left(x\right)\right) \tag{6}$$

The final classifier indicated as "T," is made up of weak classifiers. "H t (x)" stands for the output of a specific weak classifier, "t," and "alpha t" stands for the weight that the AdaBoost method gives classifier "t." The final result is produced by executing a linear combination of all the outputs from weak classifiers, and the choice is decided by evaluating the sign of this summation. In other words, the final classification is determined by the sum of the results from the weak classifiers.

4. RESULTS ANALYSIS AND DISCUSSION

This section presents the overall outcomes achieved via the aforementioned models for COVID-19 and pneumonia classification. All the outcomes are firstly assessed via a confusion matrix (CM) which is a tabular representation that is employed when evaluating the effectiveness of a classification model using ML. It compares a set of data points' real labels to the anticipated classifications of a model for each data point. A range of measures, including accuracy, precision, recall, and F1-score, may be evaluated using the matrix's insights into the number of TP, TN, FP, and FN predictions. Fig. 2 shows the CM values of each employed model represented with labels A, B, C, D, and E respectively for KNN, AB, NN (100 NH), NN (150 NH), and RF.

The performance evaluation of various ML models for categorizing X-ray pictures into the COVID-19, Pneumonia, and Normal categories is shown in Table 1. AUC, CA, F1 Score, Precision, and Recall are only a few of the important metrics used in

Figure 2. Confusion matrix achieved via each employed model

A — KNN

Actual \ Predicted	COVID19	NORMAL	PNEUMONIA	Σ
COVID19	213	11	62	286
NORMAL	3	1121	142	1266
PNEUMONIA	7	143	3268	3418
Σ	223	1275	3472	4970

B — AdaBoost

Actual \ Predicted	COVID19	NORMAL	PNEUMONIA	Σ
COVID19	209	18	59	286
NORMAL	13	992	261	1266
PNEUMONIA	69	255	3094	3418
Σ	291	1265	3414	4970

C — Neural Network (100 NH)

Actual \ Predicted	COVID19	NORMAL	PNEUMONIA	Σ
COVID19	268	4	14	286
NORMAL	1	1167	98	1266
PNEUMONIA	12	83	3323	3418
Σ	281	1254	3435	4970

D — Neural Network (150 NH)

Actual \ Predicted	COVID19	NORMAL	PNEUMONIA	Σ
COVID19	268	4	14	286
NORMAL	3	1170	93	1266
PNEUMONIA	11	77	3330	3418
Σ	282	1251	3437	4970

E — Random Forest

Actual \ Predicted	COVID19	NORMAL	PNEUMONIA	Σ
COVID19	224	9	53	286
NORMAL	5	1055	206	1266
PNEUMONIA	22	172	3224	3418
Σ	251	1236	3483	4970

A = KNN
B = AdaBoost
C = Neural Network (100 NH)
D = Neural Network (150 NH)
E = Random Forest

the assessment (Khan et al., 2022; Akhtar et al., 2022; Ahmad et al., 2022; Fatima et al., 2022; Hosni et al., 2018a, 2018b, 2019; Anwar, 2018). The NN models stand out as obvious frontrunners with great AUC, CA, and F1 scores that are 0.998, 0.994, and 0.945 respectively, highlighting their effectiveness in classification tasks. This is especially true of the model with 150 hidden neurons. These NN models show a balanced precision-recall trade-off, demonstrating their ability to recognize real positives and true negatives. Similar to its sibling with 150 hidden neurons, the NN with 100 hidden neurons exhibits parallel performance. However, while earning decent AUC and CA, AB has space for improvement, as seen by its significantly lower F1 score. Although its accuracy and recall balance out reasonably, they fall short of the levels attained by the models indicated above. Overall, NN models outperform other models in this classification challenge, especially those with 150 and 100 hidden neurons, having high AUC, CA, and F1 scores as well as a pleasing precision-recall balance. Strong capabilities are also shown by the Random Forest model, although KNN and AB have slightly lower F1 scores, suggesting possible directions for improving their precision-recall balance. The performance of each model used for the classification of pneumonia in a dataset consisting of X-Ray images of COVID-19, Pneumonia, and Normal patients is shown in Table 2. It is noteworthy that the KNN and RF models have identical AUC values, which suggests equal discriminatory strength. With continuously high AUC, CA, F1 score, accuracy, and recall, the NN models with 150 and 100 hidden neurons demonstrate remarkable performance across all parameters, highlighting their potency in differentiating pneumonia patients with 0.961 CA value and 0.990 AUC value. While the AUC, CA, and F1 scores of AB are substantially lower, its precision and recall are still closely matched, indicating balanced performance.

An evaluation of the models used to categorize common X-ray pictures within a dataset is shown in Table 3. Although the RF model's F1 score and recall are considerably lower, it still has a competitive AUC and classification accuracy, suggesting possible challenges in precisely distinguishing normal instances. The NN models with 150 and 100 hidden neurons perform remarkably well, producing consistently excellent AUC, CA, F1 score, accuracy, and recall with scores of 0.991, 0.964, 0.930, 0.935, and 0.924 respectively, reiterating their ability to differentiate between typical X-ray pictures. Even though AB's performance was inferior, it nevertheless showed consistency in detecting typical instances.

Table 4 provides a thorough analysis of each method used to divide X-ray pictures into the COVID-19, Pneumonia, and Normal categories. With comparable

Table 1. Models evaluation through each employed assessment, measure for covid-19 classification

Model	AUC	CA	F1	Precision	Recall
KNN	0.956	0.983	0.837	0.955	0.745
RF	0.987	0.982	0.834	0.892	0.783
NN (NH 100)	0.998	0.994	0.945	0.954	0.937
NN (NH 150)	0.998	0.994	0.944	0.95	0.937
AB	0.857	0.968	0.724	0.718	0.731

Table 2. Models evaluation through each employed assessment, measure for pneumonia classification

Model	AUC	CA	F1	Precision	Recall
KNN	0.962	0.929	0.949	0.941	0.956
RF	0.962	0.909	0.934	0.926	0.943
NN (NH 100)	0.990	0.958	0.970	0.967	0.977
NN (NH 150)	0.990	0.961	0.972	0.969	0.974
AB	0.850	0.870	0.906	0.906	0.905

Table 3. Models evaluation through each employed assessment, measure for normal x-rays classification

Model	AUC	CA	F1	Precision	Recall
KNN	0.968	0.940	0.882	0.879	0.885
RF	0.964	0.921	0.843	0.854	0.833
NN (NH 100)	0.991	0.963	0.926	0.931	0.922
NN (NH 150)	0.991	0.964	0.930	0.935	0.924
AB	0.855	0.890	0.784	0.784	0.784

AUC values, the KNN and RF models perform similarly, but KNN significantly surpasses RF in terms of classification accuracy, F1 score, precision, and recall. Notably, the NN models outperform all other models, especially those with 100 and 150 hidden neurons. They outperform all other models in terms of AUC (0.992), CA (0.959), F1 score (0.959), accuracy (0.959), and recall (0.959), demonstrating their exceptional skills to accurately identify COVID-19, Pneumonia, and Normal cases in X-ray pictures. The effectiveness of NN in this application can be due to their capacity to identify complicated correlations and patterns in large amounts of data, allowing them to extract useful information from X-ray pictures that would not be readily recognizable using conventional techniques. Due to their deep design, they can automatically learn hierarchical representations, which helps explain why they do better than other systems in this categorization challenge. The NN models also stand out because of their innate ability to extract subtle information from X-Ray pictures, producing exceptionally better accuracy and consistency across all assessment measures.

The models used are thoroughly compared in Table 5 based on the AUC, which shows the likelihood that the model in the row will have a higher score than the model in the column. When the AUC difference is minimal, it shows that there is

Table 4. Models evaluation through each employed assessment, measure average over disease x-rays classification

Model	AUC	CA	F1	Precision	Recall
KNN	0.964	0.926	0.925	0.926	0.926
RF	0.966	0.906	0.905	0.905	0.906
NN (NH 100)	0.991	0.957	0.957	0.957	0.957
NN (NH 150)	0.992	0.959	0.959	0.959	0.959
AB	0.853	0.864	0.864	0.864	0.864

probably not much of a performance difference between the two models. Because a model is always guaranteed to be better than itself, the diagonal components, which represent self-comparisons, produce a perfect AUC of 1.000. It is clear from looking at the offdiagonal components that the AUC values for the KNN and RF models are significantly lower than those for the NN models with 100 and 150 hidden neurons and the AB model. This demonstrates that the NN models beat KNN and RF in terms of AUC, especially those with 100 and 150 hidden neurons. But in every comparison, the AB model continuously displays the lowest AUC, indicating that it performs rather poorly in this particular situation. These results offer a clear understanding of the relative ranking of the models that were used based on their AUC values, demonstrating the superiority of the NN models and the relative underperformance of the AB models.

An effective method for assessing the effectiveness of classification models is ROC (Receiver Operating Characteristic) analysis, particularly in cases when there are numerous classes involved, as in the classification of COVID-19, Pneumonia, and Normal X-ray pictures. The trade-off between the genuine positive rate (sensitivity) and the false positive rate (1 - specificity) at different threshold values is graphically represented by the ROC curve. While the other classes are merged to form the negative class, each class is handled as the positive class in turn. In this situation, ROC analysis can provide insight into how effectively a classification model can differentiate between the various classes. The model's capacity to distinguish between each class and the total negative class is shown by the ROC curve for that class. As in the above instance NN, a successful model will have ROC curves that are around the upper left corner of the Fig., showing high sensitivity and low false positive rates across various thresholds. Fig. 3, 4, and 5 respectively show the ROC analyses for the target classes of COVID-19, pneumonia, and normal.

Table 5. Model comparison by AUC for average over classes (COVID-19, pneumonia, and normal)

Models		KNN	RF	NN NH)	(100	NN NH)	(150	AB
KNN		-	0.239	0.000		0.000		1.000
RF		0.761	-	0.000		0.000		1.000
NN	(100							
NH)		1.000	1.000	-		0.257		1.000
NN	(150							
NH)		1.000	1.000	0.743		-		1.000
AB		0.000	0.000	0.000		0.000		-

While the study provides useful insights into the use of ML models for categorizing COVID-19, Pneumonia, and Normal patients using X-ray images, several limitations must be acknowledged. One disadvantage is the exclusive emphasis on a particular Kaggle dataset, which may restrict the findings' generalizability to varied patient groups and imaging methods. Furthermore, the study uses classic ML models alongside neural networks without investigating more sophisticated deep learning architectures or ensemble approaches, which might give even greater levels of accuracy (these are included in the study's future direction). Furthermore, the evaluation focuses largely on performance indicators like AUC, CA, F1 Score, and others, possibly disregarding interpretability factors that are critical in medical applications. Despite these limitations, the study provides a firm basis and useful insights, paving the way for more research and development in the field of medical picture categorization.

5. CONCLUSION

This paper presents a comprehensive exploration of the classification of COVID-19, pneumonia, and normal X-ray images through the utilization of various ML models, accompanied by a meticulous empirical investigation of

Figure 3. ROC analyses for target class COVID-19

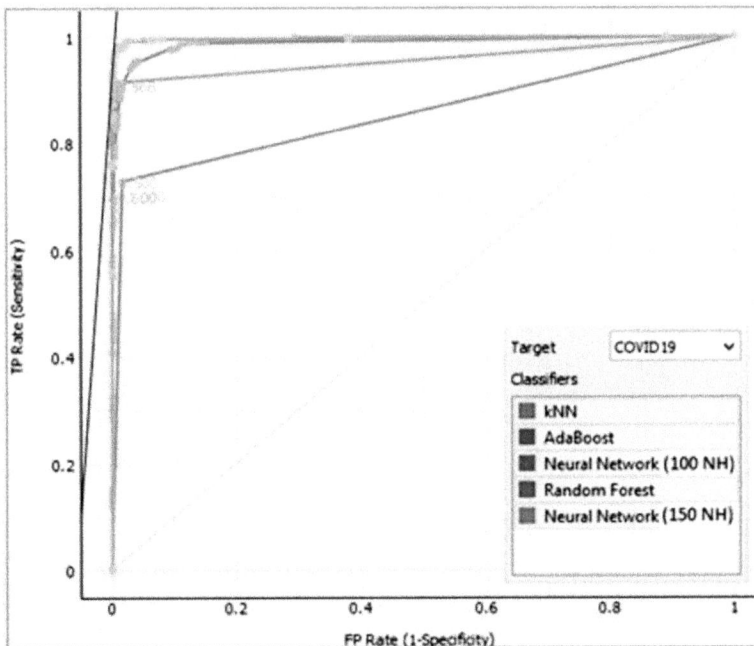

Figure 4. ROC analyses for target class pneumonia

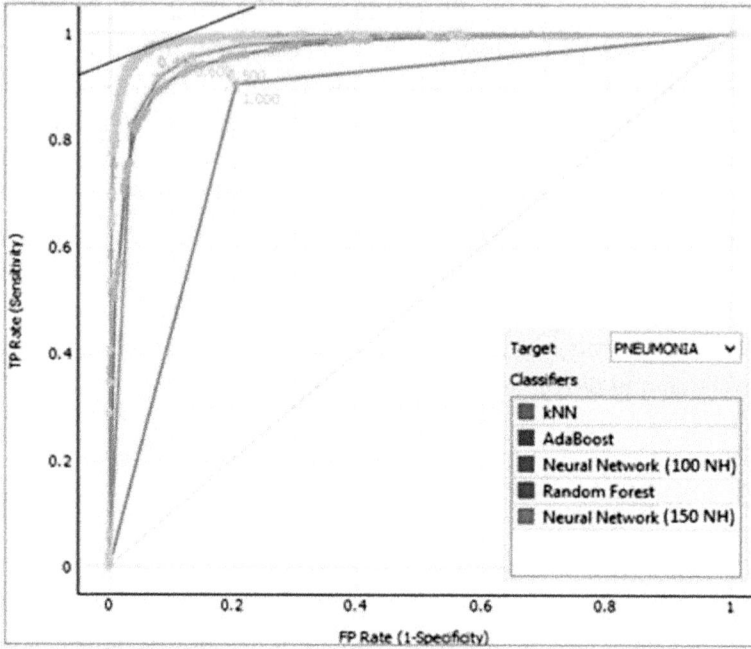

Figure 5. ROC analyses for target class normal

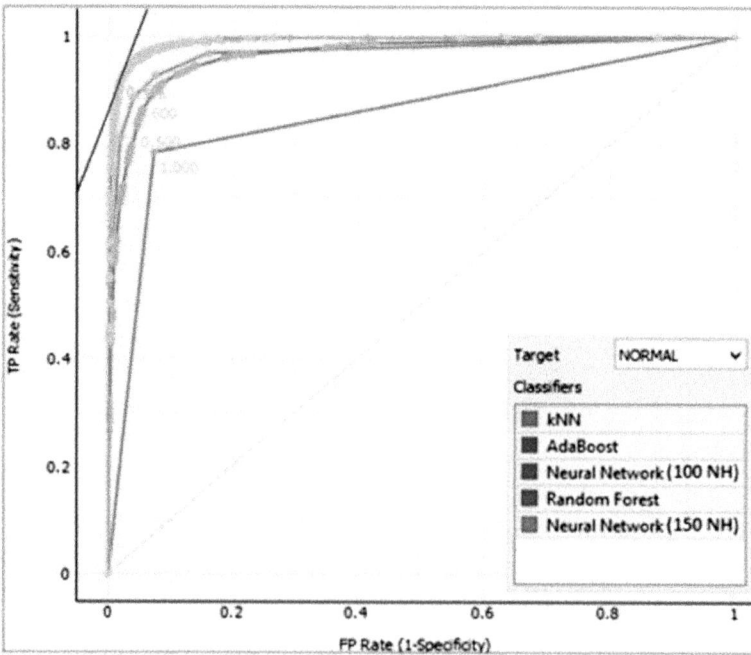

these models. The study meticulously scrutinized the performance of KNN, RF, AB, and NN architectures, employing hidden neuron counts of both 100 and 150. The efficacy of these models was subjected to a rigorous assessment encompassing an array of evaluation metrics, including AUC, CA, F1 Score, Precision, and Recall. Particularly noteworthy, the NN models incorporating 100 and 150 hidden neurons consistently outperformed their counterparts, showcasing their remarkable potential in accurately categorizing diverse medical conditions. Demonstrating commendable outcomes with elevated AUC values of 0.992, CA values of 0.959, and F1 scores, these NN models showcased enhanced performance across multiple evaluation metrics. Moreover, they demonstrated balanced precision-recall levels of 0.959 each, coupled with elevated AUC and CA values. The Receiver Operating Characteristic (ROC) analyses further validated the efficacy of the NN models in class differentiation, as their curves prominently resided in the upper left corner, substantiating their capability to achieve heightened sensitivity while minimizing false positive rates.

6. FUTURE WORK

Numerous avenues for future research endeavours can be pursued to extend and enhance the insights gleaned from this study. A pivotal direction involves delving into more intricate deep learning architectures, such as convolutional neural networks (CNNs), which hold the potential to yield even more heightened performance in tasks involving image classification, as presented in this study. The utilization of pretrained CNN models, leveraged across expansive medical imaging datasets, has the potential to augment the models' generalization capabilities through the application of transfer learning techniques. The augmentation of datasets, both in terms of diversity and size, has the potential to render the models more robust and dependable in their performance. Additionally, an exploration into ensemble methodologies, amalgamating the strengths of multiple models, could potentially elevate classification performance to a more advanced level.

REFERENCES

Abdullah, F. B., Iqbal, R., Ahmad, S., El-Affendi, M. A., & Abdullah, M. (2022). An empirical analysis of sustainable energy security for energy policy recommendations. *Sustainability (Basel)*, *14*(10), 6099. doi:10.3390/su14106099

Adhinata, F. D., Rakhmadani, D. P., Wibowo, M., & Jayadi, A. (2021). A Deep Learning Using DenseNet201 to Detect Masked or Non-masked Face. *JUITA: Jurnal Informatika, 9*(1), 115. doi:10.30595/juita.v9i1.9624

Ahmad, S., El-Affendi, M. A., Anwar, M. S., & Iqbal, R. (2022). Potential future directions in optimization of students' performance prediction system. *Computational Intelligence and Neuroscience.* doi:10.1155/2022/6864955 PMID:35619762

Ahmad, S., Li, K., Eddine, H. A. I., & Khan, M. I. (2018). A biologically inspired cognitive skills measurement approach. *Biologically Inspired Cognitive Architectures, 24,* 35-46.

Akhtar, S., Ali, A., Ahmad, S., Khan, M. I., Shah, S., & Hassan, F. (2022). The prevalence of foot ulcers in diabetic patients in Pakistan: A systematic review and meta-analysis. *Frontiers in Public Health, 10,* 1017201. doi:10.3389/fpubh.2022.1017201 PMID:36388315

Al-Areqi, F., & Konyar, M. Z. (2022). Effectiveness evaluation of different feature extraction methods for classification of covid-19 from computed tomography images: A high accuracy classification study. *Biomedical Signal Processing and Control, 76,* 103662. Advance online publication. doi:10.1016/j.bspc.2022.103662 PMID:35350595

Anwar, M. S., Wang, J., Ullah, A., Khan, W., Ahmad, S., & Li, Z. (2019, December). Impact of stalling on QoE for 360-degree virtual reality videos. In *2019 IEEE International Conference on Signal, Information and Data Processing (ICSIDP)* (pp. 1-6). IEEE. 10.1109/ICSIDP47821.2019.9173042

Anwar, M. S., Wang, J., Ullah, A., Khan, W., Li, Z., & Ahmad, S. (2018, October). User profile analysis for enhancing QoE of 360 panoramic video in virtual reality environment. In *2018 International Conference on Virtual Reality and Visualization (ICVRV)* (pp. 106-111). IEEE. 10.1109/ICVRV.2018.00030

Cheng, Z., Lu, Y., Cao, Q., Qin, L., Pan, Z., Yan, F., & Yang, W. (2020). Clinical features and chest CT manifestations of coronavirus disease 2019 (COVID-19) in a single-center study in Shanghai, China. *AJR. American Journal of Roentgenology, 215*(1), 121–126. doi:10.2214/AJR.20.22959 PMID:32174128

Cucinotta, D., & Vanelli, M. (2020). WHO declares COVID-19 a pandemic. *Acta Biomedica, 91*(1), 157–160. doi:10.23750/abm.v91i1.9397 PMID:32191675

Daniel, T. W. C., & Pardamean, B. (2023). A systematic literature review of machine learning application in COVID-19 medical image classification. *Procedia Computer Science, 216,* 749–756. doi:10.1016/j.procs.2022.12.192 PMID:36643182

Fatima, R., Samad Shaikh, N., Riaz, A., Ahmad, S., El-Affendi, M. A., Alyamani, K. A., Nabeel, M., Ali Khan, J., Yasin, A., & Latif, R. M. A. (2022). A natural language processing (NLP) evaluation on COVID-19 rumour dataset using deep learning techniques. *Computational Intelligence and Neuroscience*, *2022*, 2022. doi:10.1155/2022/6561622 PMID:36156967

Guan, W., Ni, Z., Hu, Y., Liang, W., Ou, C., He, J., Liu, L., Shan, H., Lei, C., Hui, D. S. C., Du, B., Li, L., Zeng, G., Yuen, K.-Y., Chen, R., Tang, C., Wang, T., Chen, P., Xiang, J., ... Zhong, N. (2020). Clinical Characteristics of Coronavirus Disease 2019 in China. *The New England Journal of Medicine*, *382*(18), 1708–1720. doi:10.1056/NEJMoa2002032 PMID:32109013

He, K., Zhang, X., Ren, S., & Sun, J. (2016). Deep residual learning for image recognition. *Proceedings of the IEEE Computer Society Conference on Computer Vision and Pattern Recognition*, 770–778, . (16) C.10.1109/CVPR.2016.90

Hosni, A. I. E., Li, K., & Ahmad, S. (2019, December). DARIM: Dynamic approach for rumor influence minimization in online social networks. In *International Conference on Neural Information Processing* (pp. 619-630). Cham: Springer International Publishing. 10.1007/978-3-030-36711-4_52

Hosni, A. I. E., Li, K., & Ahmed, S. (2018b). HISBmodel: A rumor diffusion model based on human individual and social behaviors in online social networks. *Neural Information Processing: 25th International Conference, ICONIP 2018, Siem Reap, Cambodia, December 13–16, 2018 Proceedings*, *25*(Part II), 14–27.

Hosni, A. I. E., Li, K., Ding, C., & Ahmed, S. (2018a, November). Least cost rumor influence minimization in multiplex social networks. In *International Conference on Neural Information Processing* (pp. 93-105). Cham: Springer International Publishing. 10.1007/978-3-030-04224-0_9

Huang, G., Liu, Z., Van Der Maaten, L., & Weinberger, K. Q. (2017). Densely connected convolutional networks. *Proceedings - 30th IEEE Conference on Computer Vision and Pattern Recognition, CVPR 2017*, 2261–2269. 10.1109/CVPR.2017.243

Hussain, L., Nguyen, T., Li, H., Abbasi, A. A., Lone, K. J., Zhao, Z., Zaib, M., Chen, A., & Duong, T. Q. (2020). Machine-learning classification of texture features of portable chest X-ray accurately classifies COVID-19 lung infection. *Biomedical Engineering Online*, *19*(1), 88. Advance online publication. doi:10.1186/s12938-020-00831-x PMID:33239006

Ismael, A. M., & Şengür, A. (2021). Deep learning approaches for COVID-19 detection based on chest X-ray images. *Expert Systems with Applications*, *164*, 114054. doi:10.1016/j.eswa.2020.114054 PMID:33013005

Jaiswal, A., Gianchandani, N., Singh, D., Kumar, V., & Kaur, M. (2021). Classification of the COVID-19 infected patients using DenseNet201 based deep transfer learning. *Journal of Biomolecular Structure & Dynamics*, *39*(15), 5682–5689. doi:10.1080/07391102.2020.1788642 PMID:32619398

Khan, B., Naseem, R., Shah, M. A., Wakil, K., Khan, A., Uddin, M. I., & Mahmoud, M. (2021). Software Defect Prediction for Healthcare Big Data: An Empirical Evaluation of Machine Learning Techniques. *Journal of Healthcare Engineering*, *2021*, 1–16. Advance online publication. doi:10.1155/2021/8899263 PMID:33815733

Khan, S., Zhang, Z., Zhu, L., Rahim, M. A., Ahmad, S., & Chen, R. (2020). SCM: Secure and accountable TLS certificate management. *International Journal of Communication Systems*, *33*(15), e4503. doi:10.1002/dac.4503

Lauer, S. A., Grantz, K. H., Bi, Q., Jones, F. K., Zheng, Q., Meredith, H. R., Azman, A. S., Reich, N. G., & Lessler, J. (2020). The incubation period of coronavirus disease 2019 (CoVID-19) from publicly reported confirmed cases: Estimation and application. *Annals of Internal Medicine*, *172*(9), 577–582. doi:10.7326/M20-0504 PMID:32150748

Morra, J. H., Tu, Z., Apostolova, L. G., Green, A. E., Toga, A. W., & Thompson, P. M. (2010). Comparison of adaboost and support vector machines for detecting alzheimer's disease through automated hippocampal segmentation. *IEEE Transactions on Medical Imaging*, *29*(1), 30–43. doi:10.1109/TMI.2009.2021941 PMID:19457748

Naseem, R., Khan, B., Shah, M. A., Wakil, K., Khan, A., Alosaimi, W., Uddin, M. I., & Alouffi, B. (2020). Performance Assessment of Classification Algorithms on Early Detection of Liver Syndrome. *Journal of Healthcare Engineering*, *2020*, 1–13. Advance online publication. doi:10.1155/2020/6680002 PMID:33489060

Ng, M. Y., Lee, E. Y. P., Yang, J., Yang, F., Li, X., Wang, H., Lui, M. M., Lo, C. S.-Y., Leung, B., Khong, P.-L., Hui, C. K.-M., Yuen, K., & Kuo, M. D. (2020). Imaging profile of the covid-19 infection: Radiologic findings and literature review. *Radiology. Cardiothoracic Imaging*, *2*(1), e200034. Advance online publication. doi:10.1148/ryct.2020200034 PMID:33778547

Nishio, M., Noguchi, S., Matsuo, H., & Murakami, T. (2020). Automatic classification between COVID-19 pneumonia, non-COVID-19 pneumonia, and the healthy on chest X-ray image: Combination of data augmentation methods. *Scientific Reports*, *10*(1), 17532. Advance online publication. doi:10.1038/s41598-020-74539-2 PMID:33067538

Noble, W. S. (2006). What is a support vector machine? *Nature Biotechnology*, *24*(12), 1565–1567. doi:10.1038/nbt1206-1565 PMID:17160063

Panwar, H., Gupta, P. K., Siddiqui, M. K., Morales-Menendez, R., Bhardwaj, P., & Singh, V. (2020). A deep learning and grad-CAM based color visualization approach for fast detection of COVID-19 cases using chest X-ray and CT-Scan images. *Chaos, Solitons, and Fractals*, *140*, 110190. Advance online publication. doi:10.1016/j.chaos.2020.110190 PMID:32836918

Pathak, Y., Shukla, P. K., Tiwari, A., Stalin, S., Singh, S., & Shukla, P. K. (2022). Deep Transfer Learning Based Classification Model for COVID-19 Disease. *Ingénierie et Recherche Biomédicale : IRBM = Biomedical Engineering and Research*, *43*(2), 87–92. doi:10.1016/j.irbm.2020.05.003 PMID:32837678

Rahim, A., Zhong, Y., Ahmad, T., Ahmad, S., Pławiak, P., & Hammad, M. (2023). Enhancing smart home security: Anomaly detection and face recognition in smart home iot devices using logit-boosted cnn models. *Sensors (Basel)*, *23*(15), 6979. doi:10.3390/s23156979 PMID:37571762

Shah, V., Keniya, R., Shridharani, A., Punjabi, M., Shah, J., & Mehendale, N. (2021). Diagnosis of COVID-19 using CT scan images and deep learning techniques. *Emergency Radiology*, *28*(3), 497–505. doi:10.1007/s10140-020-01886-y PMID:33523309

Shankar, K., & Perumal, E. (2021). A novel hand-crafted with deep learning features based fusion model for COVID-19 diagnosis and classification using chest X-ray images. *Complex & Intelligent Systems*, *7*(3), 1277–1293. doi:10.1007/s40747-020-00216-6 PMID:34777955

Simonyan, K., & Zisserman, A. (2015). Very deep convolutional networks for large-scale image recognition. *3rd International Conference on Learning Representations, ICLR 2015 - Conference Track Proceedings*.

Szegedy, S. (2017). Inception-v4, inceptionResNet and the impact of residual connections on learning. *31st AAAI Conference on Artificial Intelligence, AAAI 2017*, 4278–4284. 10.1609/aaai.v31i1.11231

Tartaglione, E., Barbano, C. A., Berzovini, C., Calandri, M., & Grangetto, M. (2020). Unveiling COVID-19 from chest x-ray with deep learning: A hurdles race with small data. *International Journal of Environmental Research and Public Health*, *17*(18), 1–17. doi:10.3390/ijerph17186933 PMID:32971995

Zamir, A., Khan, H. U., Iqbal, T., Yousaf, N., Aslam, F., Anjum, A., & Hamdani, M. (2020). Phishing web site detection using diverse machine learning algorithms. *The Electronic Library*, *38*(1), 65–80. doi:10.1108/EL-05-2019-0118

ENDNOTE

[1]   https://www.kaggle.com/datasets/amanullahasraf/covid19-pneumonia[REMOVEDHYPERLINKFIELD]normal-chest-xray-pa-dataset

Compilation of References

Abdullah, F. B., Iqbal, R., Ahmad, S., El-Affendi, M. A., & Abdullah, M. (2022b). An empirical analysis of sustainable energy security for energy policy recommendations. *Sustainability (Basel)*, *14*(10), 6099. doi:10.3390/su14106099

Abdullah, F. B., Iqbal, R., Ahmad, S., El-Affendi, M. A., & Kumar, P. (2022a). Optimization of multidimensional energy security: An index based assessment. *Energies*, *15*(11), 3929. doi:10.3390/en15113929

Adegun, A. A., Viriri, S., & Tapamo, J.-R. (2023). Review of deep learning methods for remote sensing satellite images classification: Experimental survey and comparative analysis. *Journal of Big Data*, *10*(1), 93. doi:10.1186/s40537-023-00772-x

Adhinata, F. D., Rakhmadani, D. P., Wibowo, M., & Jayadi, A. (2021). A Deep Learning Using DenseNet201 to Detect Masked or Non-masked Face. *JUITA: Jurnal Informatika*, *9*(1), 115. doi:10.30595/juita.v9i1.9624

Afchar, D., Nozick, V., Yamagishi, J., & Echizen, I. (2018, December). Mesonet: a compact facial video forgery detection network. In 2018 IEEE international workshop on information forensics and security (WIFS) (pp. 1-7). IEEE. doi:10.1109/WIFS.2018.8630761

Afshar, P., Mohammadi, A., & Plataniotis, K. N. (2018, October). Brain tumor type classification via capsule networks. In *2018 25th IEEE international conference on image processing (ICIP)* (pp. 3129-3133). IEEE. 10.1109/ICIP.2018.8451379

Afshar, P., Naderkhani, F., Oikonomou, A., Rafiee, M. J., Mohammadi, A., & Plataniotis, K. N. (2021). MIXCAPS: A capsule network-based mixture of experts for lung nodule malignancy prediction. *Pattern Recognition*, *116*, 107942. doi:10.1016/j.patcog.2021.107942

Agarwal, M., Jain, N., Kumar, M. M., & Agrawal, H. (2010). Face recognition using eigen faces and artificial neural network. *International Journal of Computer Theory and Engineering*, *2*(4), 624–629. doi:10.7763/IJCTE.2010.V2.213

AgostinelliA.DenkT. I.BorsosZ.EngelJ.VerzettiM.CaillonA.HuangQ.JansenA.RobertsA. TagliasacchiM. (2023) MusicLM: generating music from text. arXiv:2301.11325

Ahmad, Li, Eddine, & Khan. (2018). A biologically inspired cognitive skills measurement approach. *Biologically Inspired Cognitive Architectures, 24*, 35–46.

Ahmad, S., & Adnan, A. (2015, July). Machine learning based cognitive skills calculations for different emotional conditions. In *2015 IEEE 14th International Conference on Cognitive Informatics & Cognitive Computing (ICCI* CC)* (pp. 162-168). IEEE. 10.1109/ICCI-CC.2015.7259381

Ahmad, S., Anwar, M. S., Khan, M. A., Shahzad, M., Ebrahim, M., & Memon, I. (2021). Deep frustration severity network for the prediction of declined students' cognitive skills. *2021 4th international conference on Computing & Information Sciences (ICCIS)*, 1–6.

Ahmad, S., Li, K., Eddine, H. A. I., & Khan, M. I. (2018). A biologically inspired cognitive skills measurement approach. *Biologically inspired cognitive architectures, 24*, 35-46.

Ahmad, S., Li, K., Eddine, H. A. I., & Khan, M. I. (2018). A biologically inspired cognitive skills measurement approach. *Biologically Inspired Cognitive Architectures, 24*, 35-46.

Ahmad, S., Adnan, A., Khan, G., & Mehmood, N. (2015). Emotions, age, and gender based cognitive skills calculations. *International Journal of Computer Theory and Engineering, 7*(1), 76–80. doi:10.7763/IJCTE.2015.V7.934

Ahmad, S., Anwar, M. S., Ebrahim, M., Khan, W., Raza, K., Adil, S. H., & Amin, A. (2020). Deep network for the iterative estimations of students' cognitive skills. *IEEE Access : Practical Innovations, Open Solutions, 8*, 103100–103113. doi:10.1109/ACCESS.2020.2999064

Ahmad, S., Aoun, N. B., El Affendi, M. A., Anwar, M. S., Abbas, S., & Abd El Latif, A. A. (2022). Optimization of Students' Performance Prediction through an Iterative Model of Frustration Severity. *Computational Intelligence and Neuroscience, 2022*, 2022. doi:10.1155/2022/3183492 PMID:36017453

Ahmad, S., El-Affendi, M. A., Anwar, M. S., & Iqbal, R. (2022). Potential future directions in optimization of students' performance prediction system. *Computational Intelligence and Neuroscience, 2022*, 2022. doi:10.1155/2022/6864955 PMID:35619762

Ahmad, S., Li, K., Amin, A., & Khan, S. (2018). A novel technique for the evaluation of posterior probabilities of student cognitive skills. *IEEE Access : Practical Innovations, Open Solutions, 6*, 53153–53167. doi:10.1109/ACCESS.2018.2870877

Ahmed, S., Eisa, Al-Wesabi, Elsafi, Al Duhayyim, Yaseen, Hamza, & Motwakel. (2022). *Parkinson's Detection Using RNN-Graph-LSTM with Optimization Based on Speech Signals*. Tech Science Press. doi:10.32604/cmc.2022.024596

Ahonen, T., Hadid, A., & Pietikäinen, M. (2004). Face recognition with local binary patterns. In *Computer Vision-ECCV 2004: 8th European Conference on Computer Vision, Prague, Czech Republic, May 11-14, 2004*. [Springer Berlin Heidelberg.]. *Proceedings, 8*(Part I), 469–481.

Ahonen, T., Hadid, A., & Pietikainen, M. (2006). Face description with local binary patterns: Application to face recognition. *IEEE Transactions on Pattern Analysis and Machine Intelligence, 28*(12), 2037–2041. doi:10.1109/TPAMI.2006.244 PMID:17108377

Akhtar, S., Ali, A., Ahmad, S., Khan, M. I., Shah, S., & Hassan, F. (2022). The prevalence of foot ulcers in diabetic patients in Pakistan: A systematic review and meta-analysis. *Frontiers in Public Health*, *10*, 1017201. doi:10.3389/fpubh.2022.1017201 PMID:36388315

Akhtar, S., Ramzan, M., Shah, S., Ahmad, I., Khan, M. I., Ahmad, S., El-Affendi, M. A., & Qureshi, H. (2022). Forecasting exchange rate of Pakistan using time series analysis. *Mathematical Problems in Engineering*, *2022*, 2022. doi:10.1155/2022/9108580

Al-Areqi, F., & Konyar, M. Z. (2022). Effectiveness evaluation of different feature extraction methods for classification of covid-19 from computed tomography images: A high accuracy classification study. *Biomedical Signal Processing and Control*, *76*, 103662. Advance online publication. doi:10.1016/j.bspc.2022.103662 PMID:35350595

Al-Bilbisi, H. (2019). Spatial monitoring of urban expansion using satellite remote sensing images: A case study of amman city, jordan. *Sustainability (Basel)*, *11*(8), 2260. doi:10.3390/su11082260

Alhichri, H., Alswayed, A. S., Bazi, Y., Ammour, N., & Alajlan, N. A. (2021). Classification of remote sensing images using efficientnet-b3 cnn model with attention. *IEEE Access : Practical Innovations, Open Solutions*, *9*, 14078–14094. doi:10.1109/ACCESS.2021.3051085

Amer, M., & Maul, T. (2020). Path capsule networks. *Neural Processing Letters*, *52*(1), 545–559. doi:10.1007/s11063-020-10273-0

Aminuddin, J., Abdullatif, R. F., Anggraini, E. I., Gumelar, S. F., & Rahmawati, A. (2023). Development of convolutional neural network algorithm on ships detection in Natuna Islands-Indonesia using land look satellite imagery. *Remote Sensing Applications: Society and Environment*, *32*, 101025. https://doi.org/https://doi.org/10.1016/j.rsase.2023.101025

Anwar, M. S., Ullah, I., Ahmad, S., Choi, A., Ahmad, S., Wang, J., & Aurangzeb, K. (n.d.). Immersive Learning and AR/VR-Based Education: Cybersecurity Measures and Risk Management. In Cybersecurity Management in Education Technologies (pp. 1-22). CRC Press.

Anwar, M. S., Alhalabi, W., Choi, A., Ullah, I., & Alhudali, A. (2024). Internet of Metaverse Things (IoMT): Applications, Technology Challenges and Security Consideration. In *Future Communication Systems Using Artificial Intelligence, Internet of Things and Data Science* (pp. 133–158). CRC Press.

Anwar, M. S., Wang, J., Ahmad, S., Khan, W., Ullah, A., Shah, M., & Fei, Z. (2020). Impact of the impairment in 360-degree videos on users VR involvement and machine learning-based QoE predictions. *IEEE Access : Practical Innovations, Open Solutions*, *8*, 204585–204596. doi:10.1109/ACCESS.2020.3037253

Anwar, M. S., Wang, J., Khan, W., Ullah, A., Ahmad, S., & Fei, Z. (2020). Subjective QoE of 360-degree virtual reality videos and machine learning predictions. *IEEE Access : Practical Innovations, Open Solutions*, *8*, 148084–148099. doi:10.1109/ACCESS.2020.3015556

Anwar, M. S., Wang, J., Ullah, A., Khan, W., Ahmad, S., & Fei, Z. (2020). Measuring quality of experience for 360-degree videos in virtual reality. *Science China. Information Sciences*, *63*(10), 1–15. doi:10.1007/s11432-019-2734-y

Anwar, M. S., Wang, J., Ullah, A., Khan, W., Ahmad, S., & Li, Z. (2019). Impact of stalling on qoe for 360-degree virtual reality videos. *2019 IEEE International Conference on Signal, Information and Data Processing (ICSIDP)*, 1–6. 10.1109/ICSIDP47821.2019.9173042

Anwar, M. S., Wang, J., Ullah, A., Khan, W., Li, Z., & Ahmad, S. (2018, October). User profile analysis for enhancing QoE of 360 panoramic video in virtual reality environment. In *2018 International Conference on Virtual Reality and Visualization (ICVRV)* (pp. 106-111). IEEE. 10.1109/ICVRV.2018.00030

Armato, S. G. III, McLennan, G., Bidaut, L., McNitt-Gray, M. F., Meyer, C. R., Reeves, A. P., Zhao, B., Aberle, D. R., Henschke, C. I., Hoffman, E. A., Kazerooni, E. A., MacMahon, H., van Beek, E. J. R., Yankelevitz, D., Biancardi, A. M., Bland, P. H., Brown, M. S., Engelmann, R. M., Laderach, G. E., ... Clarke, L. P. (2011). The lung image database consortium (LIDC) and image database resource initiative (IDRI): A completed reference database of lung nodules on CT scans. *Medical Physics*, *38*(2), 915–931. doi:10.1118/1.3528204 PMID:21452728

Askell, A., Brundage, M., & Hadfield, G. (2019). The role of cooperation in responsible ai development. arXiv preprint arXiv:1907.04534.

Badue, C., Guidolini, R., Carneiro, R. V., Azevedo, P., Cardoso, V. B., Forechi, A., Jesus, L., Berriel, R., Paixão, T. M., Mutz, F., de Paula Veronese, L., Oliveira-Santos, T., & De Souza, A. F. (2021). Self-driving cars: A survey. *Expert Systems with Applications*, *165*, 113816. doi:10.1016/j.eswa.2020.113816

Bae, G., de La Gorce, M., Baltrušaitis, T., Hewitt, C., Chen, D., Valentin, J., Cipolla, R., & Shen, J. (2023). DigiFace-1M: 1 Million Digital Face Images for Face Recognition. In *Proceedings of the IEEE/CVF Winter Conference on Applications of Computer Vision* (pp. 3526-3535). 10.1109/WACV56688.2023.00352

Baird, A., & Maruping, L. M. (2021). The next generation of research on IS use: A theoretical framework of delegation to and from agentic IS artifacts. *Management Information Systems Quarterly*, *45*(1), 315–341. doi:10.25300/MISQ/2021/15882

Bansal, A., Nanduri, A., Castillo, C. D., Ranjan, R., & Chellappa, R. (2017, October). Umdfaces: An annotated face dataset for training deep networks. In 2017 IEEE international joint conference on biometrics (IJCB) (pp. 464-473). IEEE.

Bansal, M., Kumar, M., Sachdeva, M., & Mittal, A. (2023). Transfer learning for image classification using VGG19: Caltech-101 image data set. *Journal of Ambient Intelligence and Humanized Computing*, *14*(4), 3609–3620. doi:10.1007/s12652-021-03488-z PMID:34548886

Barman, N., Zadtootaghaj, S., Schmidt, S., Martini, M. G., & Möller, S. (2018, June). GamingVideoSET: a dataset for gaming video streaming applications. In *2018 16th Annual Workshop on Network and Systems Support for Games (NetGames)* (pp. 1-6). IEEE. 10.1109/NetGames.2018.8463362

Barocas, S., & Selbst, A. D. (2016). Big data's disparate impact. *California Law Review*, *104*, 671.

Baron-Cohen, S. (1997). *Mindblindness: an essay on autism and theory of mind.* MIT Press.

Berente, N., Gu, B., Recker, J., & Santhanam, R. (2021). Special issue editor's comments: Managing artificial intelligence. *Management Information Systems Quarterly*, *45*(3), 1433–1450.

Besiroglu, T., Bergerson, S. A., Michael, A., Heim, L., Luo, X., & Thompson, N. (2024). The Compute Divide in Machine Learning: A Threat to Academic Contribution and Scrutiny? *Computers & Society.*

Beveridge, J. R., Phillips, P. J., Bolme, D. S., Draper, B. A., Givens, G. H., Lui, Y. M., ... Cheng, S. (2013, September). The challenge of face recognition from digital point-and-shoot cameras. In *2013 IEEE Sixth International Conference on Biometrics: Theory, Applications and Systems (BTAS)* (pp. 1-8). IEEE. 10.1109/BTAS.2013.6712704

Bindemann, M., Fysh, M. C., Sage, S. S., Douglas, K., & Tummon, H. M. (2017). Person identification from aerial footage by a remote-controlled drone. *Scientific Reports*, *7*(1), 13629. doi:10.1038/s41598-017-14026-3 PMID:29051619

Bochkovskiy, A., Wang, C.-Y., & Liao, H.-Y. M. (2020). *Yolov4: Optimal speed and accuracy of object detection.* arXiv preprint arXiv:2004.10934.

Bohara, M., Patel, K., Patel, B., & Desai, J. (2021). An ai based web portal for cotton price analysis and prediction. *3rd International Conference on Integrated Intelligent Computing Communication & Security (ICIIC 2021),* 33–39. 10.2991/ahis.k.210913.005

Bommasani, Creel, Kumar, Jurafsky, & Liang. (2022). *Picking on the same person: Does algorithmic monoculture lead to outcome homogenization?* Academic Press.

Bommasani, R., Hudson, D. A., Adeli, E., Altman, R., Arora, S., von Arx, S., Bernstein, M. S., Bohg, J., Bosselut, A., Brunskill, E., Brynjolfsson, E., Buch, S., Card, D., Castellon, R., Chatterji, N. S., Chen, A. S., Creel, K. A., Davis, J., Demszky, D., . . . Liang, P. (2021) On the 123 S. Feuerriegel et al.: Generative AI, Bus Inf Syst Eng opportunities and risks of foundation models. arXiv:2108. 07258. https://10.48550/arXiv.2108.07258

Borgeaud, S., Mensch, A., Hoffmann, J., Cai, T., Rutherford, E., Millican, K., Driessche, G. B. V. D., Lespiau, J.-B., Damoc, B., & Clark, A. (2022). Improving language models by retrieving from trillions of tokens. *International conference on machine learning*, 2206–2240.

Boruah, M., & Das, R. (2024). MLCapsNet+: A multi-capsule network for the identification of the HIV ISs along important sequence positions. *Image and Vision Computing*, *145*, 104990. doi:10.1016/j.imavis.2024.104990

Bouthillier, X., Laurent, C., & Vincent, P. (2019). Unreproducible research is reproducible. *Proceedings of the International Conference on Machine Learning (PMLR).*

Boutros, F., Damer, N., Kirchbuchner, F., & Kuijper, A. (2022). Elasticface: Elastic margin loss for deep face recognition. In *Proceedings of the IEEE/CVF conference on computer vision and pattern recognition* (pp. 1578-1587). 10.1109/CVPRW56347.2022.00164

Bradley, A. P. (1997). The use of the area under the ROC curve in the evaluation of machine learning algorithms. *Pattern Recognition*, *30*(7), 1145–1159. doi:10.1016/S0031-3203(96)00142-2

Brennan, T. J., & Macauley, M. K. (1995). Remote sensing satellites and privacy: A framework for policy assessment. *Information & Communications Technology Law*, *4*(3), 233–248. doi:10.1080/13600834.1995.9965723

Budiarsa, R., Wardoyo, R., & Musdholifah, A. (2023). Face recognition for occluded face with mask region convolutional neural network and fully convolutional network: a literature review. *International Journal of Electrical & Computer Engineering, 13*(5).

Cai, W., Wei, Z., Song, Y., Li, M., & Yang, X. (2021). Residual-capsule networks with threshold convolution for segmentation of wheat plantation rows in UAV images. *Multimedia Tools and Applications*, *80*(21-23), 32131–32147. doi:10.1007/s11042-021-11203-5

Campus, K. (n.d.). *Deep Frustration Severity Network for the Prediction of Declined Students' Cognitive Skills*. Academic Press.

Cao, Q., Shen, L., Xie, W., Parkhi, O. M., & Zisserman, A. (2018, May). Vggface2: A dataset for recognising faces across pose and age. In *2018 13th IEEE international conference on automatic face & gesture recognition (FG 2018)* (pp. 67-74). IEEE. 10.1109/FG.2018.00020

Carlson, S. M., Koenig, M. A., & Harms, M. B. (2013). Theory of mind. *Wiley Interdisciplinary Reviews: Cognitive Science*, *4*(4), 391–402. doi:10.1002/wcs.1232 PMID:26304226

Caruana, R., & Niculescu-Mizil, A. (2006). An Empirical Comparison of Supervised Learning Algorithms. *Proceedings of the 23rd International Conference on Machine Learning*, 161–168. 10.1145/1143844.1143865

Celikok, M. M., Peltola, T., Daee, P., & Kaski, S. (2019) Interactive AI with a theory of mind. ACM CHI 2019 workshop: computational modeling in human-computer interaction.

Chan, A., Salganik, R., Markelius, A., Pang, C., Rajkumar, N., Krasheninnikov, D., Langosco, L., He, Z., Duan, Y., Carroll, M., Lin, M., Mayhew, A., Collins, K., Molamohammadi, M., Burden, J., Zhao, W., Rismani, S., Voudouris, K., Bhatt, U., ... Maharaj, T. (2023). Harms from increasingly agentic algorithmic systems. *2023 ACM Conference on Fairness, Accountability, and Transparency*, 651-666. arXiv:2302.10329.

Chao, H., Dong, L., Liu, Y., & Lu, B. (2019). Emotion recognition from multiband EEG signals using CapsNet. *Sensors (Basel)*, *19*(9), 2212. doi:10.3390/s19092212 PMID:31086110

Compilation of References

Charrua, A. B., Padmanaban, R., Cabral, P., Bandeira, S., & Romeiras, M. M. (2021). Impacts of the tropical cyclone idai in mozambique: A multitemporal landsat satellite imagery analysis. *Remote Sensing (Basel)*, *13*(2), 201. doi:10.3390/rs13020201

Chen, J., Kellokumpu, V., Zhao, G., & Pietikäinen, M. (2013). RLBP: Robust Local Binary Pattern. BMVC.

Chen, M., Tworek, J., Jun, H., & Yuan, Q. (2021). *Evaluating large language models trained on code*. arXiv:2107.03374

Chen, Z., & Crandall, D. (2018). *Generalized capsule networks with trainable routing procedure.* arXiv preprint arXiv:1808.08692

Cheng, G., & Han, J. (2016). A survey on object detection in optical remote sensing images. *ISPRS Journal of Photogrammetry and Remote Sensing*, *117*, 11–28. doi:10.1016/j.isprsjprs.2016.03.014

Cheng, J., Huang, W., Cao, S., Yang, R., Yang, W., Yun, Z., Wang, Z., & Feng, Q. (2015). Enhanced performance of brain tumor classification via tumor region augmentation and partition. *PLoS One*, *10*(10), e0140381. doi:10.1371/journal.pone.0140381 PMID:26447861

Cheng, J., Yang, W., Huang, M., Huang, W., Jiang, J., Zhou, Y., Yang, R., Zhao, J., Feng, Y., Feng, Q., & Chen, W. (2016). Retrieval of brain tumors by adaptive spatial pooling and fisher vector representation. *PLoS One*, *11*(6), e0157112. doi:10.1371/journal.pone.0157112 PMID:27273091

Cheng, Z., Lu, Y., Cao, Q., Qin, L., Pan, Z., Yan, F., & Yang, W. (2020). Clinical features and chest CT manifestations of coronavirus disease 2019 (COVID-19) in a single-center study in Shanghai, China. *AJR. American Journal of Roentgenology*, *215*(1), 121–126. doi:10.2214/AJR.20.22959 PMID:32174128

Chen, L., Qin, N., Dai, X., & Huang, D. (2020). Fault diagnosis of high-speed train bogie based on capsule network. *IEEE Transactions on Instrumentation and Measurement*, *69*(9), 6203–6211. doi:10.1109/TIM.2020.2968161

ChenL.ZahariaM.ZouJ. (2023) How is chatgpt's behavior changing over time? arXiv:2307.09009

Chen, T., Wang, Z., Yang, X., & Jiang, K. (2019). A deep capsule neural network with stochastic delta rule for bearing fault diagnosis on raw vibration signals. *Measurement*, *148*, 106857. doi:10.1016/j.measurement.2019.106857

Chen, X., Liu, Y., Achuthan, K., & Zhang, X. (2020). A ship movement classification based on Automatic Identification System (AIS) data using Convolutional Neural Network. *Ocean Engineering*, *218*, 108182. doi:10.1016/j.oceaneng.2020.108182

Chen, Z., Luo, P., Wang, Y., Chen, Y., & Tang, X. (2021). BlazeFace: Sub-millisecond Neural Face Detection on Mobile GPUs. *Proceedings of the IEEE Conference on Computer Vision and Pattern Recognition (CVPR)*.

Chiang, T. (2023). *ChatGPT is a blurry JPEG of the web*. https://www. newyorker.com/tech/annals-of-technology/chatgpt-is-a-blurryjpeg-of-the-web

Choi, J., Seo, H., Im, S., & Kang, M. (2019). Attention routing between capsules. *Proceedings of the IEEE/CVF international conference on computer vision workshops.*

Chui, A., Patnaik, A., Ramesh, K. & Wang, L. (2019). *Capsule Networks and Face Recognition.* Lindawangg. github. io.

Clark, K., Vendt, B., Smith, K., Freymann, J., Kirby, J., Koppel, P., Moore, S., Phillips, S., Maffitt, D., Pringle, M., Tarbox, L., & Prior, F. (2013). The Cancer Imaging Archive (TCIA): Maintaining and operating a public information repository. *Journal of Digital Imaging, 26*(6), 1045–1057. doi:10.1007/s10278-013-9622-7 PMID:23884657

Cole, M. (1987). *Remote Sensing: Principles and Interpretation.* Academic Press.

Creswell, A., Shanahan, M., & Higgins, I. (2022). Selection-inference: Exploiting large language models for interpretable logical reasoning. arXiv preprint arXiv:2205.09712.

Cucinotta, D., & Vanelli, M. (2020). WHO declares COVID-19 a pandemic. *Acta Biomedica, 91*(1), 157–160. doi:10.23750/abm.v91i1.9397 PMID:32191675

Cui, B., Chen, X., & Lu, Y. (2020). Semantic segmentation of remote sensing images using transfer learning and deep convolutional neural network with dense connection. *IEEE Access : Practical Innovations, Open Solutions, 8,* 116744–116755. doi:10.1109/ACCESS.2020.3003914

Dai, Li, Li, Tiong, Zhao, Wang, Li, Fung, & Hoi. (2023). *Instructblip: Towards generalpurpose vision-language models with instruction tuning.* Academic Press.

Daniel, T. W. C., & Pardamean, B. (2023). A systematic literature review of machine learning application in COVID-19 medical image classification. *Procedia Computer Science, 216,* 749–756. doi:10.1016/j.procs.2022.12.192 PMID:36643182

Davis, D. S., & Sanger, M. C. (2021). Ethical challenges in the practice of remote sensing and geophysical archaeology. *Archaeological Prospection, 28*(3), 271–278. doi:10.1002/arp.1837

Dellermann, D., Ebel, P., So¨llner, M., & Leimeister, J. M. (2019). Hybrid intelligence. *Business & Information Systems Engineering, 61*(5), 637–643. doi:10.1007/s12599-019-00595-2

Deng, J., Guo, J., Xue, N., & Zafeiriou, S. (2020). RetinaFace: Single-stage dense face localisation in the wild. arXiv preprint arXiv:1905.00641.

Deng, J., Guo, J., Xue, N., & Zafeiriou, S. (2019). Arcface: Additive angular margin loss for deep face recognition. In *Proceedings of the IEEE/CVF conference on computer vision and pattern recognition* (pp. 4690-4699). 10.1109/CVPR.2019.00482

Devi, K., & Muthusenthil, B. (2022). Intrusion detection framework for securing privacy attack in cloud computing environment using DCCGAN-RFOA. *Transactions on Emerging Telecommunications Technologies, 33*(9), e4561. doi:10.1002/ett.4561

DevlinJ.ChangM. W.LeeK.ToutanovaK. (2018) BERT: Pretraining of deep bidirectional transformers for language understanding. arXiv:1810.04805.

Ding, L., Zhang, J., & Bruzzone, L. (2020). Semantic segmentation of large-size vhr remote sensing images using a two-stage multiscale training architecture. *IEEE Transactions on Geoscience and Remote Sensing*, *58*(8), 5367–5376. doi:10.1109/TGRS.2020.2964675

Dong, C., Liu, J., & Xu, F. (2018). Ship Detection in Optical Remote Sensing Images Based on Saliency and a Rotation-Invariant Descriptor. *Remote Sensing (Basel)*, *10*(3), 400. Advance online publication. doi:10.3390/rs10030400

Dong, C., Liu, J., Xu, F., & Liu, C. (2019). Ship Detection from Optical Remote Sensing Images Using Multi-Scale Analysis and Fourier HOG Descriptor. *Remote Sensing (Basel)*, *11*(13), 1529. Advance online publication. doi:10.3390/rs11131529

Dong, Y., Zhang, H., Wang, C., & Wang, Y. (2019). Fine-grained ship classification based on deep residual learning for high-resolution SAR images. *Remote Sensing Letters*, *10*(11), 1095–1104. doi:10.1080/2150704X.2019.1650982

Doon, R., Rawat, T. K., & Gautam, S. (2018). *Cifar-10 classification using deep convolutional neural network. In 2018 IEEE Punecon*. IEEE.

Duan, S., Cheng, P., Wang, Z., Wang, Z., Chen, K., Sun, X., & Fu, K. (2024). MDCNet: A Multi-platform Distributed Collaborative Network for Object Detection in Remote Sensing Imagery. *IEEE Transactions on Geoscience and Remote Sensing*, *62*, 1–15. doi:10.1109/TGRS.2024.3353192

Duarte, K., Rawat, Y. S., & Shah, M. (2019). Capsulevos: Semi-supervised video object segmentation using capsule routing. In *Proceedings of the IEEE/CVF international conference on computer vision* (pp. 8480-8489). 10.1109/ICCV.2019.00857

Du, B., Ru, L., Wu, C., & Zhang, L. (2019). Unsupervised deep slow feature analysis for change detection in multi-temporal remote sensing images. *IEEE Transactions on Geoscience and Remote Sensing*, *57*(12), 9976–9992. doi:10.1109/TGRS.2019.2930682

El Emam, K., & Arbuckle, L. (2013). *Anonymizing health data: case studies and methods to get you started*. O'Reilly Media, Inc.

Eyiokur, F. I., Ekenel, H. K., & Waibel, A. (2023). Unconstrained face mask and face-hand interaction datasets: Building a computer vision system to help prevent the transmission of COVID-19. *Signal, Image and Video Processing*, *17*(4), 1027–1034. doi:10.1007/s11760-022-02308-x PMID:35910402

Faheem, M. Y., Wang, X., Ahmad, S., Azeem, M. B., & Essani, I. Y. (2019, December). Ultra-low power small size rf transceiver module for 5g/lte applications. In *2019 4th International Conference on Emerging Trends in Engineering, Sciences and Technology (ICEEST)* (pp. 1-6). IEEE.

Faisal, M., Ali, I., Khan, M. S., Kim, J., & Kim, S. M. (2020a). Cyber security and key management issues for internet of things: Techniques, requirements, and challenges. *Complexity*, *2020*, 1–9. doi:10.1155/2020/6619498

Faisal, M., Ali, I., Khan, M. S., Kim, S. M., & Kim, J. (2020b). Establishment of trust in internet of things by integrating trusted platform module: To counter cybersecurity challenges. *Complexity, 2020*, 1–9. doi:10.1155/2020/6612919

Faisal, M., Attiq-Ur-Rehman, S. N., & Perveen, Z. (2020). Security architecture of cloud network against cyber threats. *Science International (Lahore), 32*(1), 63–67.

Fatima, R., Samad Shaikh, N., Riaz, A., Ahmad, S., El-Affendi, M. A., Alyamani, K. A., Nabeel, M., Ali Khan, J., Yasin, A., & Latif, R. M. A. (2022). A natural language processing (nlp) evaluation on covid-19 rumour dataset using deep learning techniques. *Computational Intelligence and Neuroscience, 2022*, 2022. doi:10.1155/2022/6561622 PMID:36156967

Fei, H., Ji, D., Zhang, Y., & Ren, Y. (2020). Topic-enhanced capsule network for multi-label emotion classification. *IEEE/ACM Transactions on Audio, Speech, and Language Processing, 28*, 1839–1848. doi:10.1109/TASLP.2020.3001390

Feng, N., Song, Z., Yu, J., Chen, Y. P. P., Zhao, Y., He, Y., & Guan, T. (2020). SSET: A dataset for shot segmentation, event detection, player tracking in soccer videos. *Multimedia Tools and Applications, 79*(39-40), 28971–28992. doi:10.1007/s11042-020-09414-3

Fernandez-Diaz, J. C., Cohen, A. S., González, A. M., & Fisher, C. T. (2018). *Shifting perspectives and ethical concerns in the era of remote sensing.* Academic Press.

Ferrari, C., Berretti, S., & Del Bimbo, A. (2018, August). Extended youtube faces: a dataset for heterogeneous open-set face identification. In *2018 24th International Conference on Pattern Recognition (ICPR)* (pp. 3408-3413). IEEE. 10.1109/ICPR.2018.8545642

Foody, G. M. (1996). Relating the land-cover composition of mixed pixels to artificial neural network classification output. *Photogrammetric Engineering and Remote Sensing, 62*(5), 491–498.

Forde, J.Z., & Paganini, M. (2019). *The scientific method in the science of machine learning.* Academic Press.

Fu, K., Chang, Z., Zhang, Y., Xu, G., Zhang, K., & Sun, X. (2020). Rotation-aware and multi-scale convolutional neural network for object detection in remote sensing images. *ISPRS Journal of Photogrammetry and Remote Sensing, 161*, 294–308. doi:10.1016/j.isprsjprs.2020.01.025

Fu, R., Chen, C., Yan, S., Zhang, R., Wang, X., & Chen, H. (2024). FADL-Net: Frequency-Assisted Dynamic Learning Network for Oriented Object Detection in Remote Sensing Images. *IEEE Transactions on Industrial Informatics*, 1–13. doi:10.1109/TII.2024.3378841

Gabrynowicz, J. I. (1993). (in press). Remote Sensing Law: Obstacle or Opportunity for Geographic Information Systems. *National Center for Geographic Information and Analysis and the Center for the Arizona State University College of Law*, 275–278.

Gallego, A.-J., Pertusa, A., & Gil, P. (2018). Automatic ship classification from optical aerial images with convolutional neural networks. *Remote Sensing (Basel), 10*(4), 511. doi:10.3390/rs10040511

Gardiner, S. M. (2011). *A perfect moral storm: The ethical tragedy of climate change.* Oxford University Press. doi:10.1093/acprof:oso/9780195379440.001.0001

Glennan, S., & Illari, P. M. (Eds.). (2018). *The Routledge handbook of mechanisms and mechanical philosophy.* Routledge.

Gong, R., Huang, J., Zhao, Y., Geng, H., Gao, X., Wu, Q., Ai, W., Zhou, Z., Terzopoulos, D., & Zhu, S.-C. (2023). Arnold: A benchmark for language-grounded task learning with continuous states in realistic 3d scenes. *Proceedings of the IEEE/CVF International Conference on Computer Vision (ICCV).* 10.1109/ICCV51070.2023.01873

Goodfellow, I., Pouget-Abadie, J., Mirza, M., Xu, B., Warde-Farley, D., Ozair, S., Courville, A., & Bengio, Y. (2014). Generative adversarial nets. *Advances in Neural Information Processing Systems*, *27*, 2672–2680.

Gordienko, N., Kochura, Y., Taran, V., Peng, G., Gordienko, Y., & Stirenko, S. (2018). *Capsule deep neural network for recognition of historical Graffiti handwriting.* arXiv preprint arXiv:1809.06693.

Gostin, L. O., Halabi, S. F., & Wilson, K. (2018). Health data and privacy in the digital era. *Journal of the American Medical Association*, *320*(3), 233–234. doi:10.1001/jama.2018.8374 PMID:29926092

Gray, Gray, & Wegner. (2007). Dimensions of mind perception. *Sci, 315*(5812).

Gross, R., Matthews, I., Cohn, J., Kanade, T., & Baker, S. (2010). Multi-pie. *Image and Vision Computing*, *28*(5), 807–813. doi:10.1016/j.imavis.2009.08.002 PMID:20490373

Gu, J., Wu, B., & Tresp, V. (2021). Effective and efficient vote attack on capsule networks. *arXiv preprint arXiv:2102.10055.*

Guan, W., Ni, Z., Hu, Y., Liang, W., Ou, C., He, J., Liu, L., Shan, H., Lei, C., Hui, D. S. C., Du, B., Li, L., Zeng, G., Yuen, K.-Y., Chen, R., Tang, C., Wang, T., Chen, P., Xiang, J., ... Zhong, N. (2020). Clinical Characteristics of Coronavirus Disease 2019 in China. *The New England Journal of Medicine*, *382*(18), 1708–1720. doi:10.1056/NEJMoa2002032 PMID:32109013

Gui, S., Song, S., Qin, R., & Tang, Y. (2024). Remote Sensing Object Detection in the Deep Learning Era—A Review. *Remote Sensing (Basel)*, *16*(2), 327. doi:10.3390/rs16020327

Guo, Y., Zhang, L., Hu, Y., He, X., & Gao, J. (2016). Ms-celeb-1m: A dataset and benchmark for large-scale face recognition. *Computer Vision–ECCV 2016: 14th European Conference, Amsterdam, The Netherlands, October 11-14, 2016 Proceedings, 14*(Part III), 87–102.

Gupta, A., & Mason, M. (2014). Transparency and international environmental politics. In *Advances in international environmental politics* (pp. 356–380). Palgrave Macmillan UK.

Gu, Y., Wang, Y., & Li, Y. (2019). A survey on deep learning-driven remote sensing image scene understanding: Scene classification, scene retrieval and scene-guided object detection. *Applied Sciences (Basel, Switzerland)*, *9*(10), 2110. doi:10.3390/app9102110

Gyngell, C., Bowman-Smart, H., & Savulescu, J. (2019). Moral reasons to edit the human genome: Picking up from the Nuffield report. *Journal of Medical Ethics*, *45*(8), 514–523. doi:10.1136/medethics-2018-105084 PMID:30679191

Hadfield, Cuéllar, & Oreilly. (2023). *Its time to create a national registry for large ai models.* Available: https://carnegieendowment.org/2023/07/12/it-s-time-to-create-national-registry-for-large-ai-models-pub-90180]

Hammouche, R., Attia, A., Akhrouf, S., & Akhtar, Z. (2022). Gabor filter bank with deep autoencoder based face recognition system. *Expert Systems with Applications*, *197*, 116743. doi:10.1016/j.eswa.2022.116743

Hangaragi, S., Singh, T., & Neelima, N. (2023). Face detection and Recognition using Face Mesh and deep neural network. *Procedia Computer Science*, *218*, 741–749. doi:10.1016/j.procs.2023.01.054

Haq, M. U., Sethi, M. A. J., Ahmad, S., ELAffendi, M. A., & Asim, M. (2024). Automatic Player Face Detection and Recognition for Players in Cricket Games. *IEEE Access : Practical Innovations, Open Solutions*, *12*, 41219–41233. doi:10.1109/ACCESS.2024.3377564

Haq, M. U., Sethi, M. A. J., & Rehman, A. U. (2023). Capsule Network with Its Limitation, Modification, and Applications—A Survey. *Machine Learning and Knowledge Extraction*, *5*(3), 891–921. doi:10.3390/make5030047

Haq, M. U., Sethi, M. A. J., Ullah, R., Shazhad, A., Hasan, L., & Karami, G. M. (2022). COMSATS face: A dataset of face images with pose variations, its design, and aspects. *Mathematical Problems in Engineering*, *2022*, 2022. doi:10.1155/2022/4589057

Haq, M. U., Shahzad, A., Mahmood, Z., Shah, A. A., Muhammad, N., & Akram, T. (2019). Boosting the face recognition performance of ensemble based LDA for pose, non-uniform illuminations, and low-resolution images. [TIIS]. *KSII Transactions on Internet and Information Systems*, *13*(6), 3144–3164.

Haq, M. U., Shahzad, A., Mahmood, Z., Shah, A. A., Muhammad, N., & Akram, T. (2019). Boosting the face recognition performance of ensemble based lda for pose, non-uniform illuminations, and low-resolution images. *KSII Transactions on Internet and Information Systems*, *13*(6), 3144–3164.

Hartmann, J., Bergner, A., & Hildebrand, C. (2023). MindMiner: Uncovering linguistic markers of mind perception as a new lens to understand consumer-smart object relationships. *Journal of Consumer Psychology*, *33*(4), 645–667. Advance online publication. doi:10.1002/jcpy.1381

He, K., Zhang, X., Ren, S., & Sun, J. (2016). Deep residual learning for image recognition. *Proceedings of the IEEE Computer Society Conference on Computer Vision and Pattern Recognition*, 770–778, . (16) C.10.1109/CVPR.2016.90

Heidarian, S., Afshar, P., Enshaei, N., Naderkhani, F., Rafiee, M. J., Fard, F. B., Samimi, K., Atashzar, S. F., Oikonomou, A., Plataniotis, K. N., & Mohammadi, A. (2021). Covid-fact: A fully-automated capsule network-based framework for identification of covid-19 cases from chest ct scans. *Frontiers in Artificial Intelligence*, *4*, 4. doi:10.3389/frai.2021.598932 PMID:34113843

He, J., Wang, Y., & Liu, H. (2021). Ship Classification in Medium-Resolution SAR Images via Densely Connected Triplet CNNs Integrating Fisher Discrimination Regularized Metric Learning. *IEEE Transactions on Geoscience and Remote Sensing*, *59*(4), 3022–3039. doi:10.1109/TGRS.2020.3009284

Hendrycks, D., Mazeika, M., & Woodside, T. (2023). *An overview of catastrophic ai risks*. arXiv preprint arXiv:2306.12001.

Hinton, G. E., Sabour, S., & Frosst, N. (2018, May). Matrix capsules with EM routing. *International conference on learning representations*.

Hodge, J. G. Jr, Gostin, L. O., & Jacobson, P. D. (1999). Legal issues concerning electronic health information: Privacy, quality, and liability. *Journal of the American Medical Association*, *282*(15), 1466–1471. doi:10.1001/jama.282.15.1466 PMID:10535438

Hosni, A. I. E., Li, K., & Ahmad, S. (2019). Darim: Dynamic approach for rumor influence minimization in online social networks. *International Conference on Neural Information Processing*, 619–630. 10.1007/978-3-030-36711-4_52

Hosni, A. I. E., Li, K., & Ahmad, S. (2020). Minimizing rumor influence in multiplex online social networks based on human individual and social behaviors. *Information Sciences*, *512*, 1458–1480. doi:10.1016/j.ins.2019.10.063

Hosni, A. I. E., Li, K., & Ahmed, S. (2018). HISBmodel: A rumor diffusion model based on human individual and social behaviors in online social networks. *Neural Information Processing: 25th International Conference, ICONIP 2018, Siem Reap, Cambodia, December 13–16, 2018 Proceedings*, *25*(Part II), 14–27.

Hosni, A. I. E., Li, K., Ding, C., & Ahmed, S. (2018). Least cost rumor influence minimization in multiplex social networks. *International Conference on Neural Information Processing*, 93–105. 10.1007/978-3-030-04224-0_9

Hossain, M. D., & Chen, D. (2019). Segmentation for object-based image analysis (obia): A review of algorithms and challenges from remote sensing perspective. *ISPRS Journal of Photogrammetry and Remote Sensing*, *150*, 115–134. doi:10.1016/j.isprsjprs.2019.02.009

Hsu, H. J., & Chen, K. T. (2017, June). DroneFace: an open dataset for drone research. In *Proceedings of the 8th ACM on multimedia systems conference* (pp. 187-192). 10.1145/3083187.3083214

Hu, X.D. & Li, Z.H. (2022). Intrusion Detection Method Based on Capsule Network for Industrial Internet. *Acta Electonica Sinica, 50*(6), 1457.

Huang, G. B., Mattar, M., Berg, T., & Learned-Miller, E. (2008, October). Labeled faces in the wild: A database for studying face recognition in unconstrained environments. *Workshop on faces in 'Real-Life' Images: Detection, Alignment, and Recognition.*

Huang, G. B., Ramesh, M., Berg, T., & Learned-Miller, E. (2007). Labeled Faces in the Wild: A Database for Studying Face Recognition in Unconstrained Environments. *Proceedings of the Workshop on Faces in "Real-Life" Images: Detection, Alignment, and Recognition.*

Huang, G., Liu, Z., Van Der Maaten, L., & Weinberger, K. Q. (2017). Densely connected convolutional networks. *Proceedings - 30th IEEE Conference on Computer Vision and Pattern Recognition, CVPR 2017,* 2261–2269. 10.1109/CVPR.2017.243

Huang, I.-L., Lee, M.-C., Nieh, C.-Y., & Huang, J.-C. (2024). Ship Classification Based on AIS Data and Machine Learning Methods. *Electronics (Basel), 13*(1), 98. Advance online publication. doi:10.3390/electronics13010098

Huang, J., Singh, V., Kumar, S., & Jain, A. K. (2012). *Cross-Pose LFW: A database for studying cross-pose face recognition in unconstrained environments. Technical Report.* Department of Computer Science and Engineering, Michigan State University.

Huang, L., Li, W., Chen, C., Zhang, F., & Lang, H. (2018). Multiple features learning for ship classification in optical imagery. *Multimedia Tools and Applications, 77*(11), 13363–13389. doi:10.1007/s11042-017-4952-y

Huang, L., Wang, J., & Cai, D. (2021). Graph capsule network for object recognition. *IEEE Transactions on Image Processing, 30,* 1948–1961.

Huang, Y. H., & Chen, H. H. (2022). Deep face recognition for dim images. *Pattern Recognition, 126,* 108580. doi:10.1016/j.patcog.2022.108580

Hussain, L., Nguyen, T., Li, H., Abbasi, A. A., Lone, K. J., Zhao, Z., Zaib, M., Chen, A., & Duong, T. Q. (2020). Machine-learning classification of texture features of portable chest X-ray accurately classifies COVID-19 lung infection. *Biomedical Engineering Online, 19*(1), 88. Advance online publication. doi:10.1186/s12938-020-00831-x PMID:33239006

Ismael, A. M., & Şengür, A. (2021). Deep learning approaches for COVID-19 detection based on chest X-ray images. *Expert Systems with Applications, 164,* 114054. doi:10.1016/j.eswa.2020.114054 PMID:33013005

Izacard, G., Lewis, P., Lomeli, M., Hosseini, L., Petroni, F., Schick, T., Dwivedi-Yu, J., Joulin, A., Riedel, S., & Grave, E. (2022). *Few-shot learning with retrieval augmented language models.* arXiv preprint arXiv:2208.03299.

Jain, R. (2019). Improving performance and inference on audio classification tasks using capsule networks. arXiv preprint arXiv:1902.05069.

Jaiswal, A. (2018). Capsulegan: Generative adversarial capsule network. *Proceedings of the European Conference on Computer Vision (ECCV) Workshops.*

Jaiswal, A., Gianchandani, N., Singh, D., Kumar, V., & Kaur, M. (2021). Classification of the COVID-19 infected patients using DenseNet201 based deep transfer learning. *Journal of Biomolecular Structure & Dynamics*, *39*(15), 5682–5689. doi:10.1080/07391102.2020.17886 42 PMID:32619398

Janakiramaiah, B., Kalyani, G., Karuna, A., Prasad, L. N., & Krishna, M. (2023). Military object detection in defense using multi-level capsule networks. *Soft Computing*, *27*(2), 1045–1059. doi:10.1007/s00500-021-05912-0

Janiesch, C., Zschech, P., & Heinrich, K. (2021). Machine learning and deep learning. *Electronic Markets*, *31*(3), 685–695. doi:10.1007/s12525-021-00475-2

Jasanoff, S. (2016). *The ethics of invention: Technology and the human future*. WW Norton & Company.

Jayasundara, V., Roy, D., & Fernando, B. (2021). Flowcaps: Optical flow estimation with capsule networks for action recognition. In *Proceedings of the IEEE/CVF winter conference on applications of computer vision* (pp. 3409-3418). 10.1109/WACV48630.2021.00345

Jeevan, G., Zacharias, G. C., Nair, M. S., & Rajan, J. (2022). An empirical study of the impact of masks on face recognition. *Pattern Recognition*, *122*, 108308. doi:10.1016/j.patcog.2021.108308

Jiang, H., Luo, T., Peng, H., & Zhang, G. (2024). MFCANet: Multiscale Feature Context Aggregation Network for Oriented Object Detection in Remote-Sensing Images. *IEEE Access*.

Jiang, M., Yang, X., Dong, Z., Fang, S., & Meng, J. (2016). Ship Classification Based on Superstructure Scattering Features in SAR Images. *IEEE Geoscience and Remote Sensing Letters*, *13*(5), 616–620. doi:10.1109/LGRS.2016.2514482

Jia, W., Xu, S., Liang, Z., Zhao, Y., Min, H., Li, S., & Yu, Y. (2021). Real-time automatic helmet detection of motorcyclists in urban traffic using improved yolov5 detector. *IET Image Processing*, *15*(14), 3623–3637. doi:10.1049/ipr2.12295

Jobin, A., Ienca, M., & Vayena, E. (2019). The global landscape of AI ethics guidelines. *Nature Machine Intelligence*, *1*(9), 389–399. doi:10.1038/s42256-019-0088-2

Justin, S. P., Andre, J. P., Bennett, A. L., & Fabbri, D. (2017). Deep learning for brain tumor classification. *Proc. SPIE Medical Imaging 2017: Biomedical Applications in Molecular, Structural, and Functional Imaging*, 10137.

Kaggle. (n.d.). https://www.kaggle.com/datasets/andrewmvd/convid19-x-rays

Kalka, N. D., Maze, B., Duncan, J. A., O'Connor, K., Elliott, S., Hebert, K., . . . Jain, A. K. (2018, October). Ijb–s: Iarpa janus surveillance video benchmark. In *2018 IEEE 9th international conference on biometrics theory, applications and systems (BTAS)* (pp. 1-9). IEEE.

Kalra, I., Singh, M., Nagpal, S., Singh, R., Vatsa, M., & Sujit, P. B. (2019, May). Dronesurf: Benchmark dataset for drone-based face recognition. In *2019 14th IEEE International Conference on Automatic Face & Gesture Recognition (FG 2019)* (pp. 1-7). IEEE.

Kanth, R. R., & Jacob, T. P. (2023, April). Enhanced capsule generative adversarial network with Blockchain fostered Intrusion Detection System for Enhancing Cyber security in Cloud. In *2023 2nd International Conference on Smart Technologies and Systems for Next Generation Computing (ICSTSN)* (pp. 1-6). IEEE. 10.1109/ICSTSN57873.2023.10151609

Karpagam, M., Jeyavathana, R. B., Chinnappan, S. K., Kanimozhi, K. V., & Sambath, M. (2022). A novel face recognition model for fighting against human trafficking in surveillance videos and rescuing victims. *Soft Computing*, 1–16.

Karpathy, A., Toderici, G., Shetty, S., Leung, T., Sukthankar, R., & Fei-Fei, L. (2014). Large-scale video classification with convolutional neural networks. In *Proceedings of the IEEE conference on Computer Vision and Pattern Recognition* (pp. 1725-1732). 10.1109/CVPR.2014.223

Kasar, M. M., Bhattacharyya, D., & Kim, T. H. (2016). Face recognition using neural network: A review. *International Journal of Security and Its Applications*, *10*(3), 81–100. doi:10.14257/ijsia.2016.10.3.08

Kasper-Eulaers, M., Hahn, N., Berger, S., Sebulonsen, T., Myrland, Ø., & Kummervold, P. E. (2021). Detecting heavy goods vehicles in rest areas in winter conditions using yolov5. *Algorithms*, *14*(4), 114. doi:10.3390/a14040114

Kemelmacher-Shlizerman, I., Seitz, S. M., Miller, D., & Brossard, E. (2016). The megaface benchmark: 1 million faces for recognition at scale. In *Proceedings of the IEEE conference on computer vision and pattern recognition* (pp. 4873-4882). 10.1109/CVPR.2016.527

Kennicott, P. C. (1974). *Bibliographic Annual in Speech Communication 1973*. Academic Press.

Khan, M. I., Qureshi, H., Bae, S. J., Khattak, A. A., Anwar, M. S., Ahmad, S., ... Ahmad, S. (2023). *Malaria prevalence in Pakistan: A systematic review and meta-analysis (2006–2021)*. Heliyon.

Khan, M., Rahim, M., Alanzi, A. M., Ahmad, S., Fatlane, J. M., Aphane, M., & Khalifa, H. A. E. W. (2023). Dombi Aggregation Operators For p, q, r–Spherical Fuzzy Sets: Application in the Stability Assessment of Cryptocurrencies. *IEEE Access*.

Khan, W., Hua, W., Ayaz, M., Shahid Anwar, M., & Ahmad, S. (2020). A balanced energy efficient (BEE) routing scheme for underwater WSNs. In *Innovative Mobile and Internet Services in Ubiquitous Computing: Proceedings of the 13th International Conference on Innovative Mobile and Internet Services in Ubiquitous Computing (IMIS-2019)* (pp. 82-93). Springer International Publishing.

Khan, B., Naseem, R., Shah, M. A., Wakil, K., Khan, A., Uddin, M. I., & Mahmoud, M. (2021). Software Defect Prediction for Healthcare Big Data: An Empirical Evaluation of Machine Learning Techniques. *Journal of Healthcare Engineering*, *2021*, 1–16. Advance online publication. doi:10.1155/2021/8899263 PMID:33815733

Khandelwal, U., Levy, O., Jurafsky, D., Zettlemoyer, L., & Lewis, M. (2019). *Generalization through memorization: Nearest neighbor language models*. arXiv preprint arXiv:1911.00172.

Khan, F., Yu, X., Yuan, Z., & Rehman, A. U. (2023). ECG classification using 1-D convolutional deep residual neural network. *PLoS One*, *18*(4), e0284791. doi:10.1371/journal.pone.0284791 PMID:37098024

Khan, S., Kamal, A., Fazil, M., Alshara, M. A., Sejwal, V. K., Alotaibi, R. M., Baig, A. R., & Alqahtani, S. (2022). HCovBi-caps: Hate speech detection using convolutional and Bi-directional gated recurrent unit with Capsule network. *IEEE Access : Practical Innovations, Open Solutions*, *10*, 7881–7894. doi:10.1109/ACCESS.2022.3143799

Khan, S., Zhang, Z., Zhu, L., Rahim, M. A., Ahmad, S., & Chen, R. (2020). Scm: Secure and accountable tls certificate management. *International Journal of Communication Systems*, *33*(15), e4503. doi:10.1002/dac.4503

Khan, W., Wang, H., Anwar, M. S., Ayaz, M., Ahmad, S., & Ullah, I. (2019). A multi-layer cluster based energy efficient routing scheme for UWSNs. *IEEE Access : Practical Innovations, Open Solutions*, *7*, 77398–77410. doi:10.1109/ACCESS.2019.2922060

Khattab, O., Santhanam, K., Li, X. L., Hall, D., Liang, P., Potts, C., & Zaharia, M. (2022). *Demonstrate-search-predict: Composing retrieval and language models for knowledge-intensive nlp*. arXiv preprint arXiv:2212.14024.

Khayal, O. (2019). *Human Factors and Ergonomics*. Academic Press.

Kim, M., Reingold, O., & Rothblum, G. (2018). Fairness through computationally-bounded awareness. *Advances in Neural Information Processing Systems*, 31.

Kindt, E. J. (2013). Privacy and data protection issues of biometric applications. In *A Comparative Legal Analysis* (Vol. 12). Springer. doi:10.1007/978-94-007-7522-0

Kleinberg, J., & Raghavan, M. (2021). Algorithmic monoculture and social welfare. *Proceedings of the National Academy of Sciences of the United States of America*, *118*(22), e2018340118. doi:10.1073/pnas.2018340118 PMID:34035166

Klinger, R., De Clercq, O., Mohammad, S.M. & Balahur, A. (2018). *Iest: Wassa-2018 implicit emotions shared task*. arXiv preprint arXiv.

Koelstra, S., Muhl, C., Soleymani, M., Lee, J. S., Yazdani, A., Ebrahimi, T., Pun, T., Nijholt, A., & Patras, I. (2011). Deap: A database for emotion analysis; using physiological signals. *IEEE Transactions on Affective Computing*, *3*(1), 18–31. doi:10.1109/T-AFFC.2011.15

Kolyperas, D., Maglaras, G., & Sparks, L. (2019). Sport fans' roles in value co-creation. *European Sport Management Quarterly*, *19*(2), 201–220. doi:10.1080/16184742.2018.1505925

Koplow, D. A. (1992). Overflying a Country without Overlooking the Constitution: Legal Implications of Aerial Overflights. *Open Skies, Arms Control and Cooperative Security*, 93-112.

Kowalczewski, P. Ł., Siejak, P., Jarzębski, M., Jakubowicz, J., Jeżowski, P., Walkowiak, K., Smarzyński, K., Ostrowska-Ligęza, E., & Baranowska, H. M. (2023). Comparison of technological and physicochemical properties of cricket powders of different origin. *Journal of Insects as Food and Feed*, *9*(5), 637–646. doi:10.3920/JIFF2022.0030

Kraus, M., Feuerriegel, S., & Oztekin, A. (2020). Deep learning in business analytics and operations research: Models, applications and managerial implications. *European Journal of Operational Research*, *281*(3), 628–641. doi:10.1016/j.ejor.2019.09.018

Krenn, M., Pollice, R., Guo, S. Y., Aldeghi, M., Cervera-Lierta, A., Friederich, P., dos Passos Gomes, G., Häse, F., Jinich, A., Nigam, A., Yao, Z., & Aspuru-Guzik, A. (2022). On scientific understanding with artificial intelligence. *Nature Reviews. Physics*, *4*(12), 761–769. doi:10.1038/s42254-022-00518-3 PMID:36247217

Krizhevsky, A. & Hinton, G. (2009). *Learning multiple layers of features from tiny images*. Academic Press.

Krizhevsky, A., Sutskever, I., & Hinton, G. E. (2012). Imagenet classification with deep convolutional neural networks. *Advances in Neural Information Processing Systems*, 25.

Kru¨gel, S., Ostermaier, A., & Uhl, M. (2023). ChatGPT's inconsistent moral advice influences users' judgment. *Scientific Reports*, *13*(1), 4569. doi:10.1038/s41598-023-31341-0 PMID:37024502

Kshirsagar, V. P., Baviskar, M. R., & Gaikwad, M. E. (2011, March). Face recognition using Eigenfaces. In *2011 3rd International Conference on Computer Research and Development* (Vol. 2, pp. 302-306). IEEE. 10.1109/ICCRD.2011.5764137

Kumar, A., & Sachdeva, N. (2021). Multimodal cyberbullying detection using capsule network with dynamic routing and deep convolutional neural network. *Multimedia Systems*, 1–10.

Lang, H., Li, C., & Xu, J. (2022). Multisource heterogeneous transfer learning via feature augmentation for ship classification in SAR imagery. *IEEE Transactions on Geoscience and Remote Sensing*, *60*, 1–14. doi:10.1109/TGRS.2022.3178703

Lang, H., & Wu, S. (2017). Ship classification in moderate-resolution SAR image by naive geometric features-combined multiple kernel learning. *IEEE Geoscience and Remote Sensing Letters*, *14*(10), 1765–1769. doi:10.1109/LGRS.2017.2734889

LangH.YangG. (2022). *ORS Dataset*. IEEE Dataport. doi:10.21227/y4f3-wh22

Lang, H., Zhang, J., Zhang, X., & Meng, J. (2016). Ship classification in SAR image by joint feature and classifier selection. *IEEE Geoscience and Remote Sensing Letters*, *13*(2), 212–216. doi:10.1109/LGRS.2015.2506570

Laroca, R., Severo, E., Zanlorensi, L. A., Oliveira, L. S., Gon‚calves, G. R., Schwartz, W. R., & Menotti, D. (2018). *A robust real-time automatic license plate recognition based on the yolo detector. In 2018 international joint conference on neural networks (ijcnn)*. IEEE. doi:10.1109/IJCNN.2018.8489629

Larsson, G., Maire, M., & Shakhnarovich, G. (2016). Fractalnet: Ultra-deep neural networks without residuals. arXiv preprint arXiv:1605.07648

Latin, H. A., Tennehill, G. W., & White, R. E. (1976). Remote sensing evidence and environmental law. *California Law Review*, *64*(6), 1300. doi:10.2307/3480040

Lauer, S. A., Grantz, K. H., Bi, Q., Jones, F. K., Zheng, Q., Meredith, H. R., Azman, A. S., Reich, N. G., & Lessler, J. (2020). The incubation period of coronavirus disease 2019 (CoVID-19) from publicly reported confirmed cases: Estimation and application. *Annals of Internal Medicine*, *172*(9), 577–582. doi:10.7326/M20-0504 PMID:32150748

Levin, N., Kyba, C. C., Zhang, Q., de Miguel, A. S., Roma'n, M. O., Li, X., Portnov, B. A., Molthan, A. L., Jechow, A., & Miller, S. D. (2020). Remote sensing of night lights: A review and an outlook for the future. *Remote Sensing of Environment*, *237*, 111443. doi:10.1016/j.rse.2019.111443

Liang, H., Zhang, W., Li, W., Yu, J., & Xu, L. (2023). *Intergen: Diffusion-based multi-human motion generation under complex interactions*. arXiv preprint arXiv:2304.05684.

Lian, Y., Shi, X., Shen, S., & Hua, J. (2024). Multitask learning for image translation and salient object detection from multimodal remote sensing images. *The Visual Computer*, *40*(3), 1395–1414. doi:10.1007/s00371-023-02857-3

Li, J., Tian, P., Song, R., Xu, H., Li, Y., & Du, Q. (2024). PCViT: A Pyramid Convolutional Vision Transformer Detector for Object Detection in Remote Sensing Imagery. *IEEE Transactions on Geoscience and Remote Sensing*, *62*, 1–15. doi:10.1109/TGRS.2024.3360456

Li, K., Wan, G., Cheng, G., Meng, L., & Han, J. (2020). Object detection in optical remote sensing images: A survey and a new benchmark. *ISPRS Journal of Photogrammetry and Remote Sensing*, *159*, 296–307. doi:10.1016/j.isprsjprs.2019.11.023

Lillesand, T., Kiefer, R. W., & Chipman, J. (2015). *Remote sensing and image interpretation*. John Wiley & Sons.

Li, N., Shen, X., Sun, L., Xiao, Z., Ding, T., Li, T., & Li, X. (2023). Chinese face dataset for face recognition in an uncontrolled classroom environment. *IEEE Access : Practical Innovations, Open Solutions*, *11*, 86963–86976. doi:10.1109/ACCESS.2023.3302919

Lin, J., Zhao, Y., Wang, S., & Tang, Y. (2024). A robust training method for object detectors in remote sensing image. *Displays*, *81*, 102618. doi:10.1016/j.displa.2023.102618

Li, R., Zheng, S., Duan, C., Su, J., & Zhang, C. (2021). Multistage attention resunet for semantic segmentation of fine-resolution remote sensing images. *IEEE Geoscience and Remote Sensing Letters*, *19*, 1–5.

Liu, K., Yu, S., & Liu, S. (2020). An improved InceptionV3 network for obscured ship classification in remote sensing images. *IEEE Journal of Selected Topics in Applied Earth Observations and Remote Sensing*, *13*, 4738–4747. doi:10.1109/JSTARS.2020.3017676

Liu, P., Yuan, W., Fu, J., Jiang, Z., Hayashi, H., & Neubig, G. (2023). Pre-train, prompt, and predict: A systematic survey of prompting methods in natural language processing. *ACM Computing Surveys*, *55*(9), 1–35. doi:10.1145/3560815

Liu, S., Wang, Z., An, Y., Zhao, J., Zhao, Y., & Zhang, Y. D. (2023). EEG emotion recognition based on the attention mechanism and pre-trained convolution capsule network. *Knowledge-Based Systems*, *265*, 110372. doi:10.1016/j.knosys.2023.110372

Liu, X., Zhai, H., Shen, Y., Lou, B., Jiang, C., Li, T., Hussain, S. B., & Shen, G. (2020). Large-scale crop mapping from multisource remote sensing images in google earth engine. *IEEE Journal of Selected Topics in Applied Earth Observations and Remote Sensing*, *13*, 414–427. doi:10.1109/JSTARS.2019.2963539

Liu, Z., Luo, P., Wang, X., & Tang, X. (2015). Deep learning face attributes in the wild. *Proceedings of the IEEE International Conference on Computer Vision*, 3730-3738. 10.1109/ICCV.2015.425

Li, X., Liu, J., Xie, Y., Gong, P., Zhang, X., & He, H. (2024). Magdra: A multi-modal attention graph network with dynamic routing-by-agreement for multi-label emotion recognition. *Knowledge-Based Systems*, *283*, 111126. doi:10.1016/j.knosys.2023.111126

Li, Y., Choi, D., Chung, J., Kushman, N., Schrittwieser, J., Leblond, R., Eccles, T., Keeling, J., Gimeno, F., Dal Lago, A., Hubert, T., Choy, P., de Masson d'Autume, C., Babuschkin, I., Chen, X., Huang, P.-S., Welbl, J., Gowal, S., Cherepanov, A., ... Vinyals, O. (2022). Competition-level code generation with alphacode. *Science*, *378*(6624), 1092–1097. doi:10.1126/science.abq1158 PMID:36480631

Lu, J., Plataniotis, K. N., & Venetsanopoulos, A. N. (2003). Face recognition using LDA-based algorithms. *IEEE Transactions on Neural Networks*, *14*(1), 195–200. doi:10.1109/TNN.2002.806647 PMID:18238001

Luo, Y., Chen, S., & Ma, X. G. (2020). Drone lams: A drone-based face detection dataset with large angles and many scenarios. *arXiv preprint arXiv:2011.07689*.

Luque, A., Carrasco, A., Martín, A., & de las Heras, A. (2019). The impact of class imbalance in classification performance metrics based on the binary confusion matrix. *Pattern Recognition*, *91*, 216–231. doi:10.1016/j.patcog.2019.02.023

Ma, Y. J., Liang, W., Wang, G., Huang, D.-A., Bastani, O., Jayaraman, D., Zhu, Y., Fan, L., & Anandkumar, A. (2023). *Eureka: Human-level reward design via coding large language models*. arXiv preprint arXiv:2310.12931.

Ma, L., Liu, Y., Zhang, X., Ye, Y., Yin, G., & Johnson, B. A. (2019). Deep learning in remote sensing applications: A meta-analysis and review. *ISPRS Journal of Photogrammetry and Remote Sensing*, *152*, 166–177. doi:10.1016/j.isprsjprs.2019.04.015

Compilation of References

Mangalraj, P., Sivakumar, V., Karthick, S., Haribaabu, V., Ramraj, S., & Samuel, D. J. (2020). A review of multi-resolution analysis (mra) and multigeometric analysis (mga) tools used in the fusion of remote sensing images. *Circuits, Systems, and Signal Processing*, *39*(6), 3145–3172. doi:10.1007/s00034-019-01316-6

Manoharan, J. S. (2021). Capsule Network Algorithm for Performance Optimization of Text Classification. *Journal of Soft Computing Paradigm*, *3*(01), 1–9. doi:10.36548/jscp.2021.1.001

Marchisio, A., Nanfa, G., Khalid, F., Hanif, M. A., Martina, M., & Shafique, M. (2023). SeVuc: A study on the Security Vulnerabilities of Capsule Networks against adversarial attacks. *Microprocessors and Microsystems*, *96*, 104738. doi:10.1016/j.micpro.2022.104738

Ma, X., Dai, Z., He, Z., Ma, J., Wang, Y., & Wang, Y. (2017). Learning traffic as images: A deep convolutional neural network for large-scale transportation network speed prediction. *Sensors (Basel)*, *17*(4), 818. doi:10.3390/s17040818 PMID:28394270

Mazzia, V., Salvetti, F., & Chiaberge, M. (2021). Efficient-capsnet: Capsule network with self-attention routing. *Scientific Reports*, *11*(1), 1–13. doi:10.1038/s41598-021-93977-0 PMID:34282164

McIntosh, T. R., Susnjak, T., Liu, T., Watters, P., & Halgamuge, M. N. (2023). From google gemini to openai q*(q-star): A survey of reshaping the generative artificial intelligence (ai) research landscape. *arXiv preprint arXiv:2312.10868*.

Medvedev, I., Shadmand, F., & Gonçalves, N. (2023). Young Labeled Faces in the Wild (YLFW): A Dataset for Children Faces Recognition. *arXiv preprint arXiv:2301.05776*.

Mehmood, G., Khan, M. Z., Abbas, S., Faisal, M., & Rahman, H. U. (2020). An energy-efficient and cooperative fault-tolerant communication approach for wireless body area network. *IEEE Access : Practical Innovations, Open Solutions*, *8*, 69134–69147. doi:10.1109/ACCESS.2020.2986268

Mekonnen, K. A. (2023). Balanced Face Dataset: Guiding StyleGAN to Generate Labeled Synthetic Face Image Dataset for Underrepresented Group. arXiv preprint arXiv:2308.03495.

Memon, F. S., Abdullah, F. B., Iqbal, R., Ahmad, S., Hussain, I., & Abdullah, M. (2023). Addressing women's climate change awareness in Sindh, Pakistan: An empirical study of rural and urban women. *Climate and Development*, *15*(7), 565–577. doi:10.1080/17565529.2022.2125784

Miao, J., & Zhu, W. (2022). Precision–recall curve (PRC) classification trees. *Evolutionary Intelligence*, *15*(3), 1545–1569. doi:10.1007/s12065-021-00565-2

Miller, R. B., & Small, C. (2003). Cities from space: Potential applications of remote sensing in urban environmental research and policy. *Environmental Science & Policy*, *6*(2), 129–137. doi:10.1016/S1462-9011(03)00002-9

Mobiny, A., & Van Nguyen, H. (2018, September). Fast capsnet for lung cancer screening. In *International conference on medical image computing and computer-assisted intervention* (pp. 741-749). Cham: Springer International Publishing.

Moreira, A., Prats-Iraola, P., Younis, M., Krieger, G., Hajnsek, I., & Papathanassiou, K. P. (2013). A tutorial on synthetic aperture radar. *IEEE Geoscience and Remote Sensing Magazine*, *1*(1), 6–43. doi:10.1109/MGRS.2013.2248301

Morra, J. H., Tu, Z., Apostolova, L. G., Green, A. E., Toga, A. W., & Thompson, P. M. (2010). Comparison of adaboost and support vector machines for detecting alzheimer's disease through automated hippocampal segmentation. *IEEE Transactions on Medical Imaging*, *29*(1), 30–43. doi:10.1109/TMI.2009.2021941 PMID:19457748

MorrisM.R.Sohl-dicksteinJ.FiedelN.WarkentinT.DafoeA.FaustA.FarabetC.LeggS. (2023). Levels of agi: Operationalizing progress on the path to agi. arXiv:2311.02462.

Muhajir, M., Oktiana, M., Muchtar, K., Fitria, M., Akhyar, A., Pratama, M. D., & Lin, C. Y. (2023, August). USK-FEMO: A Face Emotion Dataset using Deep Learning for Effective Learning. In *2023 2nd International Conference on Computer System, Information Technology, and Electrical Engineering (COSITE)* (pp. 199-203). IEEE. 10.1109/COSITE60233.2023.10249834

Mukhometzianov, R., & Carrillo, J. (2018). *CapsNet comparative performance evaluation for image classification*. arXiv preprint arXiv:1805.11195.

Mullapudi, A., Vibhute, A. D., Mali, S., & Patil, C. H. (2023). A review of agricultural drought assessment with remote sensing data: Methods, issues, challenges and opportunities. *Applied Geomatics*, *15*(1), 1–13. doi:10.1007/s12518-022-00484-6

Munawar, F., Khan, U., Shahzad, A., Haq, M. U., Mahmood, Z., Khattak, S., & Khan, G. Z. (2019, January). An empirical study of image resolution and pose on automatic face recognition. In *2019 16th International Bhurban Conference on Applied Sciences and Technology (IBCAST)* (pp. 558-563). IEEE. 10.1109/IBCAST.2019.8667233

Munawar, F., Khan, U., Shahzad, A., Haq, M. U., Mahmood, Z., Khattak, S., & Khan, G. Z. (n.d.). *An Empirical Study of Image Resolution and Pose on Automatic Face Recognition*. Academic Press.

Munga, J. (2022). *To Close Africa's Digital Divide*. Policy Must Address the Usage Gap. Available: https://carnegieendowment.org/2022/04/26/to-close-africa-s-digital-divide-policy-must-address-usage-gap-pub-86959

Mzid, N., Pignatti, S., Huang, W., & Casa, R. (2021). An analysis of bare soil occurrence in arable croplands for remote sensing topsoil applications. *Remote Sensing (Basel)*, *13*(3), 474. doi:10.3390/rs13030474

Nair, P., Doshi, R., & Keselj, S. (2021). Pushing the limits of capsule networks. arXiv preprint arXiv:2103.08074.

Najibi, M., Samangouei, P., Chellappa, R., & Davis, L. S. (2018). SSH: Single Stage Headless Face Detector. *Proceedings of the European Conference on Computer Vision (ECCV)*.

Nan, G., Zhao, Y., Fu, L., & Ye, Q. (2024). Object Detection by Channel and Spatial Exchange for Multimodal Remote Sensing Imagery. *IEEE Journal of Selected Topics in Applied Earth Observations and Remote Sensing*, *17*, 8581–8593. doi:10.1109/JSTARS.2024.3388013

Naseem, R., Khan, B., Shah, M. A., Wakil, K., Khan, A., Alosaimi, W., Uddin, M. I., & Alouffi, B. (2020). Performance Assessment of Classification Algorithms on Early Detection of Liver Syndrome. *Journal of Healthcare Engineering*, *2020*, 1–13. Advance online publication. doi:10.1155/2020/6680002 PMID:33489060

Netzer, Y., Wang, T., Coates, A., Bissacco, A., Wu, B. & Ng, A.Y. (2011). *Reading digits in natural images with unsupervised feature learning.* Academic Press.

Ng, M. Y., Lee, E. Y. P., Yang, J., Yang, F., Li, X., Wang, H., Lui, M. M., Lo, C. S.-Y., Leung, B., Khong, P.-L., Hui, C. K.-M., Yuen, K., & Kuo, M. D. (2020). Imaging profile of the covid-19 infection: Radiologic findings and literature review. *Radiology. Cardiothoracic Imaging*, *2*(1), e200034. Advance online publication. doi:10.1148/ryct.2020200034 PMID:33778547

Nguyen, H. P., & Ribeiro, B. (2019). Advanced capsule networks via context awareness. In *Artificial Neural Networks and Machine Learning–ICANN 2019: Theoretical Neural Computation: 28th International Conference on Artificial Neural Networks, Munich, Germany, September 17–19, 2019, Proceedings, Part I 28* (pp. 166-177). Springer International Publishing.

Nishikawa, S., Ri, R., Yamada, I., Tsuruoka, Y., & Echizen, I. (2022). *Ease: Entity-aware contrastive learning of sentence embedding.* arXiv preprint arXiv:2205.04260. doi:10.18653/v1/2022.naacl-main.284

Nishio, M., Noguchi, S., Matsuo, H., & Murakami, T. (2020). Automatic classification between COVID-19 pneumonia, non-COVID-19 pneumonia, and the healthy on chest X-ray image: Combination of data augmentation methods. *Scientific Reports*, *10*(1), 17532. Advance online publication. doi:10.1038/s41598-020-74539-2 PMID:33067538

Noble, W. S. (2006). What is a support vector machine? *Nature Biotechnology*, *24*(12), 1565–1567. doi:10.1038/nbt1206-1565 PMID:17160063

O'Niel, C. (2016). *Weapons of math destruction.* Crown/Archetype.

Ohm, P. (2009). Broken promises of privacy: Responding to the surprising failure of anonymization. *UCLA l. Rev.*, *57*, 1701.

Olson, D., & Anderson, J. (2021). Review on unmanned aerial vehicles, remote sensors, imagery processing, and their applications in agriculture. *Agronomy Journal*, *113*(2), 971–992. doi:10.1002/agj2.20595

OpenA. I. (2023). *GPT-4 technical report.* arXiv:2303.0877

Osco, L. P., Junior, J. M., Ramos, A. P. M., de Castro Jorge, L. A., Fatholahi, S. N., de Andrade Silva, J., Matsubara, E. T., Pistori, H., Gonçalves, W. N., & Li, J. (2021). A review on deep learning in uav remote sensing. *International Journal of Applied Earth Observation and Geoinformation*, *102*, 102456. doi:10.1016/j.jag.2021.102456

Palmer, M. C., & King, R. H. (2006). Has salary discrimination really disappeared from Major League Baseball? *Eastern Economic Journal*, *32*(2), 285–297.

Panis, G., Lanitis, A., Tsapatsoulis, N., & Cootes, T. F. (2016). Overview of research on facial ageing using the FG-NET ageing database. *IET Biometrics*, *5*(2), 37–46. doi:10.1049/iet-bmt.2014.0053

Panwar, H., Gupta, P. K., Siddiqui, M. K., Morales-Menendez, R., Bhardwaj, P., & Singh, V. (2020). A deep learning and grad-CAM based color visualization approach for fast detection of COVID-19 cases using chest X-ray and CT-Scan images. *Chaos, Solitons, and Fractals*, *140*, 110190. Advance online publication. doi:10.1016/j.chaos.2020.110190 PMID:32836918

Pan, Z., Xu, J., Guo, Y., Hu, Y., & Wang, G. (2020). Deep learning segmentation and classification for urban village using a worldview satellite image based on u-net. *Remote Sensing (Basel)*, *12*(10), 1574. doi:10.3390/rs12101574

Pappalardo, L., Cintia, P., Rossi, A., Massucco, E., Ferragina, P., Pedreschi, D., & Giannotti, F. (2019). A public data set of spatio-temporal match events in soccer competitions. *Scientific Data*, *6*(1), 236. doi:10.1038/s41597-019-0247-7 PMID:31659162

Parkhi, O. M., Vedaldi, A., & Zisserman, A. (2015). Deep Face Recognition. *Proceedings of the British Machine Vision Conference*, 1-12. 10.5244/C.29.41

ParkJ. S.O'BrienJ. C.CaiC. J.MorrisM. R.LiangP.BernsteinM. S. (2023) *Generative agents: interactive simulacra of human behavior.* arXiv:2304.03442 doi:10.1145/3586183.3606763

Pathak, Y., Shukla, P. K., Tiwari, A., Stalin, S., Singh, S., & Shukla, P. K. (2022). Deep Transfer Learning Based Classification Model for COVID-19 Disease. *Ingénierie et Recherche Biomédicale : IRBM = Biomedical Engineering and Research*, *43*(2), 87–92. doi:10.1016/j.irbm.2020.05.003 PMID:32837678

Patrick, M. K., Adekoya, A. F., Mighty, A. A., & Edward, B. Y. (2022). Capsule networks–a survey. *Journal of King Saud University. Computer and Information Sciences*, *34*(1), 1295–1310. doi:10.1016/j.jksuci.2019.09.014

Peer, D., Stabinger, S., & Rodriguez-Sanchez, A. (2018). Training deep capsule networks. arXiv preprint arXiv:1812.09707.

Peng, C., Li, Y., Jiao, L., Chen, Y., & Shang, R. (2019). Densely based multiscale and multi-modal fully convolutional networks for high-resolution remote-sensing image semantic segmentation. *IEEE Journal of Selected Topics in Applied Earth Observations and Remote Sensing*, *12*(8), 2612–2626. doi:10.1109/JSTARS.2019.2906387

Petersen, K., Feldt, R., Mujtaba, S., & Mattsson, M. (2008). Systematic mapping studies in software engineering. *12th International Conference on Evaluation and Assessment in Software Engineering (EASE)*, *12*, 1–10. 10.14236/ewic/EASE2008.8

Phaye, R., Sikka, Dhall, & Bathula. (2018). Dense and diverse capsule networks: Making the capsules learn better. arXiv preprint arXiv:1805.04001

Phillips, P. J., Beveridge, J. R., Draper, B. A., Givens, G. H., O'Toole, A. J., Bolme, D. S., Dunlop, J., Lui, Y. M., Sahibzada, H., & Weimer, J. A. (2015). IARPA Janus Benchmark A (IJB-A) Face Dataset. *Proceedings of the IEEE Conference on Computer Vision and Pattern Recognition (CVPR)*.

Compilation of References

Phillips, P. J., Wechsler, H., Huang, J., & Rauss, P. J. (1998). The FERET database and evaluation procedure for face-recognition algorithms. *Image and Vision Computing*, *16*(5), 295–306. doi:10.1016/S0262-8856(97)00070-X

Pires de Lima, R., & Marfurt, K. (2019). Convolutional neural network for remotesensing scene classification: Transfer learning analysis. *Remote Sensing (Basel)*, *12*(1), 86. doi:10.3390/rs12010086

Quan, H., Xu, X., Zheng, T., Li, Z., Zhao, M., & Cui, X. (2021). DenseCapsNet: Detection of COVID-19 from X-ray images using a capsule neural network. *Computers in Biology and Medicine*, *133*, 104399. doi:10.1016/j.compbiomed.2021.104399 PMID:33892307

Rabinowitz, N. C., Perbet, F., Song, H. F., Zhang, C., Eslami, S. M. A., & Botvinick, M. M. (2018) Machine theory of mind. *International conference on machine learning, PMLR.* http://proceedings.mlr.press/v80/rabinowitz18a.html

Radford, A., Kim, J. W., Hallacy, C., Ramesh, A., Goh, G., Agarwal, S., Sastry, G., Askell, A., Mishkin, P., & Clark, J. (2021). Learning transferable visual models from natural language supervision. *International conference on machine learning*, 8748–8763.

Radford, A., Metz, L., & Chintala, S. (2015). Unsupervised representation learning with deep convolutional generative adversarial networks. arXiv preprint arXiv:1511.06434.

Rahim, M., Amin, F., Tag Eldin, E. M., Khalifa, A. E. W., & Ahmad, S. (2024). p, q-Spherical fuzzy sets and their aggregation operators with application to third-party logistic provider selection. *Journal of Intelligent & Fuzzy Systems*, (Preprint), 1-24.

Rahim, A., Zhong, Y., Ahmad, T., Ahmad, S., & ElAffendi, M. A. (2023). Hyper-Tuned Convolutional Neural Networks for Authorship Verification in Digital Forensic Investigations. *Computers, Materials & Continua*, *76*(2). Advance online publication. doi:10.32604/cmc.2023.039340

Rahim, A., Zhong, Y., Ahmad, T., Ahmad, S., Pławiak, P., & Hammad, M. (2023). P. P lawiak, M. Hammad, Enhancing smart home security: Anomaly detection and face recognition in smart home iot devices using logit-boosted cnn models. *Sensors (Basel)*, *23*(15), 6979. doi:10.3390/s23156979 PMID:37571762

Rajeshkumar, G., Braveen, M., Venkatesh, R., Shermila, P. J., Prabu, B. G., Veerasamy, B., ... Jeyam, A. (2023). Smart office automation via faster R-CNN based face recognition and internet of things. *Measurement. Sensors*, *27*, 100719. doi:10.1016/j.measen.2023.100719

Ram, O., Levine, Y., Dalmedigos, I., Muhlgay, D., Shashua, A., Leyton-Brown, K., & Shoham, Y. (2023). *In-context retrieval-augmented language models.* arXiv preprint arXiv:2302.00083.

Ramasinghe, S., Athuraliya, C. D., & Khan, S. H. (2018). A context-aware capsule network for multi-label classification. *Proceedings of the European Conference on Computer Vision (ECCV) Workshops.*

Ramaswamy, V., & Ozcan, K. (2018). What is co-creation? An interactional creation framework and its implications for value creation. *Journal of Business Research*, *84*, 196–205. doi:10.1016/j.jbusres.2017.11.027

Rathnayaka, P., Abeysinghe, S., Samarajeewa, C., Manchanayake, I., & Walpola, M. (2018). Sentylic at IEST 2018: Gated recurrent neural network and capsule network-based approach for implicit emotion detection. arXiv preprint arXiv:1809.01452. doi:10.18653/v1/W18-6237

Redmon, J., & Farhadi, A. (2018). *Yolov3: An incremental improvement.* arXiv preprint arXiv:1804.02767.

Reed, S., Zolna, K., Parisotto, E., Colmenarejo, S. G., Novikov, A., Barth-Maron, G., Gimenez, M., Sulsky, Y., Kay, J., & Springenberg, J. T. (2022). *A generalist agent.* arXiv preprint arXiv:2205.06175.

Rehman, A., Saba, T., Mujahid, M., Alamri, F. S., & ElHakim, N. (2023). Parkinson's disease detection using hybrid lstm-gru deep learning model. *Electronics (Basel)*, *12*(13), 2856. doi:10.3390/electronics12132856

Reisenbichler, M., Reutterer, T., Schweidel, D. A., & Dan, D. (2022). Frontiers: Supporting content marketing with natural language generation. *Marketing Science*, *41*(3), 441–452. doi:10.1287/mksc.2022.1354

Rombach, R., Blattmann, A., Lorenz, D., Esser, P., & Ommer, B. (2022). Highresolution image synthesis with latent diffusion models. IEEE/CVF conference on computer vision and pattern recognition, 10684–10695.

Rossi, A. (2019). *Legal design for the general data protection regulation. A methodology for the visualization and communication of legal concepts.* Academic Press.

Ru, L., Du, B., & Wu, C. (2020). Multi-temporal scene classification and scene change detection with correlation based fusion. *IEEE Transactions on Image Processing*, *30*, 1382–1394. doi:10.1109/TIP.2020.3039328 PMID:33237858

Sabour, S., Frosst, N., & Hinton, G. (2018, February). Matrix capsules with EM routing. In *6th international conference on learning representations, ICLR* (pp. 1-15). Academic Press.

Sabour, S., Frosst, N., & Hinton, G. E. (2017). Dynamic routing between capsules. *Advances in Neural Information Processing Systems*, 30.

Sachs, G. (2023). *Generative AI could raise global GDP by 7%.* https://www.goldmansachs.com/insights/pages/generative-aicould-raise-global-gdp-by-7-percent.html

Saenz, A. D., Harned, Z., Banerjee, O., Abràmoff, M. D., & Rajpurkar, P. (2023). "Autonomous ai systems in the face of liability, regulations and costs. *NPJ Digital Medicine*, *6*, 13. PMID:37803209

Sagar, A. S., Chen, Y., Xie, Y., & Kim, H. S. (2024). MSA R-CNN: A comprehensive approach to remote sensing object detection and scene understanding. *Expert Systems with Applications*, *241*, 122788. doi:10.1016/j.eswa.2023.122788

Sánchez Pedroche, D., Amigo, D., García, J., & Molina, J. M. (2020). Architecture for Trajectory-Based Fishing Ship Classification with AIS Data. *Sensors (Basel)*, *20*(13), 3782. Advance online publication. doi:10.3390/s20133782 PMID:32640561

Saqur, R., & Vivona, S. (2018). Capsgan: Using dynamic routing for generative adversarial networks. arXiv preprint arXiv:1806.03968.

Saqur, R., & Vivona, S. (2020). Capsgan: Using dynamic routing for generative adversarial networks. In A*dvances in Computer Vision: Proceedings of the 2019 Computer Vision Conference (CVC),* Volume 2 (pp. 511-525). Springer International Publishing. 10.1007/978-3-030-17798-0_41

Schneier, B. (2018). Artificial intelligence and the attack/defense balance. *IEEE Security and Privacy*, *16*(02), 96–96. doi:10.1109/MSP.2018.1870857

Shahid Anwar, M., Wang, J., Ahmad, S., Ullah, A., Khan, W., & Fei, Z. (2020). Evaluating the factors affecting QoE of 360-degree videos and cybersickness levels predictions in virtual reality. *Electronics (Basel)*, *9*(9), 1530. doi:10.3390/electronics9091530

Shah, V., Keniya, R., Shridharani, A., Punjabi, M., Shah, J., & Mehendale, N. (2021). Diagnosis of COVID-19 using CT scan images and deep learning techniques. *Emergency Radiology*, *28*(3), 497–505. doi:10.1007/s10140-020-01886-y PMID:33523309

Shankar, K., & Perumal, E. (2021). A novel hand-crafted with deep learning features based fusion model for COVID-19 diagnosis and classification using chest X-ray images. *Complex & Intelligent Systems*, *7*(3), 1277–1293. doi:10.1007/s40747-020-00216-6 PMID:34777955

Sheng, Z. H. A. N. G., Shanshan, L. I., Guofang, W. E. I., Xinnai, Z. H. A. N. G., & Jianwei, G. A. O. (2024). Refined multi-scale feature-oriented object detection of the remote sensing images. *National Remote Sensing Bulletin*, *26*(12), 2616–2628.

Shi, W., Min, S., Yasunaga, M., Seo, M., James, R., Lewis, M., Zettlemoyer, L., & Yih, W.-t. (2023). *Replug: Retrieval augmented black-box language models*. arXiv preprint arXiv:2301.12652.

Shi, R., Niu, L., & Zhou, R. (2022). Sparse CapsNet with explicit regularizer. *Pattern Recognition*, *124*, 108486. doi:10.1016/j.patcog.2021.108486

Siddiqui, H. U. R., Younas, F., Rustam, F., Flores, E. S., Ballester, J. B., Diez, I. D. L. T., Dudley, S., & Ashraf, I. (2023). Enhancing Cricket Performance Analysis with Human Pose Estimation and Machine Learning. *Sensors (Basel)*, *23*(15), 6839. doi:10.3390/s23156839 PMID:37571624

Signoroni, A., Savardi, M., Baronio, A., & Benini, S. (2019). Deep learning meets hyperspectral image analysis: A multidisciplinary review. *Journal of Imaging*, *5*(5), 52. doi:10.3390/jimaging5050052 PMID:34460490

Silver, M., Tiwari, A., & Karnieli, A. (2019). Identifying vegetation in arid regions using object-based image analysis with rgb-only aerial imagery. *Remote Sensing (Basel)*, *11*(19), 2308. doi:10.3390/rs11192308

Simonyan, K., & Zisserman, A. (2014). Very deep convolutional networks for large-scale image recognition. *ArXiv Preprint ArXiv:1409.1556.*

Simonyan, K., & Zisserman, A. (2015). Very deep convolutional networks for large-scale image recognition. *3rd International Conference on Learning Representations, ICLR 2015 - Conference Track Proceedings.*

Sim, T., Baker, S., & Bsat, M. (2002, May). The CMU pose, illumination, and expression (PIE) database. In *Proceedings of fifth IEEE international conference on automatic face gesture recognition* (pp. 53-58). IEEE. 10.1109/AFGR.2002.1004130

Singh, M., Nagpal, S., Singh, R., & Vatsa, M. (2019). Dual directed capsule network for very low-resolution image recognition. In *Proceedings of the IEEE/CVF International Conference on Computer Vision* (pp. 340-349). 10.1109/ICCV.2019.00043

Slonecker, E. T., Shaw, D. M., & Lillesand, T. M. (1998). Emerging legal and ethical issues in advanced remote sensing technology. *Photogrammetric Engineering and Remote Sensing, 64*(6), 589–595.

Sohail, M. Z., Zafar, T., Khan, T. A., Asim, M., Ahmad, S., Mairaj, T., & El Affendi, M. A. (2023). Prediction of Time to Failure (TTF) of Power Systems Using a Deep Learning Technique. *Journal of Hunan University Natural Sciences, 50*(12).

Sowmya, D. R., Deepa Shenoy, P., & Venugopal, K. R. (2017). Remote sensing satellite image processing techniques for image classification: A comprehensive survey. *International Journal of Computer Applications, 161*(11), 24–37. doi:10.5120/ijca2017913306

Sreekala, K., Cyril, C. P. D., Neelakandan, S., Chandrasekaran, S., Walia, R., & Martinson, E. O. (2022). Capsule network-based deep transfer learning model for face recognition. *Wireless Communications and Mobile Computing, 2022,* 1–12. doi:10.1155/2022/2086613

Sridhar, S., & Sanagavarapu, S. (2021, January). Fake news detection and analysis using multitask learning with BiLSTM CapsNet model. In *2021 11th International Conference on Cloud Computing, Data Science & Engineering (Confluence)* (pp. 905-911). IEEE. 10.1109/Confluence51648.2021.9377080

Stephen, M., Potter, K., & Mohamed, S. (2024). Ethical Considerations in Machine Learning: Balancing Innovation and Responsibility. *Computer Science.*

Sun, H., Fu, L., Li, J., Guo, Q., Meng, Z., Zhang, T., ... Yu, H. (2024). Defense against Adversarial Cloud Attack on Remote Sensing Salient Object Detection. In *Proceedings of the IEEE/CVF Winter Conference on Applications of Computer Vision* (pp. 8345-8354). 10.1109/WACV57701.2024.00816

Sun, X., Shi, A., Huang, H., & Mayer, H. (2020). Bas ^{4} net: Boundary-aware semi-supervised semantic segmentation network for very high resolution remote sensing images. *IEEE Journal of Selected Topics in Applied Earth Observations and Remote Sensing, 13,* 5398–5413. doi:10.1109/JSTARS.2020.3021098

Szegedy, C., Vanhoucke, V., Ioffe, S., Shlens, J., & Wojna, Z. (2016). Rethinking the inception architecture for computer vision. *Proceedings of the IEEE Conference on Computer Vision and Pattern Recognition*, 2818–2826. 10.1109/CVPR.2016.308

Szegedy, S. (2017). Inception-v4, inceptionResNet and the impact of residual connections on learning. *31st AAAI Conference on Artificial Intelligence, AAAI 2017*, 4278–4284. 10.1609/aaai.v31i1.11231

Tahsin, M. S., Al Karim, M., Ahmed, M. U., Tafannum, F., & Firoz, N. (2023). An integrated approach for diabetes detection using fisher score feature selection and capsule network. *Journal of Computing Science and Engineering : JCSE*, *4*(2), 61–77.

Tampubolon, H., Yang, C. L., Chan, A. S., Sutrisno, H., & Hua, K. L. (2019). Optimized capsnet for traffic jam speed prediction using mobile sensor data under urban swarming transportation. *Sensors (Basel)*, *19*(23), 5277. doi:10.3390/s19235277 PMID:31795519

Tang, S., Wu, W., Yan, J., & Zhang, X. (2018). PyramidBox: A context-assisted single shot face detector. *Proceedings of the European Conference on Computer Vision (ECCV)*. 10.1007/978-3-030-01240-3_49

Tartaglione, E., Barbano, C. A., Berzovini, C., Calandri, M., & Grangetto, M. (2020). Unveiling COVID-19 from chest x-ray with deep learning: A hurdles race with small data. *International Journal of Environmental Research and Public Health*, *17*(18), 1–17. doi:10.3390/ijerph17186933 PMID:32971995

Teto, J. K., & Xie, Y. (2019, March). Automatically Identifying of animals in the wilderness: Comparative studies between CNN and C-Capsule Network. In *Proceedings of the 2019 3rd International Conference on Compute and Data Analysis* (pp. 128-133). 10.1145/3314545.3314559

Tharwat, A., Gaber, T., Ibrahim, A., & Hassanien, A. E. (2017). Linear discriminant analysis: A detailed tutorial. *AI Communications*, *30*(2), 169–190. doi:10.3233/AIC-170729

Theckedath, D., & Sedamkar, R. R. (2020). Detecting affect states using VGG16, ResNet50 and SE-ResNet50 networks. *SN Computer Science*, *1*(2), 1–7. doi:10.1007/s42979-020-0114-9

Tiwari, S., & Jain, A. (2021). Convolutional capsule network for COVID-19 detection using radiography images. *International Journal of Imaging Systems and Technology*, *31*(2), 525–539. doi:10.1002/ima.22566 PMID:33821095

Tokarev, K., Orlova, Y. A., Rogachev, A., Chernyavsky, A., & Tokareva, Y. M. (2020). The intelligent analysis system and remote sensing images segmentation engineering by using methods of advanced machine learning and neural network modeling. *IOP Conference Series: Materials Science and Engineering*, *734*, 012124. 10.1088/1757-899X/734/1/012124

Turk, M., & Pentland, A. (1991). Eigenfaces for recognition. *Journal of Cognitive Neuroscience*, *3*(1), 71–86. doi:10.1162/jocn.1991.3.1.71 PMID:23964806

Ullah, A., Jami, A., Aziz, M. W., Naeem, F., Ahmad, S., Anwar, M. S., & Jing, W. (2019, December). Deep Facial Expression Recognition of facial variations using fusion of feature extraction with classification in end to end model. In *2019 4th International Conference on Emerging Trends in Engineering, Sciences and Technology (ICEEST)* (pp. 1-6). IEEE. 10.1109/ICEEST48626.2019.8981687

Ullah, H., Haq, M. U., Khattak, S., Khan, G. Z., & Mahmood, Z. (n.d.). *A Robust Face Recognition Method for Occluded and Low-Resolution Images.* Academic Press.

Ullah, A., Ullah, A., Ahmad, S., Haq, M. U., Shah, K., & Mlaiki, N. (2022). Series type solution of fuzzy fractional order Swift–Hohenberg equation by fuzzy hybrid Sumudu transform. *Mathematical Problems in Engineering*, 2022, 2022. doi:10.1155/2022/3864053

Ullah, F., Khan, M. Z., Faisal, M., Rehman, H. U., Abbas, S., & Mubarek, F. S. (2021). An energy efficient and reliable routing scheme to enhance the stability period in wireless body area networks. *Computer Communications*, 165, 20–32. doi:10.1016/j.comcom.2020.10.017

Ullah, H., Haq, M. U., Khattak, S., Khan, G. Z., & Mahmood, Z. (2019, August). A robust face recognition method for occluded and low-resolution images. In *2019 International Conference on Applied and Engineering Mathematics (ICAEM)* (pp. 86-91). IEEE. 10.1109/ICAEM.2019.8853753

Van Calster, B., McLernon, D. J., Van Smeden, M., Wynants, L., Steyerberg, E. W., & Collins, P. B. G. S. (2019). Calibration: The Achilles heel of predictive analytics. *BMC Medicine*, 17(1), 230. doi:10.1186/s12916-019-1466-7 PMID:31842878

Varoquaux, G., & Cheplygina, V. (2022). Machine learning for medical imaging: Methodological failures and recommendations for the future. *NPJ Digital Medicine*, 5(1), 48. doi:10.1038/s41746-022-00592-y PMID:35413988

Vogels, M. F., de Jong, S. M., Sterk, G., & Addink, E. A. (2019). Mapping irrigated agriculture in complex landscapes using spot6 imagery and object-based image analysis–a case study in the central rift valley, ethiopia–. *International Journal of Applied Earth Observation and Geoinformation*, 75, 118–129. doi:10.1016/j.jag.2018.07.019

Wahab, H., Mehmood, I., Ugail, H., Sangaiah, A. K., & Muhammad, K. (2023). Machine learning based small bowel video capsule endoscopy analysis: Challenges and opportunities. *Future Generation Computer Systems*, 143, 191–214. doi:10.1016/j.future.2023.01.011

Wang, Y., Li, P., Sun, M., & Liu, Y. (2023b). *Self-knowledge guided retrieval augmentation for large language models.* arXiv preprint arXiv:2310.05002. doi:10.18653/v1/2023.findings-emnlp.691

Wang, Z., Cai, S., Liu, A., Ma, X., & Liang, Y. (2023). *Describe, explain, plan and select: Interactive planning with large language models enables open-world multi-task agents.* arXiv preprint arXiv:2302.01560.

Wang, C., Liu, B., Liu, L., Zhu, Y., Hou, J., Liu, P., & Li, X. (2021). A review of deep learning used in the hyperspectral image analysis for agriculture. *Artificial Intelligence Review*, 54(7), 5205–5253. doi:10.1007/s10462-021-10018-y

Wang, P., Sun, X., Diao, W., & Fu, K. (2019). Fmssd: Feature-merged single-shot detection for multiscale objects in large-scale remote sensing imagery. *IEEE Transactions on Geoscience and Remote Sensing*, *58*(5), 3377–3390. doi:10.1109/TGRS.2019.2954328

Wang, Q. W., Shen, F., Cheng, L., Jiang, J., He, G., Sheng, W., Jing, N., & Zhigang, M. (2021). Ship detection based on fused features and rebuilt YOLOv3 networks in optical remote-sensing images. *International Journal of Remote Sensing*, *42*(2), 520–536. doi:10.1080/01431161.2020.1811422

Wang, W., Liu, K., Tang, R., & Wang, S. (2019). Remote sensing image-based analysis of the urban heat island effect in shenzhen, china. *Physics and Chemistry of the Earth Parts A/B/C*, *110*, 168–175. doi:10.1016/j.pce.2019.01.002

Wang, X., Wang, Y., Guo, S., Kong, L., & Cui, G. (2023). Capsule Network With Multiscale Feature Fusion for Hidden Human Activity Classification. *IEEE Transactions on Instrumentation and Measurement*, *72*, 1–12. doi:10.1109/TIM.2023.3238749

Wang, Y., Ning, D., & Feng, S. (2020). A novel capsule network based on wide convolution and multi-scale convolution for fault diagnosis. *Applied Sciences (Basel, Switzerland)*, *10*(10), 3659. doi:10.3390/app10103659

Wang, Z., Bovik, A. C., Sheikh, H. R., & Simoncelli, E. P. (2004). Image quality assessment: From error visibility to structural similarity. *IEEE Transactions on Image Processing*, *13*(4), 600–612. doi:10.1109/TIP.2003.819861 PMID:15376593

Wang, Z., Huang, B., Wang, G., Yi, P., & Jiang, K. (2023). Masked face recognition dataset and application. *IEEE Transactions on Biometrics, Behavior, and Identity Science*, *5*(2), 298–304. doi:10.1109/TBIOM.2023.3242085

Wang, Z., Zheng, L., Du, W., Cai, W., Zhou, J., Wang, J., Han, X., & He, G. (2019). A novel method for intelligent fault diagnosis of bearing based on capsule neural network. *Complexity*, *2019*, 2019. doi:10.1155/2019/6943234

Waweru, L. W., Kipyego, B. T., & Muchangi, D. M. (2021). Classification of plant leaf diseases based on capsule network-support vector machine model. *Int J Electr Eng Technol*, *12*, 188–199.

Wen, D., Huang, X., Bovolo, F., Li, J., Ke, X., Zhang, A., & Benediktsson, J. A. (2021). Change detection from very-high-spatial-resolution optical remote sensing images: Methods, applications, and future directions. *IEEE Geoscience and Remote Sensing Magazine*, *9*(4), 68–101. doi:10.1109/MGRS.2021.3063465

Weng, Z., Meng, F., Liu, S., Zhang, Y., Zheng, Z., & Gong, C. (2022). Cattle face recognition based on a Two-Branch convolutional neural network. *Computers and Electronics in Agriculture*, *196*, 106871. doi:10.1016/j.compag.2022.106871

Wickramasinghe, I. (2022). Applications of machine learning in cricket: A systematic review. *Machine Learning with Applications*, *10*, 100435. doi:10.1016/j.mlwa.2022.100435

Wu, H., Mao, J., Sun, W., Zheng, B., Zhang, H., Chen, Z., & Wang, W. (2016, August). Probabilistic robust route recovery with spatio-temporal dynamics. In *Proceedings of the 22nd ACM SIGKDD International Conference on Knowledge Discovery and Data Mining* (pp. 1915-1924) 10.1145/2939672.2939843

Wu, J., Zhu, Y., Wang, Z., Song, Z., Liu, X., Wang, W., Zhang, Z., Yu, Y., Xu, Z., Zhang, T., & Zhou, J. (2017). A novel ship classification approach for high resolution SAR images based on the BDA-KELM classification model. *International Journal of Remote Sensing*, *38*(23), 6457–6476. doi:10.1080/01431161.2017.1356487

Wu, Z., Huang, Y., Zhang, L., & Wang, L. (2021). FaceDet: A Lightweight Anchor-Free Face Detector for Mobile Devices. *Proceedings of the IEEE Conference on Computer Vision and Pattern Recognition (CVPR)*.

Xiao, S., Zhang, Y., & Chang, X. (2022). Ship Detection Based on Compressive Sensing Measurements of Optical Remote Sensing Scenes. *IEEE Journal of Selected Topics in Applied Earth Observations and Remote Sensing*, *15*, 8632–8649. doi:10.1109/JSTARS.2022.3209024

Xie, Y., Tian, J., & Zhu, X. X. (2020). Linking points with labels in 3d: A review of point cloud semantic segmentation. *IEEE Geoscience and Remote Sensing Magazine*, *8*(4), 38–59. doi:10.1109/MGRS.2019.2937630

Xu, M., Huang, P., Yu, W., Liu, S., Zhang, X., Niu, Y., Zhang, T., Xia, F., Tan, J., & Zhao, D. (2023). *Creative robot tool use with large language models*. arXiv preprint arXiv:2310.13065.

Xu, Y., Du, B., Zhang, L., Cerra, D., Pato, M., Carmona, E., Prasad, S., Yokoya, N., Ha¨nsch, R., & Le Saux, B. (2019). Advanced multi-sensor optical remote sensing for urban land use and land cover classification: Outcome of the 2018 ieee grss data fusion contest. *IEEE Journal of Selected Topics in Applied Earth Observations and Remote Sensing*, *12*(6), 1709–1724. doi:10.1109/JSTARS.2019.2911113

Yang, B., Bao, W., & Wang, J. (2022). Active disease-related compound identification based on capsule network. *Briefings in Bioinformatics*, *23*(1), bbab462. doi:10.1093/bib/bbab462 PMID:35057581

Yang, S., Luo, P., Loy, C. C., & Tang, X. (2016). Wider face: A face detection benchmark. In *Proceedings of the IEEE conference on computer vision and pattern recognition* (pp. 5525-5533). IEEE.

Yan, Z., Song, X., Zhong, H., Yang, L., & Wang, Y. (2022). Ship Classification and Anomaly Detection Based on Spaceborne AIS Data Considering Behavior Characteristics. *Sensors (Basel)*, *22*(20), 7713. Advance online publication. doi:10.3390/s22207713 PMID:36298063

Yi, D., Su, J., & Chen, W.-H. (2021). Probabilistic faster r-cnn with stochastic region proposing: Towards object detection and recognition in remote sensing imagery. *Neurocomputing*, *459*, 290–301. doi:10.1016/j.neucom.2021.06.072

Yin, S., Li, H., Teng, L., Jiang, M., & Karim, S. (2020). An optimised multi-scale fusion method for airport detection in large-scale optical remote sensing images. *International Journal of Image and Data Fusion*, *11*(2), 201–214. doi:10.1080/19479832.2020.1727573

Yu, J., Wang, X., Tu, S., Cao, S., Zhang-Li, D., Lv, X., Peng, H., Yao, Z., Zhang, X., & Li, H. (2023). *Kola: Carefully benchmarking world knowledge of large language models*. arXiv preprint arXiv:2306.09296.

Yu, H., & Yang, J. (2001). A direct LDA algorithm for high-dimensional data—With application to face recognition. *Pattern Recognition*, *34*(10), 2067–2070. doi:10.1016/S0031-3203(00)00162-X

Yu, S., Jiang, Y., Lu, J., & Zhou, J. (2019). CenterFace: Joint Face Detection and Alignment Using Face as Point. *Proceedings of the IEEE Conference on Computer Vision and Pattern Recognition (CVPR)*.

Zamir, A., Khan, H. U., Iqbal, T., Yousaf, N., Aslam, F., Anjum, A., & Hamdani, M. (2020). Phishing web site detection using diverse machine learning algorithms. *The Electronic Library*, *38*(1), 65–80. doi:10.1108/EL-05-2019-0118

Zhang, X. (2021). The AlexNet, LeNet-5 and VGG NET applied to CIFAR-10. In *2021 2nd International Conference on Big Data & Artificial Intelligence & Software Engineering (ICBASE)* (pp. 414-419). IEEE. 10.1109/ICBASE53849.2021.00083

Zhang, Y., Liu, T., Yu, P., Wang, S., & Tao, R. (2024). SFSANet: Multi-scale Object Detection In Remote Sensing Image Based on Semantic Fusion and Scale Adaptability. *IEEE Transactions on Geoscience and Remote Sensing*.

Zhang, C., Yue, P., Tapete, D., Jiang, L., Shangguan, B., Huang, L., & Liu, G. (2020). A deeply supervised image fusion network for change detection in high resolution bi-temporal remote sensing images. *ISPRS Journal of Photogrammetry and Remote Sensing*, *166*, 183–200. doi:10.1016/j.isprsjprs.2020.06.003

Zhang, G., Yu, W., & Hou, R. (2024). MFIL-FCOS: A Multi-Scale Fusion and Interactive Learning Method for 2D Object Detection and Remote Sensing Image Detection. *Remote Sensing (Basel)*, *16*(6), 936. doi:10.3390/rs16060936

Zhang, H., Tian, M., Shao, G., Cheng, J., & Liu, J. (2022). Target detection of forward-looking sonar image based on improved yolov5. *IEEE Access : Practical Innovations, Open Solutions*, *10*, 18023–18034. doi:10.1109/ACCESS.2022.3150339

Zhang, J., Shan, S., Kan, M., & Chen, X. (2014). WebFace: A Scalable Face Image Dataset with Varying Pose and Age. *Proceedings of the IEEE Conference on Computer Vision and Pattern Recognition (CVPR)*.

Zhang, L., Edraki, M., & Qi, G.-J. (2018). Cappronet: Deep feature learning via orthogonal projections onto capsule subspaces. *Advances in Neural Information Processing Systems*, 31.

Zhang, S., Wu, R., Xu, K., Wang, J., & Sun, W. (2019). R-cnn-based ship detection from high resolution remote sensing imagery. *Remote Sensing (Basel)*, *11*(6), 631. doi:10.3390/rs11060631

Zhang, T., Zhang, X., Zhu, X., Wang, G., Han, X., Tang, X., & Jiao, L. (2024). Multistage Enhancement Network for Tiny Object Detection in Remote Sensing Images. *IEEE Transactions on Geoscience and Remote Sensing*, *62*, 1–12. doi:10.1109/TGRS.2024.3396874

Zhang, Y., Sheng, W., Jiang, J., Jing, N., Wang, Q., & Mao, Z. (2020). Priority Branches for Ship Detection in Optical Remote Sensing Images. *Remote Sensing (Basel)*, *12*(7), 1196. Advance online publication. doi:10.3390/rs12071196

Zhao, Z., Kleinhans, A., Sandhu, G., Patel, I., & Unnikrishnan, K. P. (2019). *Capsule networks with max-min normalization.* arXiv preprint arXiv:1903.09662

Zheng, X., Fan, Y., Wu, B., Zhang, Y., Wang, J., & Pan, S. (2023). Robust physical-world attacks on face recognition. *Pattern Recognition*, *133*, 109009. doi:10.1016/j.patcog.2022.109009

Zhou, Y., Daamen, W., Vellinga, T., & Hoogendoorn, S. P. (2019). Ship classification based on ship behavior clustering from AIS data. *Ocean Engineering*, *175*, 176–187. doi:10.1016/j.oceaneng.2019.02.005

Zhu, X., Lyu, S., Wang, X., & Zhao, Q. (2021). Tph-yolov5: Improved yolov5 based on transformer prediction head for object detection on drone-captured scenarios. *Proceedings of the IEEE/CVF international conference on computer vision*, 2778–2788.

Zhuang, Y., Qi, B., Chen, H., Bi, F., Li, L., & Xie, Y. (2018). Locally Oriented Scene Complexity Analysis Real-Time Ocean Ship Detection from Optical Remote Sensing Images. *Sensors (Basel)*, *18*(11), 3799. Advance online publication. doi:10.3390/s18113799 PMID:30404224

ZieglerD. M.StiennonN.WuJ.BrownT. B.RadfordA.AmodeiD.ChristianoP.IrvingG. (2019). *Fine-tuning language models from human preferences.* arXiv:1909.08593

Zuboff, S. (2023). The age of surveillance capitalism. In *Social theory re-wired* (pp. 203–213). Routledge. doi:10.4324/9781003320609-27

Related References

To continue our tradition of advancing academic research, we have compiled a list of recommended IGI Global readings. These references will provide additional information and guidance to further enrich your knowledge and assist you with your own research and future publications.

Abbasnejad, B., Moeinzadeh, S., Ahankoob, A., & Wong, P. S. (2021). The Role of Collaboration in the Implementation of BIM-Enabled Projects. In J. Underwood & M. Shelbourn (Eds.), *Handbook of Research on Driving Transformational Change in the Digital Built Environment* (pp. 27–62). IGI Global. https://doi.org/10.4018/978-1-7998-6600-8.ch002

Abdulrahman, K. O., Mahamood, R. M., & Akinlabi, E. T. (2022). Additive Manufacturing (AM): Processing Technique for Lightweight Alloys and Composite Material. In K. Kumar, B. Babu, & J. Davim (Ed.), *Handbook of Research on Advancements in the Processing, Characterization, and Application of Lightweight Materials* (pp. 27-48). IGI Global. https://doi.org/10.4018/978-1-7998-7864-3.ch002

Agrawal, R., Sharma, P., & Saxena, A. (2021). A Diamond Cut Leather Substrate Antenna for BAN (Body Area Network) Application. In V. Singh, V. Dubey, A. Saxena, R. Tiwari, & H. Sharma (Eds.), *Emerging Materials and Advanced Designs for Wearable Antennas* (pp. 54–59). IGI Global. https://doi.org/10.4018/978-1-7998-7611-3.ch004

Ahmad, F., Al-Ammar, E. A., & Alsaidan, I. (2022). Battery Swapping Station: A Potential Solution to Address the Limitations of EV Charging Infrastructure. In M. Alam, R. Pillai, & N. Murugesan (Eds.), *Developing Charging Infrastructure and Technologies for Electric Vehicles* (pp. 195–207). IGI Global. doi:10.4018/978-1-7998-6858-3.ch010

Aikhuele, D. (2018). A Study of Product Development Engineering and Design Reliability Concerns. *International Journal of Applied Industrial Engineering*, 5(1), 79–89. doi:10.4018/IJAIE.2018010105

Al-Khatri, H., & Al-Atrash, F. (2021). Occupants' Habits and Natural Ventilation in a Hot Arid Climate. In R. González-Lezcano (Ed.), *Advancements in Sustainable Architecture and Energy Efficiency* (pp. 146–168). IGI Global. https://doi.org/10.4018/978-1-7998-7023-4.ch007

Al-Shebeeb, O. A., Rangaswamy, S., Gopalakrishan, B., & Devaru, D. G. (2017). Evaluation and Indexing of Process Plans Based on Electrical Demand and Energy Consumption. *International Journal of Manufacturing, Materials, and Mechanical Engineering*, 7(3), 1–19. doi:10.4018/IJMMME.2017070101

Amuda, M. O., Lawal, T. F., & Akinlabi, E. T. (2017). Research Progress on Rheological Behavior of AA7075 Aluminum Alloy During Hot Deformation. *International Journal of Materials Forming and Machining Processes*, 4(1), 53–96. doi:10.4018/IJMFMP.2017010104

Amuda, M. O., Lawal, T. F., & Mridha, S. (2021). Microstructure and Mechanical Properties of Silicon Carbide-Treated Ferritic Stainless Steel Welds. In L. Burstein (Ed.), *Handbook of Research on Advancements in Manufacturing, Materials, and Mechanical Engineering* (pp. 395–411). IGI Global. https://doi.org/10.4018/978-1-7998-4939-1.ch019

Anikeev, V., Gasem, K. A., & Fan, M. (2021). Application of Supercritical Technologies in Clean Energy Production: A Review. In L. Chen (Ed.), *Handbook of Research on Advancements in Supercritical Fluids Applications for Sustainable Energy Systems* (pp. 792–821). IGI Global. https://doi.org/10.4018/978-1-7998-5796-9.ch022

Arafat, M. Y., Saleem, I., & Devi, T. P. (2022). Drivers of EV Charging Infrastructure Entrepreneurship in India. In M. Alam, R. Pillai, & N. Murugesan (Eds.), *Developing Charging Infrastructure and Technologies for Electric Vehicles* (pp. 208–219). IGI Global. https://doi.org/10.4018/978-1-7998-6858-3.ch011

Araujo, A., & Manninen, H. (2022). Contribution of Project-Based Learning on Social Skills Development: An Industrial Engineer Perspective. In A. Alves & N. van Hattum-Janssen (Eds.), *Training Engineering Students for Modern Technological Advancement* (pp. 119–145). IGI Global. https://doi.org/10.4018/978-1-7998-8816-1.ch006

Related References

Armutlu, H. (2018). Intelligent Biomedical Engineering Operations by Cloud Computing Technologies. In U. Kose, G. Guraksin, & O. Deperlioglu (Eds.), *Nature-Inspired Intelligent Techniques for Solving Biomedical Engineering Problems* (pp. 297–317). Hershey, PA: IGI Global. doi:10.4018/978-1-5225-4769-3.ch015

Atik, M., Sadek, M., & Shahrour, I. (2017). Single-Run Adaptive Pushover Procedure for Shear Wall Structures. In V. Plevris, G. Kremmyda, & Y. Fahjan (Eds.), *Performance-Based Seismic Design of Concrete Structures and Infrastructures* (pp. 59–83). Hershey, PA: IGI Global. doi:10.4018/978-1-5225-2089-4.ch003

Attia, H. (2021). Smart Power Microgrid Impact on Sustainable Building. In R. González-Lezcano (Ed.), *Advancements in Sustainable Architecture and Energy Efficiency* (pp. 169–194). IGI Global. https://doi.org/10.4018/978-1-7998-7023-4.ch008

Aydin, A., Akyol, E., Gungor, M., Kaya, A., & Tasdelen, S. (2018). Geophysical Surveys in Engineering Geology Investigations With Field Examples. In N. Ceryan (Ed.), *Handbook of Research on Trends and Digital Advances in Engineering Geology* (pp. 257–280). Hershey, PA: IGI Global. doi:10.4018/978-1-5225-2709-1.ch007

Ayoobkhan, M. U. D., Y., A., J., Easwaran, B., & R., T. (2021). Smart Connected Digital Products and IoT Platform With the Digital Twin. In P. Vasant, G. Weber, & W. Punurai (Ed.), Research Advancements in Smart Technology, Optimization, and Renewable Energy (pp. 330-350). IGI Global. https://doi.org/ doi:10.4018/978-1-7998-3970-5.ch016

Baeza Moyano, D., & González Lezcano, R. A. (2021). The Importance of Light in Our Lives: Towards New Lighting in Schools. In R. González-Lezcano (Ed.), *Advancements in Sustainable Architecture and Energy Efficiency* (pp. 239–256). IGI Global. https://doi.org/10.4018/978-1-7998-7023-4.ch011

Bagdadee, A. H. (2021). A Brief Assessment of the Energy Sector of Bangladesh. *International Journal of Energy Optimization and Engineering, 10*(1), 36–55. doi:10.4018/IJEOE.2021010103

Baklezos, A. T., & Hadjigeorgiou, N. G. (2021). Magnetic Sensors for Space Applications and Magnetic Cleanliness Considerations. In C. Nikolopoulos (Ed.), *Recent Trends on Electromagnetic Environmental Effects for Aeronautics and Space Applications* (pp. 147–185). IGI Global. https://doi.org/10.4018/978-1-7998-4879-0.ch006

Bas, T. G. (2017). Nutraceutical Industry with the Collaboration of Biotechnology and Nutrigenomics Engineering: The Significance of Intellectual Property in the Entrepreneurship and Scientific Research Ecosystems. In T. Bas & J. Zhao (Eds.), *Comparative Approaches to Biotechnology Development and Use in Developed and Emerging Nations* (pp. 1–17). Hershey, PA: IGI Global. doi:10.4018/978-1-5225-1040-6.ch001

Bazeer Ahamed, B., & Periakaruppan, S. (2021). Taxonomy of Influence Maximization Techniques in Unknown Social Networks. In P. Vasant, G. Weber, & W. Punurai (Eds.), *Research Advancements in Smart Technology, Optimization, and Renewable Energy* (pp. 351-363). IGI Global. https://doi.org/10.4018/978-1-7998-3970-5.ch017

Beale, R., & André, J. (2017). *Design Solutions and Innovations in Temporary Structures*. Hershey, PA: IGI Global. doi:10.4018/978-1-5225-2199-0

Behnam, B. (2017). Simulating Post-Earthquake Fire Loading in Conventional RC Structures. In P. Samui, S. Chakraborty, & D. Kim (Eds.), *Modeling and Simulation Techniques in Structural Engineering* (pp. 425–444). Hershey, PA: IGI Global. doi:10.4018/978-1-5225-0588-4.ch015

Ben Hamida, I., Salah, S. B., Msahli, F., & Mimouni, M. F. (2018). Distribution Network Reconfiguration Using SPEA2 for Power Loss Minimization and Reliability Improvement. *International Journal of Energy Optimization and Engineering*, 7(1), 50–65. doi:10.4018/IJEOE.2018010103

Bentarzi, H. (2021). Fault Tree-Based Root Cause Analysis Used to Study Mal-Operation of a Protective Relay in a Smart Grid. In A. Recioui & H. Bentarzi (Eds.), *Optimizing and Measuring Smart Grid Operation and Control* (pp. 289–308). IGI Global. https://doi.org/10.4018/978-1-7998-4027-5.ch012

Beysens, D. A., Garrabos, Y., & Zappoli, B. (2021). Thermal Effects in Near-Critical Fluids: Piston Effect and Related Phenomena. In L. Chen (Ed.), *Handbook of Research on Advancements in Supercritical Fluids Applications for Sustainable Energy Systems* (pp. 1–31). IGI Global. https://doi.org/10.4018/978-1-7998-5796-9.ch001

Bhaskar, S. V., & Kudal, H. N. (2017). Effect of TiCN and AlCrN Coating on Tribological Behaviour of Plasma-nitrided AISI 4140 Steel. *International Journal of Surface Engineering and Interdisciplinary Materials Science*, 5(2), 1–17. doi:10.4018/IJSEIMS.2017070101

Bhuyan, D. (2018). Designing of a Twin Tube Shock Absorber: A Study in Reverse Engineering. In K. Kumar & J. Davim (Eds.), *Design and Optimization of Mechanical Engineering Products* (pp. 83–104). Hershey, PA: IGI Global. doi:10.4018/978-1-5225-3401-3.ch005

Blumberg, G. (2021). Blockchains for Use in Construction and Engineering Projects. In J. Underwood & M. Shelbourn (Eds.), *Handbook of Research on Driving Transformational Change in the Digital Built Environment* (pp. 179–208). IGI Global. https://doi.org/10.4018/978-1-7998-6600-8.ch008

Bolboaca, A. M. (2021). Considerations Regarding the Use of Fuel Cells in Combined Heat and Power for Stationary Applications. In G. Badea, R. Felseghi, & I. Aşchilean (Eds.), *Hydrogen Fuel Cell Technology for Stationary Applications* (pp. 239–275). IGI Global. https://doi.org/10.4018/978-1-7998-4945-2.ch010

Burstein, L. (2021). Simulation Tool for Cable Design. In L. Burstein (Ed.), *Handbook of Research on Advancements in Manufacturing, Materials, and Mechanical Engineering* (pp. 54–74). IGI Global. https://doi.org/10.4018/978-1-7998-4939-1.ch003

Calderon, F. A., Giolo, E. G., Frau, C. D., Rengel, M. G., Rodriguez, H., Tornello, M., ... Gallucci, R. (2018). Seismic Microzonation and Site Effects Detection Through Microtremors Measures: A Review. In N. Ceryan (Ed.), *Handbook of Research on Trends and Digital Advances in Engineering Geology* (pp. 326–349). Hershey, PA: IGI Global. doi:10.4018/978-1-5225-2709-1.ch009

Ceryan, N., & Can, N. K. (2018). Prediction of The Uniaxial Compressive Strength of Rocks Materials. In N. Ceryan (Ed.), *Handbook of Research on Trends and Digital Advances in Engineering Geology* (pp. 31–96). Hershey, PA: IGI Global. doi:10.4018/978-1-5225-2709-1.ch002

Ceryan, S. (2018). Weathering Indices Used in Evaluation of the Weathering State of Rock Material. In N. Ceryan (Ed.), *Handbook of Research on Trends and Digital Advances in Engineering Geology* (pp. 132–186). Hershey, PA: IGI Global. doi:10.4018/978-1-5225-2709-1.ch004

Chen, H., Padilla, R. V., & Besarati, S. (2017). Supercritical Fluids and Their Applications in Power Generation. In L. Chen & Y. Iwamoto (Eds.), *Advanced Applications of Supercritical Fluids in Energy Systems* (pp. 369–402). Hershey, PA: IGI Global. doi:10.4018/978-1-5225-2047-4.ch012

Chen, H., Padilla, R. V., & Besarati, S. (2021). Supercritical Fluids and Their Applications in Power Generation. In L. Chen (Ed.), *Handbook of Research on Advancements in Supercritical Fluids Applications for Sustainable Energy Systems* (pp. 566–599). IGI Global. https://doi.org/10.4018/978-1-7998-5796-9.ch016

Chen, L. (2017). Principles, Experiments, and Numerical Studies of Supercritical Fluid Natural Circulation System. In L. Chen & Y. Iwamoto (Eds.), *Advanced Applications of Supercritical Fluids in Energy Systems* (pp. 136–187). Hershey, PA: IGI Global. doi:10.4018/978-1-5225-2047-4.ch005

Chen, L. (2021). Principles, Experiments, and Numerical Studies of Supercritical Fluid Natural Circulation System. In L. Chen (Ed.), *Handbook of Research on Advancements in Supercritical Fluids Applications for Sustainable Energy Systems* (pp. 219–269). IGI Global. https://doi.org/10.4018/978-1-7998-5796-9.ch007

Chiba, Y., Marif, Y., Henini, N., & Tlemcani, A. (2021). Modeling of Magnetic Refrigeration Device by Using Artificial Neural Networks Approach. *International Journal of Energy Optimization and Engineering*, *10*(4), 68–76. https://doi.org/10.4018/IJEOE.2021100105

Clementi, F., Di Sciascio, G., Di Sciascio, S., & Lenci, S. (2017). Influence of the Shear-Bending Interaction on the Global Capacity of Reinforced Concrete Frames: A Brief Overview of the New Perspectives. In V. Plevris, G. Kremmyda, & Y. Fahjan (Eds.), *Performance-Based Seismic Design of Concrete Structures and Infrastructures* (pp. 84–111). Hershey, PA: IGI Global. doi:10.4018/978-1-5225-2089-4.ch004

Codinhoto, R., Fialho, B. C., Pinti, L., & Fabricio, M. M. (2021). BIM and IoT for Facilities Management: Understanding Key Maintenance Issues. In J. Underwood & M. Shelbourn (Eds.), *Handbook of Research on Driving Transformational Change in the Digital Built Environment* (pp. 209–231). IGI Global. doi:10.4018/978-1-7998-6600-8.ch009

Cortés-Polo, D., Calle-Cancho, J., Carmona-Murillo, J., & González-Sánchez, J. (2017). Future Trends in Mobile-Fixed Integration for Next Generation Networks: Classification and Analysis. *International Journal of Vehicular Telematics and Infotainment Systems*, *1*(1), 33–53. doi:10.4018/IJVTIS.2017010103

Costa, H. G., Sheremetieff, F. H., & Araújo, E. A. (2022). Influence of Game-Based Methods in Developing Engineering Competences. In A. Alves & N. van Hattum-Janssen (Eds.), *Training Engineering Students for Modern Technological Advancement* (pp. 69–88). IGI Global. https://doi.org/10.4018/978-1-7998-8816-1.ch004

Related References

Cui, X., Zeng, S., Li, Z., Zheng, Q., Yu, X., & Han, B. (2018). Advanced Composites for Civil Engineering Infrastructures. In K. Kumar & J. Davim (Eds.), *Composites and Advanced Materials for Industrial Applications* (pp. 212–248). Hershey, PA: IGI Global. doi:10.4018/978-1-5225-5216-1.ch010

Dalgıç, S., & Kuşku, İ. (2018). Geological and Geotechnical Investigations in Tunneling. In N. Ceryan (Ed.), *Handbook of Research on Trends and Digital Advances in Engineering Geology* (pp. 482–529). Hershey, PA: IGI Global. doi:10.4018/978-1-5225-2709-1.ch014

Dang, C., & Hihara, E. (2021). Study on Cooling Heat Transfer of Supercritical Carbon Dioxide Applied to Transcritical Carbon Dioxide Heat Pump. In L. Chen (Ed.), *Handbook of Research on Advancements in Supercritical Fluids Applications for Sustainable Energy Systems* (pp. 451–493). IGI Global. https://doi.org/10.4018/978-1-7998-5796-9.ch013

Daus, Y., Kharchenko, V., & Yudaev, I. (2021). Research of Solar Energy Potential of Photovoltaic Installations on Enclosing Structures of Buildings. *International Journal of Energy Optimization and Engineering*, *10*(4), 18–34. https://doi.org/10.4018/IJEOE.2021100102

Daus, Y., Kharchenko, V., & Yudaev, I. (2021). Optimizing Layout of Distributed Generation Sources of Power Supply System of Agricultural Object. *International Journal of Energy Optimization and Engineering*, *10*(3), 70–84. https://doi.org/10.4018/IJEOE.2021070104

de la Varga, D., Soto, M., Arias, C. A., van Oirschot, D., Kilian, R., Pascual, A., & Álvarez, J. A. (2017). Constructed Wetlands for Industrial Wastewater Treatment and Removal of Nutrients. In Á. Val del Río, J. Campos Gómez, & A. Mosquera Corral (Eds.), *Technologies for the Treatment and Recovery of Nutrients from Industrial Wastewater* (pp. 202–230). Hershey, PA: IGI Global. doi:10.4018/978-1-5225-1037-6.ch008

Deb, S., Ammar, E. A., AlRajhi, H., Alsaidan, I., & Shariff, S. M. (2022). V2G Pilot Projects: Review and Lessons Learnt. In M. Alam, R. Pillai, & N. Murugesan (Eds.), *Developing Charging Infrastructure and Technologies for Electric Vehicles* (pp. 252–267). IGI Global. https://doi.org/10.4018/978-1-7998-6858-3.ch014

Dekhandji, F. Z., & Rais, M. C. (2021). A Comparative Study of Power Quality Monitoring Using Various Techniques. In A. Recioui & H. Bentarzi (Eds.), *Optimizing and Measuring Smart Grid Operation and Control* (pp. 259–288). IGI Global. https://doi.org/10.4018/978-1-7998-4027-5.ch011

Deperlioglu, O. (2018). Intelligent Techniques Inspired by Nature and Used in Biomedical Engineering. In U. Kose, G. Guraksin, & O. Deperlioglu (Eds.), *Nature-Inspired Intelligent Techniques for Solving Biomedical Engineering Problems* (pp. 51–77). Hershey, PA: IGI Global. doi:10.4018/978-1-5225-4769-3.ch003

Dhurpate, P. R., & Tang, H. (2021). Quantitative Analysis of the Impact of Inter-Line Conveyor Capacity for Throughput of Manufacturing Systems. *International Journal of Manufacturing, Materials, and Mechanical Engineering*, *11*(1), 1–17. https://doi.org/10.4018/IJMMME.2021010101

Dinkar, S., & Deep, K. (2021). A Survey of Recent Variants and Applications of Antlion Optimizer. *International Journal of Energy Optimization and Engineering*, *10*(2), 48–73. doi:10.4018/IJEOE.2021040103

Dixit, A. (2018). Application of Silica-Gel-Reinforced Aluminium Composite on the Piston of Internal Combustion Engine: Comparative Study of Silica-Gel-Reinforced Aluminium Composite Piston With Aluminium Alloy Piston. In K. Kumar & J. Davim (Eds.), *Composites and Advanced Materials for Industrial Applications* (pp. 63–98). Hershey, PA: IGI Global. doi:10.4018/978-1-5225-5216-1.ch004

Drabecki, M. P., & Kułak, K. B. (2021). Global Pandemics on European Electrical Energy Markets: Lessons Learned From the COVID-19 Outbreak. *International Journal of Energy Optimization and Engineering*, *10*(3), 24–46. https://doi.org/10.4018/IJEOE.2021070102

Dutta, M. M. (2021). Nanomaterials for Food and Agriculture. In M. Bhat, I. Wani, & S. Ashraf (Eds.), *Applications of Nanomaterials in Agriculture, Food Science, and Medicine* (pp. 75–97). IGI Global. doi:10.4018/978-1-7998-5563-7.ch004

Dutta, M. M., & Goswami, M. (2021). Coating Materials: Nano-Materials. In S. Roy & G. Bose (Eds.), *Advanced Surface Coating Techniques for Modern Industrial Applications* (pp. 1–30). IGI Global. doi:10.4018/978-1-7998-4870-7.ch001

Elsayed, A. M., Dakkama, H. J., Mahmoud, S., Al-Dadah, R., & Kaialy, W. (2017). Sustainable Cooling Research Using Activated Carbon Adsorbents and Their Environmental Impact. In T. Kobayashi (Ed.), *Applied Environmental Materials Science for Sustainability* (pp. 186–221). Hershey, PA: IGI Global. doi:10.4018/978-1-5225-1971-3.ch009

Ercanoglu, M., & Sonmez, H. (2018). General Trends and New Perspectives on Landslide Mapping and Assessment Methods. In N. Ceryan (Ed.), *Handbook of Research on Trends and Digital Advances in Engineering Geology* (pp. 350–379). Hershey, PA: IGI Global. doi:10.4018/978-1-5225-2709-1.ch010

Related References

Faroz, S. A., Pujari, N. N., Rastogi, R., & Ghosh, S. (2017). Risk Analysis of Structural Engineering Systems Using Bayesian Inference. In P. Samui, S. Chakraborty, & D. Kim (Eds.), *Modeling and Simulation Techniques in Structural Engineering* (pp. 390–424). Hershey, PA: IGI Global. doi:10.4018/978-1-5225-0588-4.ch014

Fekik, A., Hamida, M. L., Denoun, H., Azar, A. T., Kamal, N. A., Vaidyanathan, S., Bousbaine, A., & Benamrouche, N. (2022). Multilevel Inverter for Hybrid Fuel Cell/PV Energy Conversion System. In A. Fekik & N. Benamrouche (Eds.), *Modeling and Control of Static Converters for Hybrid Storage Systems* (pp. 233–270). IGI Global. https://doi.org/10.4018/978-1-7998-7447-8.ch009

Fekik, A., Hamida, M. L., Houassine, H., Azar, A. T., Kamal, N. A., Denoun, H., Vaidyanathan, S., & Sambas, A. (2022). Power Quality Improvement for Grid-Connected Photovoltaic Panels Using Direct Power Control. In A. Fekik & N. Benamrouche (Eds.), *Modeling and Control of Static Converters for Hybrid Storage Systems* (pp. 107–142). IGI Global. https://doi.org/10.4018/978-1-7998-7447-8.ch005

Fernando, P. R., Hamigah, T., Disne, S., Wickramasingha, G. G., & Sutharshan, A. (2018). The Evaluation of Engineering Properties of Low Cost Concrete Blocks by Partial Doping of Sand with Sawdust: Low Cost Sawdust Concrete Block. *International Journal of Strategic Engineering*, *1*(2), 26–42. doi:10.4018/IJoSE.2018070103

Ferro, G., Minciardi, R., Parodi, L., & Robba, M. (2022). Optimal Charging Management of Microgrid-Integrated Electric Vehicles. In M. Alam, R. Pillai, & N. Murugesan (Eds.), *Developing Charging Infrastructure and Technologies for Electric Vehicles* (pp. 133–155). IGI Global. https://doi.org/10.4018/978-1-7998-6858-3.ch007

Flumerfelt, S., & Green, C. (2022). Graduate Lean Leadership Education: A Case Study of a Program. In A. Alves & N. van Hattum-Janssen (Eds.), *Training Engineering Students for Modern Technological Advancement* (pp. 202–224). IGI Global. https://doi.org/10.4018/978-1-7998-8816-1.ch010

Galli, B. J. (2021). Implications of Economic Decision Making to the Project Manager. *International Journal of Strategic Engineering*, *4*(1), 19–32. https://doi.org/10.4018/IJoSE.2021010102

Gento, A. M., Pimentel, C., & Pascual, J. A. (2022). Teaching Circular Economy and Lean Management in a Learning Factory. In A. Alves & N. van Hattum-Janssen (Eds.), *Training Engineering Students for Modern Technological Advancement* (pp. 183–201). IGI Global. https://doi.org/10.4018/978-1-7998-8816-1.ch009

Ghosh, S., Mitra, S., Ghosh, S., & Chakraborty, S. (2017). Seismic Reliability Analysis in the Framework of Metamodelling Based Monte Carlo Simulation. In P. Samui, S. Chakraborty, & D. Kim (Eds.), *Modeling and Simulation Techniques in Structural Engineering* (pp. 192–208). Hershey, PA: IGI Global. doi:10.4018/978-1-5225-0588-4.ch006

Gil, M., & Otero, B. (2017). Learning Engineering Skills through Creativity and Collaboration: A Game-Based Proposal. In R. Alexandre Peixoto de Queirós & M. Pinto (Eds.), *Gamification-Based E-Learning Strategies for Computer Programming Education* (pp. 14–29). Hershey, PA: IGI Global. doi:10.4018/978-1-5225-1034-5.ch002

Gill, J., Ayre, M., & Mills, J. (2017). Revisioning the Engineering Profession: How to Make It Happen! In M. Gray & K. Thomas (Eds.), *Strategies for Increasing Diversity in Engineering Majors and Careers* (pp. 156–175). Hershey, PA: IGI Global. doi:10.4018/978-1-5225-2212-6.ch008

Godzhaev, Z., Senkevich, S., Kuzmin, V., & Melikov, I. (2021). Use of the Neural Network Controller of Sprung Mass to Reduce Vibrations From Road Irregularities. In P. Vasant, G. Weber, & W. Punurai (Ed.), *Research Advancements in Smart Technology, Optimization, and Renewable Energy* (pp. 69-87). IGI Global. https://doi.org/10.4018/978-1-7998-3970-5.ch005

Gomes de Gusmão, C. M. (2022). Digital Competencies and Transformation in Higher Education: Upskilling With Extension Actions. In A. Alves & N. van Hattum-Janssen (Eds.), *Training Engineering Students for Modern Technological Advancement* (pp. 313–328). IGI Global. https://doi.org/10.4018/978-1-7998-8816-1.ch015A

Goyal, N., Ram, M., & Kumar, P. (2017). Welding Process under Fault Coverage Approach for Reliability and MTTF. In M. Ram & J. Davim (Eds.), *Mathematical Concepts and Applications in Mechanical Engineering and Mechatronics* (pp. 222–245). Hershey, PA: IGI Global. doi:10.4018/978-1-5225-1639-2.ch011

Gray, M., & Lundy, C. (2017). Engineering Study Abroad: High Impact Strategy for Increasing Access. In M. Gray & K. Thomas (Eds.), *Strategies for Increasing Diversity in Engineering Majors and Careers* (pp. 42–59). Hershey, PA: IGI Global. doi:10.4018/978-1-5225-2212-6.ch003

Güler, O., & Varol, T. (2021). Fabrication of Functionally Graded Metal and Ceramic Powders Synthesized by Electroless Deposition. In S. Roy & G. Bose (Eds.), *Advanced Surface Coating Techniques for Modern Industrial Applications* (pp. 150–187). IGI Global. https://doi.org/10.4018/978-1-7998-4870-7.ch007

Related References

Guraksin, G. E. (2018). Internet of Things and Nature-Inspired Intelligent Techniques for the Future of Biomedical Engineering. In U. Kose, G. Guraksin, & O. Deperlioglu (Eds.), *Nature-Inspired Intelligent Techniques for Solving Biomedical Engineering Problems* (pp. 263–282). Hershey, PA: IGI Global. doi:10.4018/978-1-5225-4769-3.ch013

Hamida, M. L., Fekik, A., Denoun, H., Ardjal, A., & Bokhtache, A. A. (2022). Flying Capacitor Inverter Integration in a Renewable Energy System. In A. Fekik & N. Benamrouche (Eds.), *Modeling and Control of Static Converters for Hybrid Storage Systems* (pp. 287–306). IGI Global. https://doi.org/10.4018/978-1-7998-7447-8.ch011

Hasegawa, N., & Takahashi, Y. (2021). Control of Soap Bubble Ejection Robot Using Facial Expressions. *International Journal of Manufacturing, Materials, and Mechanical Engineering*, *11*(2), 1–16. https://doi.org/10.4018/IJMMME.2021040101

Hejazi, T., & Akbari, L. (2017). A Multiresponse Optimization Model for Statistical Design of Processes with Discrete Variables. In M. Ram & J. Davim (Eds.), *Mathematical Concepts and Applications in Mechanical Engineering and Mechatronics* (pp. 17–37). Hershey, PA: IGI Global. doi:10.4018/978-1-5225-1639-2.ch002

Hejazi, T., & Hejazi, A. (2017). Monte Carlo Simulation for Reliability-Based Design of Automotive Complex Subsystems. In M. Ram & J. Davim (Eds.), *Mathematical Concepts and Applications in Mechanical Engineering and Mechatronics* (pp. 177–200). Hershey, PA: IGI Global. doi:10.4018/978-1-5225-1639-2.ch009

Hejazi, T., & Poursabbagh, H. (2017). Reliability Analysis of Engineering Systems: An Accelerated Life Testing for Boiler Tubes. In M. Ram & J. Davim (Eds.), *Mathematical Concepts and Applications in Mechanical Engineering and Mechatronics* (pp. 154–176). Hershey, PA: IGI Global. doi:10.4018/978-1-5225-1639-2.ch008

Henao, J., Poblano-Salas, C. A., Vargas, F., Giraldo-Betancur, A. L., Corona-Castuera, J., & Sotelo-Mazón, O. (2021). Principles and Applications of Thermal Spray Coatings. In S. Roy & G. Bose (Eds.), *Advanced Surface Coating Techniques for Modern Industrial Applications* (pp. 31–70). IGI Global. https://doi.org/10.4018/978-1-7998-4870-7.ch002

Henao, J., & Sotelo, O. (2018). Surface Engineering at High Temperature: Thermal Cycling and Corrosion Resistance. In A. Pakseresht (Ed.), *Production, Properties, and Applications of High Temperature Coatings* (pp. 131–159). Hershey, PA: IGI Global. doi:10.4018/978-1-5225-4194-3.ch006

Hrnčič, M. K., Cör, D., & Knez, Ž. (2021). Supercritical Fluids as a Tool for Green Energy and Chemicals. In L. Chen (Ed.), *Handbook of Research on Advancements in Supercritical Fluids Applications for Sustainable Energy Systems* (pp. 761–791). IGI Global. doi:10.4018/978-1-7998-5796-9.ch021

Ibrahim, O., Erdem, S., & Gurbuz, E. (2021). Studying Physical and Chemical Properties of Graphene Oxide and Reduced Graphene Oxide and Their Applications in Sustainable Building Materials. In R. González-Lezcano (Ed.), *Advancements in Sustainable Architecture and Energy Efficiency* (pp. 221–238). IGI Global. https://doi.org/10.4018/978-1-7998-7023-4.ch010

Ihianle, I. K., Islam, S., Naeem, U., & Ebenuwa, S. H. (2021). Exploiting Patterns of Object Use for Human Activity Recognition. In A. Nwajana & I. Ihianle (Eds.), *Handbook of Research on 5G Networks and Advancements in Computing, Electronics, and Electrical Engineering* (pp. 382–401). IGI Global. https://doi.org/10.4018/978-1-7998-6992-4.ch015

Ijemaru, G. K., Ngharamike, E. T., Oleka, E. U., & Nwajana, A. O. (2021). An Energy-Efficient Model for Opportunistic Data Collection in IoV-Enabled SC Waste Management. In A. Nwajana & I. Ihianle (Eds.), *Handbook of Research on 5G Networks and Advancements in Computing, Electronics, and Electrical Engineering* (pp. 1–19). IGI Global. https://doi.org/10.4018/978-1-7998-6992-4.ch001

Ilori, O. O., Adetan, D. A., & Umoru, L. E. (2017). Effect of Cutting Parameters on the Surface Residual Stress of Face-Milled Pearlitic Ductile Iron. *International Journal of Materials Forming and Machining Processes*, 4(1), 38–52. doi:10.4018/IJMFMP.2017010103

Imam, M. H., Tasadduq, I. A., Ahmad, A., Aldosari, F., & Khan, H. (2017). Automated Generation of Course Improvement Plans Using Expert System. *International Journal of Quality Assurance in Engineering and Technology Education*, 6(1), 1–12. doi:10.4018/IJQAETE.2017010101

Injeti, S. K., & Kumar, T. V. (2018). A WDO Framework for Optimal Deployment of DGs and DSCs in a Radial Distribution System Under Daily Load Pattern to Improve Techno-Economic Benefits. *International Journal of Energy Optimization and Engineering*, 7(2), 1–38. doi:10.4018/IJEOE.2018040101

Ishii, N., Anami, K., & Knisely, C. W. (2018). *Dynamic Stability of Hydraulic Gates and Engineering for Flood Prevention*. Hershey, PA: IGI Global. doi:10.4018/978-1-5225-3079-4

Iwamoto, Y., & Yamaguchi, H. (2021). Application of Supercritical Carbon Dioxide for Solar Water Heater. In L. Chen (Ed.), *Handbook of Research on Advancements in Supercritical Fluids Applications for Sustainable Energy Systems* (pp. 370–387). IGI Global. https://doi.org/10.4018/978-1-7998-5796-9.ch010

Jayapalan, S. (2018). A Review of Chemical Treatments on Natural Fibers-Based Hybrid Composites for Engineering Applications. In K. Kumar & J. Davim (Eds.), *Composites and Advanced Materials for Industrial Applications* (pp. 16–37). Hershey, PA: IGI Global. doi:10.4018/978-1-5225-5216-1.ch002

Kapetanakis, T. N., Vardiambasis, I. O., Ioannidou, M. P., & Konstantaras, A. I. (2021). Modeling Antenna Radiation Using Artificial Intelligence Techniques: The Case of a Circular Loop Antenna. In C. Nikolopoulos (Ed.), *Recent Trends on Electromagnetic Environmental Effects for Aeronautics and Space Applications* (pp. 186–225). IGI Global. https://doi.org/10.4018/978-1-7998-4879-0.ch007

Karkalos, N. E., Markopoulos, A. P., & Dossis, M. F. (2017). Optimal Model Parameters of Inverse Kinematics Solution of a 3R Robotic Manipulator Using ANN Models. *International Journal of Manufacturing, Materials, and Mechanical Engineering*, 7(3), 20–40. doi:10.4018/IJMMME.2017070102

Kelly, M., Costello, M., Nicholson, G., & O'Connor, J. (2021). The Evolving Integration of BIM Into Built Environment Programmes in a Higher Education Institute. In J. Underwood & M. Shelbourn (Eds.), *Handbook of Research on Driving Transformational Change in the Digital Built Environment* (pp. 294–326). IGI Global. https://doi.org/10.4018/978-1-7998-6600-8.ch012

Kesimal, A., Karaman, K., Cihangir, F., & Ercikdi, B. (2018). Excavatability Assessment of Rock Masses for Geotechnical Studies. In N. Ceryan (Ed.), *Handbook of Research on Trends and Digital Advances in Engineering Geology* (pp. 231–256). Hershey, PA: IGI Global. doi:10.4018/978-1-5225-2709-1.ch006

Knoflacher, H. (2017). The Role of Engineers and Their Tools in the Transport Sector after Paradigm Change: From Assumptions and Extrapolations to Science. In H. Knoflacher & E. Ocalir-Akunal (Eds.), *Engineering Tools and Solutions for Sustainable Transportation Planning* (pp. 1–29). Hershey, PA: IGI Global. doi:10.4018/978-1-5225-2116-7.ch001

Kose, U. (2018). Towards an Intelligent Biomedical Engineering With Nature-Inspired Artificial Intelligence Techniques. In U. Kose, G. Guraksin, & O. Deperlioglu (Eds.), *Nature-Inspired Intelligent Techniques for Solving Biomedical Engineering Problems* (pp. 1–26). Hershey, PA: IGI Global. doi:10.4018/978-1-5225-4769-3.ch001

Kostić, S. (2018). A Review on Enhanced Stability Analyses of Soil Slopes Using Statistical Design. In N. Ceryan (Ed.), *Handbook of Research on Trends and Digital Advances in Engineering Geology* (pp. 446–481). Hershey, PA: IGI Global. doi:10.4018/978-1-5225-2709-1.ch013

Kumar, A., Patil, P. P., & Prajapati, Y. K. (2018). *Advanced Numerical Simulations in Mechanical Engineering.* Hershey, PA: IGI Global. doi:10.4018/978-1-5225-3722-9

Kumar, G. R., Rajyalakshmi, G., & Manupati, V. K. (2017). Surface Micro Patterning of Aluminium Reinforced Composite through Laser Peening. *International Journal of Manufacturing, Materials, and Mechanical Engineering, 7*(4), 15–27. doi:10.4018/IJMMME.2017100102

Kumar, N., Basu, D. N., & Chen, L. (2021). Effect of Flow Acceleration and Buoyancy on Thermalhydraulics of sCO2 in Mini/Micro-Channel. In L. Chen (Ed.), *Handbook of Research on Advancements in Supercritical Fluids Applications for Sustainable Energy Systems* (pp. 161–182). IGI Global. doi:10.4018/978-1-7998-5796-9.ch005

Kumari, N., & Kumar, K. (2018). Fabrication of Orthotic Calipers With Epoxy-Based Green Composite. In K. Kumar & J. Davim (Eds.), *Composites and Advanced Materials for Industrial Applications* (pp. 157–176). Hershey, PA: IGI Global. doi:10.4018/978-1-5225-5216-1.ch008

Kuppusamy, R. R. (2018). Development of Aerospace Composite Structures Through Vacuum-Enhanced Resin Transfer Moulding Technology (VERTMTy): Vacuum-Enhanced Resin Transfer Moulding. In K. Kumar & J. Davim (Eds.), *Composites and Advanced Materials for Industrial Applications* (pp. 99–111). Hershey, PA: IGI Global. doi:10.4018/978-1-5225-5216-1.ch005

Kurganov, V. A., Zeigarnik, Y. A., & Maslakova, I. V. (2021). Normal and Deteriorated Heat Transfer Under Heating Turbulent Supercritical Pressure Coolants Flows in Round Tubes. In L. Chen (Ed.), *Handbook of Research on Advancements in Supercritical Fluids Applications for Sustainable Energy Systems* (pp. 494–532). IGI Global. https://doi.org/10.4018/978-1-7998-5796-9.ch014

Li, H., & Zhang, Y. (2021). Heat Transfer and Fluid Flow Modeling for Supercritical Fluids in Advanced Energy Systems. In L. Chen (Ed.), *Handbook of Research on Advancements in Supercritical Fluids Applications for Sustainable Energy Systems* (pp. 388–422). IGI Global. https://doi.org/10.4018/978-1-7998-5796-9.ch011

Loy, J., Howell, S., & Cooper, R. (2017). Engineering Teams: Supporting Diversity in Engineering Education. In M. Gray & K. Thomas (Eds.), *Strategies for Increasing Diversity in Engineering Majors and Careers* (pp. 106–129). Hershey, PA: IGI Global. doi:10.4018/978-1-5225-2212-6.ch006

Related References

Macher, G., Armengaud, E., Kreiner, C., Brenner, E., Schmittner, C., Ma, Z., ... Krammer, M. (2018). Integration of Security in the Development Lifecycle of Dependable Automotive CPS. In N. Druml, A. Genser, A. Krieg, M. Menghin, & A. Hoeller (Eds.), *Solutions for Cyber-Physical Systems Ubiquity* (pp. 383–423). Hershey, PA: IGI Global. doi:10.4018/978-1-5225-2845-6.ch015

Madhu, M. N., Singh, J. G., Mohan, V., & Ongsakul, W. (2021). Transmission Risk Optimization in Interconnected Systems: Risk-Adjusted Available Transfer Capability. In P. Vasant, G. Weber, & W. Punurai (Ed.), *Research Advancements in Smart Technology, Optimization, and Renewable Energy* (pp. 183-199). IGI Global. https://doi.org/10.4018/978-1-7998-3970-5.ch010

Mahendramani, G., & Lakshmana Swamy, N. (2018). Effect of Weld Groove Area on Distortion of Butt Welded Joints in Submerged Arc Welding. *International Journal of Manufacturing, Materials, and Mechanical Engineering*, 8(2), 33–44. doi:10.4018/IJMMME.2018040103

Makropoulos, G., Koumaras, H., Setaki, F., Filis, K., Lutz, T., Montowtt, P., Tomaszewski, L., Dybiec, P., & Järvet, T. (2021). 5G and Unmanned Aerial Vehicles (UAVs) Use Cases: Analysis of the Ecosystem, Architecture, and Applications. In A. Nwajana & I. Ihianle (Eds.), *Handbook of Research on 5G Networks and Advancements in Computing, Electronics, and Electrical Engineering* (pp. 36–69). IGI Global. https://doi.org/10.4018/978-1-7998-6992-4.ch003

Meric, E. M., Erdem, S., & Gurbuz, E. (2021). Application of Phase Change Materials in Construction Materials for Thermal Energy Storage Systems in Buildings. In R. González-Lezcano (Ed.), *Advancements in Sustainable Architecture and Energy Efficiency* (pp. 1–20). IGI Global. https://doi.org/10.4018/978-1-7998-7023-4.ch001

Mihret, E. T., & Yitayih, K. A. (2021). Operation of VANET Communications: The Convergence of UAV System With LTE/4G and WAVE Technologies. *International Journal of Smart Vehicles and Smart Transportation*, 4(1), 29–51. https://doi.org/10.4018/IJSVST.2021010103

Mir, M. A., Bhat, B. A., Sheikh, B. A., Rather, G. A., Mehraj, S., & Mir, W. R. (2021). Nanomedicine in Human Health Therapeutics and Drug Delivery: Nanobiotechnology and Nanobiomedicine. In M. Bhat, I. Wani, & S. Ashraf (Eds.), *Applications of Nanomaterials in Agriculture, Food Science, and Medicine* (pp. 229–251). IGI Global. doi:10.4018/978-1-7998-5563-7.ch013

Mohammadzadeh, S., & Kim, Y. (2017). Nonlinear System Identification of Smart Buildings. In P. Samui, S. Chakraborty, & D. Kim (Eds.), *Modeling and Simulation Techniques in Structural Engineering* (pp. 328–347). Hershey, PA: IGI Global. doi:10.4018/978-1-5225-0588-4.ch011

Molina, G. J., Aktaruzzaman, F., Soloiu, V., & Rahman, M. (2017). Design and Testing of a Jet-Impingement Instrument to Study Surface-Modification Effects by Nanofluids. *International Journal of Surface Engineering and Interdisciplinary Materials Science*, 5(2), 43–61. doi:10.4018/IJSEIMS.2017070104

Moreno-Rangel, A., & Carrillo, G. (2021). Energy-Efficient Homes: A Heaven for Respiratory Illnesses. In R. González-Lezcano (Ed.), *Advancements in Sustainable Architecture and Energy Efficiency* (pp. 49–71). IGI Global. https://doi.org/10.4018/978-1-7998-7023-4.ch003

Msomi, V., & Jantjies, B. T. (2021). Correlative Analysis Between Tensile Properties and Tool Rotational Speeds of Friction Stir Welded Similar Aluminium Alloy Joints. *International Journal of Surface Engineering and Interdisciplinary Materials Science*, 9(2), 58–78. https://doi.org/10.4018/IJSEIMS.2021070104

Muigai, M. N., Mwema, F. M., Akinlabi, E. T., & Obiko, J. O. (2021). Surface Engineering of Materials Through Weld-Based Technologies: An Overview. In S. Roy & G. Bose (Eds.), *Advanced Surface Coating Techniques for Modern Industrial Applications* (pp. 247–260). IGI Global. doi:10.4018/978-1-7998-4870-7.ch011

Mukherjee, A., Saeed, R. A., Dutta, S., & Naskar, M. K. (2017). Fault Tracking Framework for Software-Defined Networking (SDN). In C. Singhal & S. De (Eds.), *Resource Allocation in Next-Generation Broadband Wireless Access Networks* (pp. 247–272). Hershey, PA: IGI Global. doi:10.4018/978-1-5225-2023-8.ch011

Mukhopadhyay, A., Barman, T. K., & Sahoo, P. (2018). Electroless Nickel Coatings for High Temperature Applications. In K. Kumar & J. Davim (Eds.), *Composites and Advanced Materials for Industrial Applications* (pp. 297–331). Hershey, PA: IGI Global. doi:10.4018/978-1-5225-5216-1.ch013

Mwema, F. M., & Wambua, J. M. (2022). Machining of Poly Methyl Methacrylate (PMMA) and Other Olymeric Materials: A Review. In K. Kumar, B. Babu, & J. Davim (Eds.), *Handbook of Research on Advancements in the Processing, Characterization, and Application of Lightweight Materials* (pp. 363–379). IGI Global. https://doi.org/10.4018/978-1-7998-7864-3.ch016

Mykhailyshyn, R., Savkiv, V., Boyko, I., Prada, E., & Virgala, I. (2021). Substantiation of Parameters of Friction Elements of Bernoulli Grippers With a Cylindrical Nozzle. *International Journal of Manufacturing, Materials, and Mechanical Engineering*, *11*(2), 17–39. https://doi.org/10.4018/IJMMME.2021040102

Náprstek, J., & Fischer, C. (2017). Dynamic Stability and Post-Critical Processes of Slender Auto-Parametric Systems. In V. Plevris, G. Kremmyda, & Y. Fahjan (Eds.), *Performance-Based Seismic Design of Concrete Structures and Infrastructures* (pp. 128–171). Hershey, PA: IGI Global. doi:10.4018/978-1-5225-2089-4.ch006

Nautiyal, L., Shivach, P., & Ram, M. (2018). Optimal Designs by Means of Genetic Algorithms. In M. Ram & J. Davim (Eds.), *Soft Computing Techniques and Applications in Mechanical Engineering* (pp. 151–161). Hershey, PA: IGI Global. doi:10.4018/978-1-5225-3035-0.ch007

Nazir, R. (2017). Advanced Nanomaterials for Water Engineering and Treatment: Nano-Metal Oxides and Their Nanocomposites. In T. Saleh (Ed.), *Advanced Nanomaterials for Water Engineering, Treatment, and Hydraulics* (pp. 84–126). Hershey, PA: IGI Global. doi:10.4018/978-1-5225-2136-5.ch005

Nikolopoulos, C. D. (2021). Recent Advances on Measuring and Modeling ELF-Radiated Emissions for Space Applications. In C. Nikolopoulos (Ed.), *Recent Trends on Electromagnetic Environmental Effects for Aeronautics and Space Applications* (pp. 1–38). IGI Global. https://doi.org/10.4018/978-1-7998-4879-0.ch001

Nogueira, A. F., Ribeiro, J. C., Fernández de Vega, F., & Zenha-Rela, M. A. (2018). Evolutionary Approaches to Test Data Generation for Object-Oriented Software: Overview of Techniques and Tools. In M. Khosrow-Pour, D.B.A. (Ed.), Incorporating Nature-Inspired Paradigms in Computational Applications (pp. 162-194). Hershey, PA: IGI Global. https://doi.org/ doi:10.4018/978-1-5225-5020-4.ch006

Nwajana, A. O., Obi, E. R., Ijemaru, G. K., Oleka, E. U., & Anthony, D. C. (2021). Fundamentals of RF/Microwave Bandpass Filter Design. In A. Nwajana & I. Ihianle (Eds.), *Handbook of Research on 5G Networks and Advancements in Computing, Electronics, and Electrical Engineering* (pp. 149–164). IGI Global. https://doi.org/10.4018/978-1-7998-6992-4.ch005

Ogbodo, E. A. (2021). Comparative Study of Transmission Line Junction vs. Asynchronously Coupled Junction Diplexers. In A. Nwajana & I. Ihianle (Eds.), *Handbook of Research on 5G Networks and Advancements in Computing, Electronics, and Electrical Engineering* (pp. 326–336). IGI Global. https://doi.org/10.4018/978-1-7998-6992-4.ch013

Orosa, J. A., Vergara, D., Fraguela, F., & Masdías-Bonome, A. (2021). Statistical Understanding and Optimization of Building Energy Consumption and Climate Change Consequences. In R. González-Lezcano (Ed.), *Advancements in Sustainable Architecture and Energy Efficiency* (pp. 195–220). IGI Global. https://doi.org/10.4018/978-1-7998-7023-4.ch009

Osho, M. B. (2018). Industrial Enzyme Technology: Potential Applications. In S. Bharati & P. Chaurasia (Eds.), *Research Advancements in Pharmaceutical, Nutritional, and Industrial Enzymology* (pp. 375–394). Hershey, PA: IGI Global. doi:10.4018/978-1-5225-5237-6.ch017

Ouadi, A., & Zitouni, A. (2021). Phasor Measurement Improvement Using Digital Filter in a Smart Grid. In A. Recioui & H. Bentarzi (Eds.), *Optimizing and Measuring Smart Grid Operation and Control* (pp. 100–117). IGI Global. https://doi.org/10.4018/978-1-7998-4027-5.ch005

Padmaja, P., & Marutheswar, G. (2017). Certain Investigation on Secured Data Transmission in Wireless Sensor Networks. *International Journal of Mobile Computing and Multimedia Communications*, 8(1), 48–61. doi:10.4018/IJMCMC.2017010104

Palmer, S., & Hall, W. (2017). An Evaluation of Group Work in First-Year Engineering Design Education. In R. Tucker (Ed.), *Collaboration and Student Engagement in Design Education* (pp. 145–168). Hershey, PA: IGI Global. doi:10.4018/978-1-5225-0726-0.ch007

Panchenko, V. (2021). Prospects for Energy Supply of the Arctic Zone Objects of Russia Using Frost-Resistant Solar Modules. In P. Vasant, G. Weber, & W. Punurai (Eds.), *Research Advancements in Smart Technology, Optimization, and Renewable Energy* (pp. 149-169). IGI Global. https://doi.org/10.4018/978-1-7998-3970-5.ch008

Panchenko, V. (2021). Photovoltaic Thermal Module With Paraboloid Type Solar Concentrators. *International Journal of Energy Optimization and Engineering*, 10(2), 1–23. https://doi.org/10.4018/IJEOE.2021040101

Pandey, K., & Datta, S. (2021). Dry Machining of Inconel 825 Superalloys: Performance of Tool Inserts (Carbide, Cermet, and SiAlON). *International Journal of Manufacturing, Materials, and Mechanical Engineering*, 11(4), 26–39. doi:10.4018/IJMMME.2021100102

Panneer, R. (2017). Effect of Composition of Fibers on Properties of Hybrid Composites. *International Journal of Manufacturing, Materials, and Mechanical Engineering*, 7(4), 28–43. doi:10.4018/IJMMME.2017100103

Related References

Pany, C. (2021). Estimation of Correct Long-Seam Mismatch Using FEA to Compare the Measured Strain in a Non-Destructive Testing of a Pressurant Tank: A Reverse Problem. *International Journal of Smart Vehicles and Smart Transportation*, 4(1), 16–28. doi:10.4018/IJSVST.2021010102

Paul, S., & Roy, P. (2018). Optimal Design of Power System Stabilizer Using a Novel Evolutionary Algorithm. *International Journal of Energy Optimization and Engineering*, 7(3), 24–46. doi:10.4018/IJEOE.2018070102

Paul, S., & Roy, P. K. (2021). Oppositional Differential Search Algorithm for the Optimal Tuning of Both Single Input and Dual Input Power System Stabilizer. In P. Vasant, G. Weber, & W. Punurai (Eds.), *Research Advancements in Smart Technology, Optimization, and Renewable Energy* (pp. 256-282). IGI Global. https://doi.org/10.4018/978-1-7998-3970-5.ch013

Pavaloiu, A. (2018). Artificial Intelligence Ethics in Biomedical-Engineering-Oriented Problems. In U. Kose, G. Guraksin, & O. Deperlioglu (Eds.), *Nature-Inspired Intelligent Techniques for Solving Biomedical Engineering Problems* (pp. 219–231). Hershey, PA: IGI Global. doi:10.4018/978-1-5225-4769-3.ch010

Pioro, I., Mahdi, M., & Popov, R. (2017). Application of Supercritical Pressures in Power Engineering. In L. Chen & Y. Iwamoto (Eds.), *Advanced Applications of Supercritical Fluids in Energy Systems* (pp. 404–457). Hershey, PA: IGI Global. doi:10.4018/978-1-5225-2047-4.ch013

Plaksina, T., & Gildin, E. (2017). Rigorous Integrated Evolutionary Workflow for Optimal Exploitation of Unconventional Gas Assets. *International Journal of Energy Optimization and Engineering*, 6(1), 101–122. doi:10.4018/IJEOE.2017010106

Popat, J., Kakadiya, H., Tak, L., Singh, N. K., Majeed, M. A., & Mahajan, V. (2021). Reliability of Smart Grid Including Cyber Impact: A Case Study. In R. Singh, A. Singh, A. Dwivedi, & P. Nagabhushan (Eds.), *Computational Methodologies for Electrical and Electronics Engineers* (pp. 163–174). IGI Global. https://doi.org/10.4018/978-1-7998-3327-7.ch013

Quiza, R., La Fé-Perdomo, I., Rivas, M., & Ramtahalsing, V. (2021). Triple Bottom Line-Focused Optimization of Oblique Turning Processes Based on Hybrid Modeling: A Study Case on AISI 1045 Steel Turning. In L. Burstein (Ed.), *Handbook of Research on Advancements in Manufacturing, Materials, and Mechanical Engineering* (pp. 215–241). IGI Global. https://doi.org/10.4018/978-1-7998-4939-1.ch010

Rahmani, M. K. (2022). Blockchain Technology: Principles and Algorithms. In S. Khan, M. Syed, R. Hammad, & A. Bushager (Eds.), *Blockchain Technology and Computational Excellence for Society 5.0* (pp. 16–27). IGI Global. https://doi.org/10.4018/978-1-7998-8382-1.ch002

Ramdani, N., & Azibi, M. (2018). Polymer Composite Materials for Microelectronics Packaging Applications: Composites for Microelectronics Packaging. In K. Kumar & J. Davim (Eds.), *Composites and Advanced Materials for Industrial Applications* (pp. 177–211). Hershey, PA: IGI Global. doi:10.4018/978-1-5225-5216-1.ch009

Ramesh, M., Garg, R., & Subrahmanyam, G. V. (2017). Investigation of Influence of Quenching and Annealing on the Plane Fracture Toughness and Brittle to Ductile Transition Temperature of the Zinc Coated Structural Steel Materials. *International Journal of Surface Engineering and Interdisciplinary Materials Science*, *5*(2), 33–42. doi:10.4018/IJSEIMS.2017070103

Robinson, J., & Beneroso, D. (2022). Project-Based Learning in Chemical Engineering: Curriculum and Assessment, Culture and Learning Spaces. In A. Alves & N. van Hattum-Janssen (Eds.), *Training Engineering Students for Modern Technological Advancement* (pp. 1–19). IGI Global. https://doi.org/10.4018/978-1-7998-8816-1.ch001

Rondon, B. (2021). Experimental Characterization of Admittance Meter With Crude Oil Emulsions. *International Journal of Electronics, Communications, and Measurement Engineering*, *10*(2), 51–59. https://doi.org/10.4018/IJECME.2021070104

Rudolf, S., Biryuk, V. V., & Volov, V. (2018). Vortex Effect, Vortex Power: Technology of Vortex Power Engineering. In V. Kharchenko & P. Vasant (Eds.), *Handbook of Research on Renewable Energy and Electric Resources for Sustainable Rural Development* (pp. 500–533). Hershey, PA: IGI Global. doi:10.4018/978-1-5225-3867-7.ch021

Sah, A., Bhadula, S. J., Dumka, A., & Rawat, S. (2018). A Software Engineering Perspective for Development of Enterprise Applications. In A. Elçi (Ed.), *Handbook of Research on Contemporary Perspectives on Web-Based Systems* (pp. 1–23). Hershey, PA: IGI Global. doi:10.4018/978-1-5225-5384-7.ch001

Sahli, Y., Zitouni, B., & Hocine, B. M. (2021). Three-Dimensional Numerical Study of Overheating of Two Intermediate Temperature P-AS-SOFC Geometrical Configurations. In G. Badea, R. Felseghi, & I. Aşchilean (Eds.), *Hydrogen Fuel Cell Technology for Stationary Applications* (pp. 186–222). IGI Global. https://doi.org/10.4018/978-1-7998-4945-2.ch008

Sahoo, P., & Roy, S. (2017). Tribological Behavior of Electroless Ni-P, Ni-P-W and Ni-P-Cu Coatings: A Comparison. *International Journal of Surface Engineering and Interdisciplinary Materials Science, 5*(1), 1–15. doi:10.4018/IJSEIMS.2017010101

Sahoo, S. (2018). Laminated Composite Hypar Shells as Roofing Units: Static and Dynamic Behavior. In K. Kumar & J. Davim (Eds.), *Composites and Advanced Materials for Industrial Applications* (pp. 249–269). Hershey, PA: IGI Global. doi:10.4018/978-1-5225-5216-1.ch011

Sahu, H., & Hungyo, M. (2018). Introduction to SDN and NFV. In A. Dumka (Ed.), *Innovations in Software-Defined Networking and Network Functions Virtualization* (pp. 1–25). Hershey, PA: IGI Global. doi:10.4018/978-1-5225-3640-6.ch001

Salem, A. M., & Shmelova, T. (2018). Intelligent Expert Decision Support Systems: Methodologies, Applications, and Challenges. In T. Shmelova, Y. Sikirda, N. Rizun, A. Salem, & Y. Kovalyov (Eds.), *Socio-Technical Decision Support in Air Navigation Systems: Emerging Research and Opportunities* (pp. 215–242). Hershey, PA: IGI Global. doi:10.4018/978-1-5225-3108-1.ch007

Samal, M. (2017). FE Analysis and Experimental Investigation of Cracked and Un-Cracked Thin-Walled Tubular Components to Evaluate Mechanical and Fracture Properties. In P. Samui, S. Chakraborty, & D. Kim (Eds.), *Modeling and Simulation Techniques in Structural Engineering* (pp. 266–293). Hershey, PA: IGI Global. doi:10.4018/978-1-5225-0588-4.ch009

Samal, M., & Balakrishnan, K. (2017). Experiments on a Ring Tension Setup and FE Analysis to Evaluate Transverse Mechanical Properties of Tubular Components. In P. Samui, S. Chakraborty, & D. Kim (Eds.), *Modeling and Simulation Techniques in Structural Engineering* (pp. 91–115). Hershey, PA: IGI Global. doi:10.4018/978-1-5225-0588-4.ch004

Samarasinghe, D. A., & Wood, E. (2021). Innovative Digital Technologies. In J. Underwood & M. Shelbourn (Eds.), *Handbook of Research on Driving Transformational Change in the Digital Built Environment* (pp. 142–163). IGI Global. https://doi.org/10.4018/978-1-7998-6600-8.ch006

Sawant, S. (2018). Deep Learning and Biomedical Engineering. In U. Kose, G. Guraksin, & O. Deperlioglu (Eds.), *Nature-Inspired Intelligent Techniques for Solving Biomedical Engineering Problems* (pp. 283–296). Hershey, PA: IGI Global. doi:10.4018/978-1-5225-4769-3.ch014

Schulenberg, T. (2021). Energy Conversion Using the Supercritical Steam Cycle. In L. Chen (Ed.), *Handbook of Research on Advancements in Supercritical Fluids Applications for Sustainable Energy Systems* (pp. 659–681). IGI Global. doi:10.4018/978-1-7998-5796-9.ch018

Sezgin, H., & Berkalp, O. B. (2018). Textile-Reinforced Composites for the Automotive Industry. In K. Kumar & J. Davim (Eds.), *Composites and Advanced Materials for Industrial Applications* (pp. 129–156). Hershey, PA: IGI Global. doi:10.4018/978-1-5225-5216-1.ch007

Shaaban, A. A., & Shehata, O. M. (2021). Combining Response Surface Method and Metaheuristic Algorithms for Optimizing SPIF Process. *International Journal of Manufacturing, Materials, and Mechanical Engineering, 11*(4), 1–25. https://doi.org/10.4018/IJMMME.2021100101

Shafaati Shemami, M., & Sefid, M. (2022). Implementation and Demonstration of Electric Vehicle-to-Home (V2H) Application: A Case Study. In M. Alam, R. Pillai, & N. Murugesan (Eds.), *Developing Charging Infrastructure and Technologies for Electric Vehicles* (pp. 268–293). IGI Global. https://doi.org/10.4018/978-1-7998-6858-3.ch015

Shah, M. Z., Gazder, U., Bhatti, M. S., & Hussain, M. (2018). Comparative Performance Evaluation of Effects of Modifier in Asphaltic Concrete Mix. *International Journal of Strategic Engineering, 1*(2), 13–25. doi:10.4018/IJoSE.2018070102

Sharma, N., & Kumar, K. (2018). Fabrication of Porous NiTi Alloy Using Organic Binders. In K. Kumar & J. Davim (Eds.), *Composites and Advanced Materials for Industrial Applications* (pp. 38–62). Hershey, PA: IGI Global. doi:10.4018/978-1-5225-5216-1.ch003

Shivach, P., Nautiyal, L., & Ram, M. (2018). Applying Multi-Objective Optimization Algorithms to Mechanical Engineering. In M. Ram & J. Davim (Eds.), *Soft Computing Techniques and Applications in Mechanical Engineering* (pp. 287–301). Hershey, PA: IGI Global. doi:10.4018/978-1-5225-3035-0.ch014

Shmelova, T. (2018). Stochastic Methods for Estimation and Problem Solving in Engineering: Stochastic Methods of Decision Making in Aviation. In S. Kadry (Ed.), *Stochastic Methods for Estimation and Problem Solving in Engineering* (pp. 139–160). Hershey, PA: IGI Global. doi:10.4018/978-1-5225-5045-7.ch006

Siero González, L. R., & Romo Vázquez, A. (2017). Didactic Sequences Teaching Mathematics for Engineers With Focus on Differential Equations. In M. Ramírez-Montoya (Ed.), *Handbook of Research on Driving STEM Learning With Educational Technologies* (pp. 129–151). Hershey, PA: IGI Global. doi:10.4018/978-1-5225-2026-9.ch007

Sim, M. S., You, K. Y., Esa, F., & Chan, Y. L. (2021). Nanostructured Electromagnetic Metamaterials for Sensing Applications. In M. Bhat, I. Wani, & S. Ashraf (Eds.), *Applications of Nanomaterials in Agriculture, Food Science, and Medicine* (pp. 141–164). IGI Global. https://doi.org/10.4018/978-1-7998-5563-7.ch009

Singh, R., & Dutta, S. (2018). Visible Light Active Nanocomposites for Photocatalytic Applications. In K. Kumar & J. Davim (Eds.), *Composites and Advanced Materials for Industrial Applications* (pp. 270–296). Hershey, PA: IGI Global. doi:10.4018/978-1-5225-5216-1.ch012

Skripov, P. V., Yampol'skiy, A. D., & Rutin, S. B. (2021). High-Power Heat Transfer in Supercritical Fluids: Microscale Times and Sizes. In L. Chen (Ed.), *Handbook of Research on Advancements in Supercritical Fluids Applications for Sustainable Energy Systems* (pp. 424–450). IGI Global. https://doi.org/10.4018/978-1-7998-5796-9.ch012

Sözbilir, H., Özkaymak, Ç., Uzel, B., & Sümer, Ö. (2018). Criteria for Surface Rupture Microzonation of Active Faults for Earthquake Hazards in Urban Areas. In N. Ceryan (Ed.), *Handbook of Research on Trends and Digital Advances in Engineering Geology* (pp. 187–230). Hershey, PA: IGI Global. doi:10.4018/978-1-5225-2709-1.ch005

Stanciu, I. (2018). Stochastic Methods in Microsystems Engineering. In S. Kadry (Ed.), *Stochastic Methods for Estimation and Problem Solving in Engineering* (pp. 161–176). Hershey, PA: IGI Global. doi:10.4018/978-1-5225-5045-7.ch007

Strebkov, D., Nekrasov, A., Trubnikov, V., & Nekrasov, A. (2018). Single-Wire Resonant Electric Power Systems for Renewable-Based Electric Grid. In V. Kharchenko & P. Vasant (Eds.), *Handbook of Research on Renewable Energy and Electric Resources for Sustainable Rural Development* (pp. 449–474). Hershey, PA: IGI Global. doi:10.4018/978-1-5225-3867-7.ch019

Sukhyy, K., Belyanovskaya, E., & Sukhyy, M. (2021). *Basic Principles for Substantiation of Working Pair Choice.* IGI Global. doi:10.4018/978-1-7998-4432-7.ch002

Suri, M. S., & Kaliyaperumal, D. (2022). Extension of Aspiration Level Model for Optimal Planning of Fast Charging Stations. In A. Fekik & N. Benamrouche (Eds.), *Modeling and Control of Static Converters for Hybrid Storage Systems* (pp. 91–106). IGI Global. https://doi.org/10.4018/978-1-7998-7447-8.ch004

Tallet, E., Gledson, B., Rogage, K., Thompson, A., & Wiggett, D. (2021). Digitally-Enabled Design Management. In J. Underwood & M. Shelbourn (Eds.), *Handbook of Research on Driving Transformational Change in the Digital Built Environment* (pp. 63–89). IGI Global. https://doi.org/10.4018/978-1-7998-6600-8.ch003

Terki, A., & Boubertakh, H. (2021). A New Hybrid Binary-Real Coded Cuckoo Search and Tabu Search Algorithm for Solving the Unit-Commitment Problem. *International Journal of Energy Optimization and Engineering*, 10(2), 104–119. https://doi.org/10.4018/IJEOE.2021040105

Tüdeş, Ş., Kumlu, K. B., & Ceryan, S. (2018). Integration Between Urban Planning and Natural Hazards For Resilient City. In N. Ceryan (Ed.), *Handbook of Research on Trends and Digital Advances in Engineering Geology* (pp. 591–630). Hershey, PA: IGI Global. doi:10.4018/978-1-5225-2709-1.ch017

Ulamis, K. (2018). Soil Liquefaction Assessment by Anisotropic Cyclic Triaxial Test. In N. Ceryan (Ed.), *Handbook of Research on Trends and Digital Advances in Engineering Geology* (pp. 631–664). Hershey, PA: IGI Global. doi:10.4018/978-1-5225-2709-1.ch018

Valente, M., & Milani, G. (2017). Seismic Assessment and Retrofitting of an Under-Designed RC Frame Through a Displacement-Based Approach. In V. Plevris, G. Kremmyda, & Y. Fahjan (Eds.), *Performance-Based Seismic Design of Concrete Structures and Infrastructures* (pp. 36–58). Hershey, PA: IGI Global. doi:10.4018/978-1-5225-2089-4.ch002

Vargas-Bernal, R. (2021). Advances in Electromagnetic Environmental Shielding for Aeronautics and Space Applications. In C. Nikolopoulos (Ed.), *Recent Trends on Electromagnetic Environmental Effects for Aeronautics and Space Applications* (pp. 80–96). IGI Global. https://doi.org/10.4018/978-1-7998-4879-0.ch003

Vasant, P. (2018). A General Medical Diagnosis System Formed by Artificial Neural Networks and Swarm Intelligence Techniques. In U. Kose, G. Guraksin, & O. Deperlioglu (Eds.), *Nature-Inspired Intelligent Techniques for Solving Biomedical Engineering Problems* (pp. 130–145). Hershey, PA: IGI Global. doi:10.4018/978-1-5225-4769-3.ch006

Verner, C. M., & Sarwar, D. (2021). Avoiding Project Failure and Achieving Project Success in NHS IT System Projects in the United Kingdom. *International Journal of Strategic Engineering*, *4*(1), 33–54. https://doi.org/10.4018/IJoSE.2021010103

Verrollot, J., Tolonen, A., Harkonen, J., & Haapasalo, H. J. (2018). Challenges and Enablers for Rapid Product Development. *International Journal of Applied Industrial Engineering*, *5*(1), 25–49. doi:10.4018/IJAIE.2018010102

Wan, A. C., Zulu, S. L., & Khosrow-Shahi, F. (2021). Industry Views on BIM for Site Safety in Hong Kong. In J. Underwood & M. Shelbourn (Eds.), *Handbook of Research on Driving Transformational Change in the Digital Built Environment* (pp. 120–140). IGI Global. https://doi.org/10.4018/978-1-7998-6600-8.ch005

Yardimci, A. G., & Karpuz, C. (2018). Fuzzy Rock Mass Rating: Soft-Computing-Aided Preliminary Stability Analysis of Weak Rock Slopes. In N. Ceryan (Ed.), *Handbook of Research on Trends and Digital Advances in Engineering Geology* (pp. 97–131). Hershey, PA: IGI Global. doi:10.4018/978-1-5225-2709-1.ch003

You, K. Y. (2021). Development Electronic Design Automation for RF/Microwave Antenna Using MATLAB GUI. In A. Nwajana & I. Ihianle (Eds.), *Handbook of Research on 5G Networks and Advancements in Computing, Electronics, and Electrical Engineering* (pp. 70–148). IGI Global. https://doi.org/10.4018/978-1-7998-6992-4.ch004

Yousefi, Y., Gratton, P., & Sarwar, D. (2021). Investigating the Opportunities to Improve the Thermal Performance of a Case Study Building in London. *International Journal of Strategic Engineering*, *4*(1), 1–18. https://doi.org/10.4018/IJoSE.2021010101

Zindani, D., & Kumar, K. (2018). Industrial Applications of Polymer Composite Materials. In K. Kumar & J. Davim (Eds.), *Composites and Advanced Materials for Industrial Applications* (pp. 1–15). Hershey, PA: IGI Global. doi:10.4018/978-1-5225-5216-1.ch001

Zindani, D., Maity, S. R., & Bhowmik, S. (2018). A Decision-Making Approach for Material Selection of Polymeric Composite Bumper Beam. In K. Kumar & J. Davim (Eds.), *Composites and Advanced Materials for Industrial Applications* (pp. 112–128). Hershey, PA: IGI Global. doi:10.4018/978-1-5225-5216-1.ch006

About the Contributors

Sadique Ahmad is the CEO of KnowledgeShare IU Private Limited. Also, he is a Senior Assistant Professor at the Department of Computer Sciences Bahria University Karachi Campus Pakistan (on leave). Currently, he is working as a Postdoc Fellow at Prince Sultan University, Riyadh KSA. Dr. Sadique Ahmad achieved his Ph.D. degree (2019) from the Department of Computer Sciences and Technology, Beijing Institute of Technology China while a Master's degree (2015) from the Department of Computer Sciences from IMSciences University Peshawar Pakistan. He has achieved over 50 research articles in peer-review journals and conferences. His research interests include Artificial Intelligence and more specifically Deep Learning, Behavior and Emotions Modelling, Cognitive Computing, Cognitive Modeling, Object Detection, Data Analytics, remote sensing image analysis. He has reviewed over 180 scientific research articles for various well-known journals, including Information Sciences, IEEE Access Journal, Knowledge-Based Systems, Education and Information Technologies, Information Technology and Management, ICEEST Conference, and ICONIP Conference. Dr. Sadique Ahmad has many collaborative scientific activities with international teams in different research projects in the following international universities. 1. Shenzhen University, China 2. Gachon University South Korea 3. COMSAT University Pakistan 4. Bahria University Pakistan 5. Beijing Institute of Technology China 6. University of Haripur, Pakistan 7. Szabist University Karachi Pakistan 8. Sfax (ENIS), University Tunisia 9. AL-BAHA University, Saudi Arabia 10. Menoufia University, Shebin El-Koom, Egypt 11. Prince Sultan University Saudi Arabia.

Muhammad Shahid Anwar is currently working as an Assistant Professor in the Department of AI & Software at Gachon University, Seongnam, South Korea. He received his Ph.D. degree in Information and Communication Engineering from the School of Information and Electronics, Beijing Institute of Technology, Beijing, China in 2021 and his M.Sc. in Telecommunications Technology from Aston University, Birmingham, U.K., in 2012. Dr. Shahid has authored and co-authored more than 50 publications including IEEE, Springer, IET, Hindawi, MDPI, Frontiers journals, and flagship conference papers. He has been honored with the "Outstanding Scholar of the Year 2020 Award" from the CSC Scholarship Council under the Ministry of Education China. Dr. Shahid also received the "Excellent Student of the Year 2020 Award" from the Beijing Institute of Technology, China. He has been serving as an editorial board member and guest editor in several Journals and a reviewer of quite a few Journals including ACM and IEEE Transactions. His research interests include 360-degree videos, immersive Media (Virtual Reality, AR), Cognitive Computing, Object Detection, Metaverse, and Quality of Experience (QoE) evaluations of VR telemedicine and healthcare systems. He is focusing on deep learning-based VR video evaluations and developed several Machine Learning based QoE Prediction models.

Ala Saleh Alluhaidan received the B.Sc. degree in computer science from Princess Nourah bint Abdulrahman University, Riyadh, Saudi Arabia, the M.Sc. degree in computer information systems from Grand Valley State University, MI, USA, and the Ph.D. degree in information systems and technology from Claremont Graduate University, CA, USA. She is currently an Associate Professor with the Department of Information Systems, Princess Nourah bint Abdulrahman University. Her current research interests include big data analytics, Cognitive Computing, health informatics, and Artificial Intelligence. Also, she achieved 48 research articles in peer-reviewed journals and conferences. Few research articles are under review in different prestigious journals.

Mohammed A. El-Affendi is currently a Professor of computer science with the Department of Computer Science, Prince Sultan University, the Former Dean of CCIS, AIDE, the Rector, the Founder, and the Director of the Data Science Laboratory (EIAS), and the Founder and the Director of the Center of Excellence in CyberSecurity. His current research interests include data science, intelligent and cognitive systems, machine learning, and natural language processing.

* * *

Muhammad Arshad is working as Associate Professor. His filed of interests are software engineering, web engineering, and networking.

Zahid Farid completed both his B.S and MS degrees in electrical engineering at the University of Engineering & Technology, Peshawar. Subsequently, he earned his doctoral degree in Electrical Engineering from the National University Malaysia (UKM) in 2017. His research interests are broad, encompassing artificial intelligence, optimization algorithms, positioning and tracking technologies, Wireless Sensor networks, and the Internet of Things. With over 18 years of experience in teaching and research within the engineering domain, he presently holds the position of Associate Professor at Abasyn University in Peshawar, KPK, Pakistan rephrase.

Mahmood Ul Haq received a B.S. degree in Electrical and Electronics Engineering in 2016 and MS degree in Electrical Engineering in 2018 from COMSATS University Islamabad (CUI), Abbottabad Campus, Pakistan. Currently, he is Student of Ph.D. Computer System Engineering, in University of Engineering and Technology Peshawar, Pakistan. Mahmood's research expertise encompasses topics, such as object recognition, face recognition, image segmentation and natural language processing, Mahmood, has published numerous manuscripts in reputable Journals, and Conferences. He is the recipient of Prime Minister Fee Refunding Scheme, Government of Pakistan scholarship award for MS studies.

Bll Khan is a distinguished researcher with expertise in machine learning, data science, natural language processing, and bioinformatics. Renowned for his prolific contributions to these fields, Bilal has authored numerous articles that have significantly advanced the understanding and application of data-driven technologies. His impactful work has earned him recognition as a reviewer for esteemed journals, where he continues to contribute to the scholarly discourse and shape the future of these interdisciplinary domains.

S A. Sadik is a researcher in the field of computer science and telecommunication. He holds a MSc degree in Electrical and Electronics Engineering from Kutahya Dumlupinar University and a PhD degree in Electrical and Electronics Engineering from Kutahya Dumlupinar University. Sadik's research focuses on optical fiber technology and applying artificial intelligence techniques to solve real-world problems, with a particular interest in machine learning applications in optical fiber systems.

Muhammad Athar Javed Sethi is an Assistant Professor at the Department of Computer Systems Engineering, University of Engineering and Technology (UET), Peshawar, Pakistan. He received the Bachelor of Sciences in Computer Information Systems Engineering (honors) from the UET Peshawar, Pakistan in 2004. He had done Master of Sciences in Computer Systems Engineering from the same university in 2008. He completed his Ph.D. at Universiti Teknologi PETRONAS (UTP), Malaysia in Department of Electrical & Electronic Engineering in 2016. He was able to get different scholarships to pursue and continue his studies. Dr. Sethi got IEEE student best paper award in the year 2013 and a most downloaded paper award from Elsevier in consecutive two years 2016-2017 for two different papers. His research interests include network on chip (NoC), interconnection networks, computer architecture, embedded systems and field programmable gate array (FPGA's). Dr. Sethi, has published numerous manuscripts in reputable Journals, Conferences and Books. He also wrote a book "Bio-Inspired Fault-Tolerant Algorithms for Network-on-Chip". Currently, he is actively involved in technical program committees of various international conferences. He is serving as Associate Editor at EAI Endorsed Transactions on Context-aware Systems & Applications and at EAI Endorsed Transactions on Bioengineering and Bioinformatics.

Sba Ullah completed his PhD in Mathematics from University of Malakand. His research area is Fluid Dynamics and mathematical modeling. He has published several papers in international journals. Abdullah has completed his BS mathematics degree from University of Malakand in 2018. He has received his M. Phil. Mathematics degree in Applied Mathematics from University of Malakand in 2021. His area of research are ODEs, PDEs, Integral transform methods, fractional and Fuzzy ODEs, PDEs. His impactful work has earned him recognition as a reviewer for esteemed journals, where he continues to contribute to the scholarly discourse and shape the future of these interdisciplinary domains.

Index

A

Applications 4, 14, 23, 25-28, 30-31, 39, 51-53, 56, 61, 63-64, 66, 68-69, 74, 86-89, 93-94, 102-103, 106-108, 113-117, 124, 131, 133, 135, 137-138, 143, 146, 150, 152-154, 168-169, 173-174, 176-177, 181, 192-194, 197, 211, 215

Artificial Intelligence 27, 69, 79, 82, 87, 107, 112-118, 122, 126-133, 139, 143-144, 146, 148, 192, 217

C

Capsule Network 8, 23, 55, 60-61, 64-65, 67-68, 88-90, 92-93, 95, 102-104, 107-112, 169, 171, 194

CNN 8-9, 11, 21, 26, 31, 34, 36, 41, 59, 67, 88-91, 93-95, 103, 110-111, 120, 149, 196, 213, 217

Cognitive Modeling 30, 38, 70-71, 76-78, 84

Convolutional Neural Network (CNN) 120, 149

D

Data Privacy 126

Datasets 31, 33, 35-36, 38, 42-45, 50, 55-57, 62-63, 78, 94, 100, 108, 141, 152-154, 160, 168, 173-177, 200, 202, 213, 218

Deep Learning 1-9, 11-12, 14-15, 20, 22, 25-27, 30-31, 34, 36, 38, 40-42, 45, 49-50, 56, 59, 65, 68-69, 89, 93, 107-108, 111, 113-115, 131, 136, 139-140, 146, 160, 168, 170-171, 180, 196, 199, 201, 211, 213-215, 217

Digital Privacy 82-83

Drones 55-57, 61-62, 69, 77, 117

E

Environmental Sustainability 71, 82, 84

Ethical Considerations 70-71, 76, 84, 115, 149

F

Face Recognition 23, 25-27, 55-57, 60-68, 107, 109-112, 151-154, 160, 162, 165-177, 179-181, 183-187, 190-191, 193-197, 217

Fairness 71, 79-82, 86, 128, 145

I

Image Classification 34, 40-41, 51, 53, 61, 67, 88, 109, 125, 198, 200, 213-214

Image Processing 1, 11, 23, 26, 53, 55, 89, 105, 108, 178, 197-198

Inceptionv3 30, 41, 45-46, 48-50, 52

L

Long-Lasting Memories (LLMs) 125, 149

M

Machine Learning 8, 23, 27, 31-32, 51-52, 56-57, 67, 69, 79, 82-83, 88-89, 105,

107, 110, 112-114, 117, 127-128, 138-142, 144-146, 148-150, 152, 167, 169, 171, 187, 192, 197-201, 204-205, 214, 216-217

Medical Images 130

N

Natural Language Processing (NLP) 22, 65, 107, 115-116, 118, 150, 168, 215

O

Object Detection 1-6, 9, 12, 14, 19, 21-29, 41, 51-52, 60-61, 66, 69, 108, 198

Object Recognition 1, 3-6, 8, 11, 14, 32, 88, 94, 103, 108

Open Source Software (OSS) 127-129, 131, 150

Optical Remote Sensing 28, 30-31, 39, 45, 51-54

R

Remote Sensing 1-6, 8-9, 11-12, 14, 19-31, 39, 45, 51-54, 70-78, 84-87

Remote Sensing Images 1-6, 9, 11-12, 14, 19, 21-22, 24-29, 51-52, 54

Retrieve, Analyze, Generate (RAG) 133, 150

S

Satellite Imagery 21, 30-31, 38, 51, 74

Ship Classification 30-38, 51-54

T

Technologies 1-2, 14, 21, 32, 56, 66, 73, 76-77, 79, 82-84, 88, 106, 113, 115, 123, 127-130, 132-133, 143, 192, 201

Transfer Learning 9, 21, 26, 34, 38-39, 51-52, 67, 202, 213, 216-217

Trends 1-3, 5, 12, 19, 27, 82, 115, 117, 193

V

VGG16 40-41, 46-50, 53, 202

VGG19 41, 46, 48-51, 202

Visual Language (VL) 121, 150

Visual Language Models (VLMs) 125, 150

X

X-Rays Image Processing 198

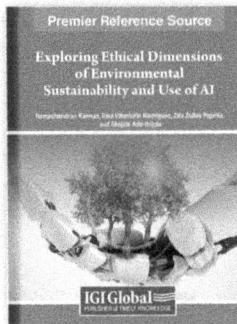

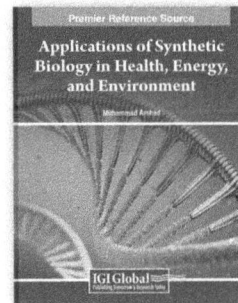

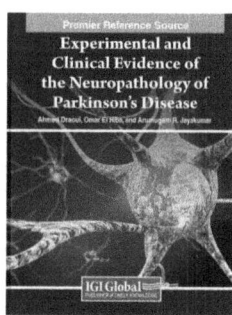

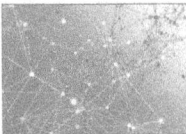
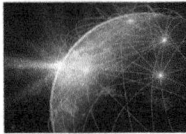

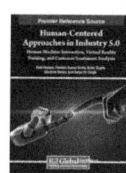

9 798369 329139